Other Titles in This Series

Volume

9   Bruce C. Berndt and Robert A. Rankin
    Ramanujan: Letters and commentary
    1995

8   Karen Hunger Parshall and David E. Rowe
    The emergence of the American mathematical research community, 1876–1900: J. J. Sylvester, Felix Klein, and E. H. Moore
    1994

7   Henk J. M. Bos
    Lectures in the history of mathematics
    1993

6   Smilka Zdravkovska and Peter L. Duren, Editors
    Golden years of Moscow mathematics
    1993

5   George W. Mackey
    The scope and history of commutative and noncommutative harmonic analysis
    1992

4   Charles W. McArthur
    Operations analysis in the U.S. Army Eighth Air Force in World War II
    1990

3   Peter L. Duren, editor, et al.
    A century of mathematics in America, part III
    1989

2   Peter L. Duren, editor, et al.
    A century of mathematics in America, part II
    1989

1   Peter L. Duren, editor, et al.
    A century of mathematics in America, part I
    1988

SRINIVASA RAMANUJAN
1887–1920

History of
Mathematics
Volume 9

RAMANUJAN
LETTERS and COMMENTARY

BRUCE C. BERNDT *and* ROBERT A. RANKIN

American Mathematical Society
London Mathematical Society

1991 *Mathematics Subject Classification.* Primary 01A32, 01A60, 01A70; Secondary 11-03, 40-03, 41-03.

Photograph on front cover: India postage stamp honoring Ramanujan. Back cover: Ramanujan with friends at Trinity College. Background photo: Letter from Ramanujan to his father.
A list of photograph credits is included at the beginning of this volume.

Library of Congress Cataloging-in-Publication Data
Ramanujan Aiyangar, Srinivasa, 1887–1920.
  [Correspondence, Selections]
  Ramanujan : letters and commentary / Bruce C. Berndt, Robert A. Rankin.
    p. cm. — (History of mathematics, ISSN 0899-2428; v. 9)
  Includes bibliographical references (p.   -   ) and index.
  ISBN 0-8218-0281-9 (hardcover) (acid-free)
  ISBN 0-8218-0470-7 (softcover) (acid-free)
    1. Ramanujan Aiyangar, Srinivasa, 1887–1920—Correspondence.   2. Mathematicians—India—Correspondence.   3. Number theory.   I. Berndt, Bruce C., 1939–   .   II. Rankin, Robert A. (Robert Alexander), 1915–   .   III. Title.   IV. Series.
  QA29.R3A4   1995
  510′.92—dc20   95-5254

Copying and reprinting. Individual readers of this publication, and nonprofit libraries acting for them, are permitted to make fair use of the material, such as to copy a chapter for use in teaching or research. Permission is granted to quote brief passages from this publication in reviews, provided the customary acknowledgment of the source is given.

Republication, systematic copying, or multiple reproduction of any material in this publication (including abstracts) is permitted only under license from the American Mathematical Society. Requests for such permission should be addressed to the Manager of Editorial Services, American Mathematical Society, P.O. Box 6248, Providence, Rhode Island 02940-6248. Requests can also be made by e-mail to reprint-permission@math.ams.org.

© Copyright 1995 by the American Mathematical Society. All rights reserved.
Printed in the United States of America.

The American Mathematical Society retains all rights
except those granted to the United States Government.
◎ The paper used in this book is acid-free and falls within the guidelines
established to ensure permanence and durability.

The London Mathematical Society is incorporated under Royal Charter
and is registered with the Charity Commissioners.

To our wives,
Helen and Mary

Contents

Preface	ix
Photograph Credits	xiii
A Brief Biography of Ramanujan	1
Chapter 1. Ramanujan in Madras	7
Chapter 2. Ramanujan's First Two Letters to Hardy and Hardy's Response	21
Chapter 3. Preparing to Go to England	73
Chapter 4. Ramanujan at Cambridge	109
Chapter 5. Ramanujan is Ill	145
Chapter 6. Ramanujan Returns to India	203
Chapter 7. After his Death	227
Chapter 8. Ramanujan's Papers and Manuscripts	259
Chapter 9. Family Histories	313
Provenance of the Letters	317
References	323
Index	335

Preface

Ramanujan's letter of 16 January 1913 to G.H. Hardy is one of the most famous letters in the history of mathematics. After introducing himself "as a clerk in the Accounts Department of the Port Trust Office at Madras," Ramanujan began to relate some of his mathematical discoveries. In this and the following letter of 27 February 1913, Ramanujan set forth over one hundred of his theorems. The day on which Hardy received Ramanujan's first letter marks the dawning of the recognition of Ramanujan's remarkable mathematical talents outside his native south India. Later, in the late 1920s, after Ramanujan's death, these letters spawned several papers by G.N. Watson, C.T. Preece, and others. Indeed, the influence of these letters still pervades contemporary research in mathematics.

Correspondence between Ramanujan and Hardy occurred not only before Ramanujan journeyed to England in March 1914, but also for two years while Ramanujan was confined to various nursing homes in 1917–1919. After returning to India, Ramanujan wrote to Hardy only once. These letters contain much interesting mathematics as well as updates on Ramanujan's illness. Unfortunately, some of the correspondence between these two great mathematicians has been lost.

Of course, Ramanujan wrote many letters to his family and friends from Cambridge. Although most of these letters have been lost, several have been preserved. We owe a great debt to P.K. Srinivasan, who collected many of these letters and made them available to the general public [263]. Other letters are secured in Cambridge University Library and in the library of Trinity College, Cambridge; photocopies of some of these are published in Ramanujan's "lost notebook" [227] and elsewhere. Further letters, mostly from the files of the Madras Port Trust Office, are safeguarded in the National Archives in Delhi.

The present authors have collected as many letters to, from, and about Ramanujan as has been possible. A list of the letters, together with their sources, is given in an appendix. There are several motivations for collecting these letters and writing extensive commentary on them.

First, as indicated above, the correspondence between Ramanujan and

Hardy contains much fascinating mathematics. We have examined each theorem quoted by Ramanujan in these letters and have traced its history and influence. It is hoped that this scholarship will be helpful to mathematicians who have been influenced by Ramanujan's mathematics.

Second, many of the people with whom Ramanujan corresponded or are mentioned in this correspondence are known to most of us today only by name. Thus, short biographies of most of the correspondents have been prepared. For example, while searching for information about Sir Gilbert Walker, Head of the Meteorological Observatory in Madras and one of the first to recognize Ramanujan's mathematical gifts, we discovered that he is well known to meteorologists today as one of the founders of modern weather forecasting.

Third, many parts of Ramanujan's personal letters need to be explained, especially to readers in the West unfamiliar with Indian places, culture, and foods.

Fourth, P.K. Srinivasan's collection [263] has been out of print for several years, and very few copies are available outside India. Thus, we are anxious that these letters and others not in Srinivasan's book be made accessible to those both within and outside India who have an interest in Ramanujan and his mathematics.

It should be emphasized that this work is not meant to be a full biography of Ramanujan. Only those parts of Ramanujan's life and work arising from the letters that we have collected are discussed. Those desiring a more complete and detailed account of Ramanujan's life are advised to read Robert Kanigel's book [142]. Less complete biographies have been written by S. Ram [195] and S.R. Ranganathan [229]. Prior to Kanigel's book, the most commonly read account of Ramanujan's life was that of P.V. Seshu Aiyar and R. Ramachandra Rao, found at the beginning of Ramanujan's *Collected Papers* [226]. This is an amalgam of earlier obituaries written by them individually [1], [243]. Hardy's obituary [98], [112, pp. 702–720], [226, pp. xxix–xxvi] focuses more on Ramanujan's mathematical contributions. Almost all of the many brief biographies of Ramanujan that have been written since Ramanujan's death are based on the articles of Seshu Aiyar, Ramachandra Rao, and Hardy. Berndt [36] has written a brief account of Ramanujan's life and mathematics for nonspecialists. Anecdotes and personal reminiscences about Ramanujan can be found in the collections of Srinivasan [263] and R. Bharathi [45].

In addition to the letters printed here, we also include some newspaper articles published during Ramanujan's lifetime, some office memos from the Madras Port Trust Office, and jottings from a family diary. In most cases, we follow each item with commentary.

The authors are extremely grateful to many people who helped secure letters and other material for this volume and who helped us in writing our commentaries.

PREFACE

The conception for this work began shortly after S. Chandrasekhar, through the cooperation of the late Prime Minister Rajiv Gandhi, obtained for one of us (BCB) copies of all the Ramanujan correspondence in the National Archives in Delhi. Since that time, Chandrasekhar has offered considerable valuable advice and obtained further information for us.

We again express our gratitude to P.K. Srinivasan who long ago realized that letters of Ramanujan would be lost unless he made efforts to collect them.

K. Srinivasa Rao has aided us in several ways. Through his efforts, we obtained copies of the letters of Ramanujan to his boyhood friend, E. Vinayaka Row. Srinivasa Rao has provided considerable information about names, places, and customs in India and has offered many useful suggestions, after a very careful reading of our manuscript.

We are thankful that Kanigel [142] has brought Ramanujan's story to the general public, and information that he gathered has been useful to us.

We are also pleased to thank George Andrews for providing commentary on Ramanujan's last letter to Hardy; Richard Askey for many helpful remarks; Krishnaswami Alladi and S. Bhargava for information about India and its customs; Paul Bialek for comments on Ramanujan's letters to Hardy written from nursing homes; John Brillhart for writing the biography of D.H. Lehmer; Dan Grayson for answering hundreds of queries about T_EX; V.S. Ramachandran, Curator of Visvesvaraya Industrial and Technological Museum, Bangalore, for several of the photographs appearing in this volume; A. Ranganathan for information about his grandfather, Dr. P.S. Chandrasekar, who tended Ramanujan in the final year of his life; T.V. Rangaswami for a copy of Ramanujan's passport; Ranga and Shantha Rao for much knowledge about Indian foods and culture; C.A. Reddi for information about his father-in-law, A.S. Ramalingam, who was a close friend of Ramanujan, and for obtaining for us documents from the Tamil Nadu Archives; K. Venkatachaliengar for several helpful comments; V. Viswanathan for information about his grandfather, S. Narayana Aiyar; K.C. Wali for helpful correspondence; and Wayne Wendland for information on Sir Gilbert Walker and weather forecasting.

Many people in several organizations have kindly helped us in uncovering information about those whose names appear in the correspondence detailed in this book. Thus, we offer our sincere thanks to librarians at the University of Illinois, in particular, Nancy Anderson and Mary Berney; Glasgow University Library; Mitchell Library, Glasgow; Trinity College, Cambridge, in particular, the Master and Fellows of Trinity, the Librarians, the College Office, and Rosemary Graham, the Manuscript Cataloguer; the University Archives at the Cambridge University Library; the Cambridge Scientific Periodicals Library; the Cambridge Philosophical Society; the Biographical Assistant and Archivist at St. John's College, Cambridge; the Archivist at Merton College, Oxford; the Archivist at Christ Church, Oxford; J. Vernon

Armitage, College of St. Hild and St. Bede, Durham; Pacelli Clanchy, Valuation Office, Dublin; David H. Fowler, University of Warwick; Donald Glynn, Rutland Centre, Dublin; Wilbur R. Knorr, Stanford University; Walter Ledermann, University of Sussex; the late William McLean, formerly Registrar of the University of Madras; Peter Neumann, the Queen's College, Oxford; David J. Simms, Trinity College, Dublin; Wolfgang Schwarz, University of Frankfurt am Main; the late Gertrude K. Stanley, formerly of Westfield College, London; M.G. Underwood, Archivist at St. John's College, Cambridge; the Librarian of the Royal Society; the Administrator of the London Mathematical Society; the India Office Library; and the Office Manager at Barclays Bank.

We note, in conclusion, that the copyright of all Ramanujan's writings, of every kind, was, after his death, purchased from his widow by the University of Madras.

This project could not have been begun if it were not for a grant from the Alfred P. Sloan Foundation, to which we offer our sincere appreciation.

Photograph Credits

The American Mathematical Society gratefully acknowledges the kindness of these individuals and institutions in granting the following permissions:

V. S. Ramachandran, Curator, Visvesvaraya Industrial and Technological Museum, Bangalore, India

> Reproduction of the cover, G. S. Carr, *Synopsis of Elementary Results Pure Mathematics*, Francis Hodgson, London, 1880, p. 5; courtesy of V. S. Ramachandran
>
> Photo of Pachaiyappa's College, Madras, p. 6; courtesy of V. S. Ramachandran
>
> Photo of R. Ramachandra Rao, p. 6; courtesy of V. S. Ramachandran
>
> Reproduction of Ramanujan's letter to Madras Port Trust, p. 9; courtesy of V. S. Ramachandran
>
> Photo of the old Port Trust Office at Madras, p. 10; courtesy of V. S. Ramachandran
>
> Photo of S. Narayana Aiyar, p. 74; courtesy of V. S. Ramachandran
>
> Photo of E. H. Neville, p. 95; courtesy of V. S. Ramachandran
>
> Photo of Ramanujan with friends at Trinity College, p. 115 and cover; courtesy of V. S. Ramachandran
>
> Reproduction of Ramanujan's letter to his father, p. 120; courtesy of V. S. Ramachandran
>
> Photo of Ramanujan with western clothing and haircut, p. 127; courtesy of V. S. Ramachandran
>
> Photo of Ramanujan and fellow graduates, p. 137; courtesy of V. S. Ramachandran
>
> Photo of A. S. Ramalingam, p. 163; courtesy of V. S. Ramachandran

Photo of Gometra, Madras, p. 231; courtesy of V. S. Ramachandran

Photo of Komalatammal, p. 236; courtesy of V. S. Ramachandran

Photo of Janaki Ammal, p. 239; courtesy of V. S. Ramachandran

Reproduction of postage stamp, p. 314 and cover; courtesy of V. S. Ramachandran

Oxford University Press

Photo of G. H. Hardy, p. 31; from G. H. Hardy, *Collected Papers*, Vol. 1, Clarendon Press, 1966; by permission of Oxford University Press

University of Dundee

Photo of B. M. Wilson, p. 268; copyright University of Dundee; by permission of University of Dundee

The Royal Society

Photo of G. N. Watson, p. 282; by permission of The Royal Society, London

S. Chandrasekhar

Photo of S. Ramanujan, frontispiece; by permission of S. Chandrasekhar

Town High School, Kumbakonam, India

Reproduction of certificate awarded to Ramanujan, p. 4

All other photos were provided by the authors.

A Brief Biography of Ramanujan

Srinivasa Ramanujan was born on 22 December 1887 in his maternal grandmother's house in Erode, a small town about 250 miles southwest of Madras. A year later, his mother Komalatammal returned with him to her home in Kumbakonam (approximately 160 miles south-southwest of Madras), where her husband, Kuppuswamy Srinivasa Aiyangar, was a clerk in a cloth merchant's shop.

In October 1892, Ramanujan enrolled at a local primary school in Kumbakonam and attended several different primary schools in the following years. In January 1898, having passed his primary examinations with distinction, he entered the Town High School of Kumbakonam. It was in his later years at this school that he borrowed a copy of G. S. Carr's *Synopsis of Elementary Results Pure Mathematics* [59], which, by its concise style, strongly influenced his mathematical writing.

In 1904, Ramanujan entered the Government College, Kumbakonam, with a scholarship. Up to this point, he had done well in all his school subjects, but gradually he became more and more absorbed in mathematics and neglected his other studies, so that his scholarship was not renewed. In 1905, being in monetary difficulties and depressed, he ran away to Vizagapatnam, 400 miles north of Madras, without informing his parents.

By the following year, he had returned. He enrolled in Pachaiyappa's College, Madras, and attended lectures there for three months in an effort to pass the First Arts examination, which would enable him to enter the University of Madras. However, he failed in all his subjects other than mathematics.

In 1909, his mother arranged a marriage between him and S. Janaki Ammal, the nine-year-old daughter of a distant relative, but she did not come to live with him until 1912. By that time, her husband had become renowned in the Madras area for his mathematical abilities, partly through mathematical problems that he posed and solved in the recently founded *Journal of the Indian Mathematical Society*. For six weeks, probably in 1911, Ramanujan held a temporary job in the Accountant General's office in Madras. Through the generosity of R. Ramachandra Rao, the Collector of Nellore, who was an uncle of Ramanujan's friend, R. Krishna Rao, he received some regular financial support for approximately one year.

The year 1912 was a turning point for Ramanujan. He applied successfully for a clerkship in the accounts section of the Madras Port Trust at a salary of 30 rupees per month. The post was obtained through the influence of R. Ramachandra Rao and provided Ramanujan some leisure time to pursue mathematics. The Chief Accountant of the Madras Port Trust, S. Narayana Aiyar, had been trained as a mathematician, and he and the Port Trust Chairman, Sir Francis Spring, took a warm interest in Ramanujan and his mathematical abilities.

Ramanujan was now well known to professors in various colleges in Madras, and they were anxious to help him to further his studies and foster his genius in a more suitable environment. The first step was taken by C.L.T. Griffith, a civil engineering professor at Madras Engineering College. He was a graduate of University College, London, and wrote to his former mathematics professor, M.J.M. Hill, enclosing some of Ramanujan's work. Hill's reply was meant to be encouraging but was not very helpful, possibly because he was put off by Ramanujan's unexplained summation of divergent series.

However, by this time, other more important figures, such as Sir Francis Spring and Sir Gilbert Walker, who was a former Senior Wrangler, added their weight, and three well-known Cambridge mathematicians were approached. They were Professor H.F. Baker, Professor E.W. Hobson, and Mr. G.H. Hardy, the Cayley Lecturer in Mathematics and a Fellow of Trinity College. The first two evidently offered no positive response, but Hardy's reply was favorable and encouraging. It should be remarked that it is not known whether Baker and Hobson received the same wealth of material that Ramanujan sent to Hardy. Hardy at once established a rapport with Ramanujan, and several letters passed between them. Madras University was persuaded to grant Ramanujan a scholarship of 75 rupees per month for two years, and Hardy took steps to arrange for Ramanujan to be brought to Cambridge and receive suitable financial support.

Various difficulties had to be overcome. To certain orthodox Brahmins, crossing the ocean from India to England makes one an outcaste, but eventually agreement was reached between Ramanujan and his mother that he should go. In reaching this decision, the role played by E.H. Neville, a young Fellow of Trinity who had been invited to lecture in Madras, was an important one, as he obtained Ramanujan's confidence. Therefore, on 17 March 1914, Ramanujan set sail from Madras for England, supported by scholarships of £250 a year from the University of Madras and £60 a year from Trinity College. This was adequate, not only to support him in Cambridge, but also for the family he left behind in India.

Soon after his arrival in London, Ramanujan moved into Neville's house in Cambridge. Some weeks later, he took up his rooms in Trinity College. Being a strict vegetarian, he never ate in the College hall, but cooked in his rooms. As it was wartime, there were very few students in residence, and gradually many of the senior Cambridge mathematicians, such as Hardy's colleague J.E. Littlewood, departed for war service. Also, vegetarian food, which even in normal times was hard to obtain outside London, became even scarcer. However, Ramanujan met Hardy nearly every day, and the discussions and collaboration which ensued were of immense value to both men. It was a very fruitful period.

Unfortunately, early in 1917, Ramanujan fell ill, and during his remaining time in England he was rarely out of nursing homes, first in Cambridge, then near Wells, and for a long period in Matlock in Derbyshire. Latterly, when he seemed to have improved slightly in health, he was a patient at Fitzroy House in London and at a sanatorium in Putney; he may have resided in other nursing homes of which we have no record. Improvements in health, such as they were, were probably due to his elections in 1918 to Fellowships of the Royal Society and Trinity College.

On 13 March 1919, Ramanujan returned to India, but he was clearly a sick man. He was attended in several different places and received the best possible medical treatment that could be provided in India, but his health continued to decline; he died on 26 April 1920. Ramanujan was still working on his mathematics four days before the end. The nature of his illness has not been definitely determined. Tuberculosis has been the favored diagnosis, but a vitamin deficiency may have been an important factor. It has recently [298] been suggested that he may have suffered from hepatic amoebiasis.

RAMANUJAN'S HOME, KUMBAKONAM

KUMBAKONAM TOWN HIGH SCHOOL

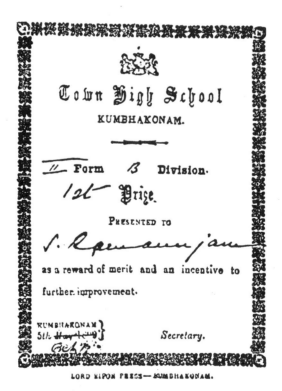

CERTIFICATE PRESENTED TO RAMANUJAN BY TOWN HIGH SCHOOL

A SYNOPSIS

OF

ELEMENTARY RESULTS

PURE MATHEMATICS:

CONTAINING

PROPOSITIONS, FORMULÆ, AND METHODS OF ANALYSIS,

WITH

ABRIDGED DEMONSTRATIONS.

SUPPLEMENTED BY AN INDEX TO THE PAPERS ON PURE MATHEMATICS WHICH ARE TO BE FOUND IN THE PRINCIPAL JOURNALS AND TRANSACTIONS OF LEARNED SOCIETIES, BOTH ENGLISH AND FOREIGN, OF THE PRESENT CENTURY.

BY

G. S. CARR, M.A.

LONDON:
FRANCIS HODGSON, 89 FARRINGDON STREET, E.C.
CAMBRIDGE; MACMILLAN & BOWES.

1886.

(All Rights reserved.)

THE PRIMARY SOURCE OF RAMANUJAN'S EARLY LEARNING OF MATHEMATICS

GOVERNMENT COLLEGE, KUMBAKONAM

Pachaiyappa's College, Madras

R. Ramachandra Rao

Chapter 1
Ramanujan in Madras

Commentary. The following letter of recommendation was attached to Ramanujan's application for a position at the Madras Port Trust.

E. W. Middlemast — Letter of Recommendation 21 September 1911

Madras

I can strongly recommend the applicant. He is a young man of <u>quite exceptional capacity in Mathematics</u> and especially in work relating to numbers. He has a natural aptitude for computation and is very quick at figure work. Though he has had no experience of statistical work I am confident that he can pick up the details in a very short time.

> E. W. Middlemast
> Ag. Principal and
> Professor of Mathematics
> The Presidency College

Commentary. Edward William Middlemast was born in 1864. He was a graduate of Cambridge University (St. John's College), being classed as 10th Wrangler in Part I of the Mathematical Tripos in 1886.

During the nineteenth century, the British government established two Presidency Colleges in India. Located in Madras and Calcutta, they were modeled after the University of London. They are now operated by the Indian government and have always been ranked among the very best Indian colleges and universities.

The Mathematical Tripos. Since so many of the mathematicians whose names appear in this book were graduates of the University of Cambridge, it seems appropriate to describe how the subject was taught and examined there. The reputation of Cambridge as a centre for research in mathematics goes back to the foundation of the Lucasian Professorship of Mathematics in 1663. The first holder of this chair was Isaac Barrow (1630–1677), an able mathematician, who resigned in 1669 in favor of Isaac Newton (1642–1727), so that he might devote more time to the study of theology. Newton held the chair for over 30 years and gave the study of mathematics a unique position in the university.

When the first honours examination in the university came into being in 1748, it was primarily mathematical. It was called the *Tripos*, after the three-legged stool used at disputations, and the candidates placed in the first class were known as *Wranglers*, from the style of argument used. Those in the second and third classes were known as *Senior* and *Junior Optimes*, respectively, and these terms are still used today. A detailed order of merit was published, with the top man being given the title of *Senior Wrangler*. This was an honour of some distinction, not only in Cambridge but throughout the country at large, and several well-known anecdotes confirm the interest that was taken in it.

In 1882 the examination was divided into two parts, and later a third part was added. Women were admitted as students to the university only after the first women's college, Girton, was founded in 1869, and women's names first appeared in the Tripos lists in 1882. Women's names were listed separately and not numbered among the men in the order of merit, with their positions being indicated by notes, such as "between 58 and 59." Thus, in 1890 male pride suffered a blow when Phillipa Fawcett was listed as "above the Senior Wrangler."

Other triposes were established in due course, beginning with the Classical Tripos in 1824. By the end of the nineteenth century, Part I of the Mathematical Tripos was the only examination in which an order of merit within each class remained. This was finally abolished after the examination in 1909, largely through the efforts of G.H. Hardy. The chief reason for this was that teaching of mathematics had passed into the hands of coaches, who drilled their pupils in the skills of answering examination questions as quickly as possible to amass the maximum number of marks. Many of the questions had become incredibly tricky and tortuous, with no attention being paid to the meaning or development of the subject. A reference to this is made in the biography of Hardy following Ramanujan's letter to him dated 16 January 1913.

S. Ramanujan to Madras Port Trust Office   **9 February 1912**

Triplicane

From

    S. Ramanujan
       7, Summer House
          Triplicane

To

    The Chief Accountant
       Port Trust
          Madras

Sir,

    I understand there is a clerkship vacant in your office, and I beg to apply for the same. I have passed the Matriculation Examination and studied up

Chapter 1. Ramanujan in Madras

> Triplicane
> 9th February 1912
>
> From
>   S. Ramanujan
>   7, Summer House
>   Triplicane.
>
> To
>   The Chief Accountant
>   Port Trust
>   Madras
>
> Sir,
>   I understand there is a clerkship vacant in your office, and I beg to apply for the same. I have passed the Matriculation Examination and studied up to the F.A. but was prevented from pursuing my studies further owing to several untoward circumstances. I have, however, been devoting all my time to Mathematics and developing the subject. I can say I am quite confident I can do justice to my work if I am appointed to the post. I therefore beg to request that you will be good enough to confer the appointment on me.
>
>   I beg to remain,
>   Sir,
>   Your most obedient Servant
>   S. Ramanujan

RAMANUJAN'S LETTER TO PORT TRUST

to the F.A. but was prevented from pursuing my studies further owing to several untoward circumstances. I have, however, been devoting all my time to Mathematics and developing the subject. I can say I am quite confident I can do justice to my work if I am appointed to the post. I therefore beg to request that you will be good enough to confer the appointment on me.

      I beg to remain,
         Sir,
         Your most obedient Servant
           S. Ramanujan

THE OLD PORT TRUST OFFICE AT MADRAS

Commentary. Ramanujan's residence at 7, Summer House in Triplicane, to which he had moved in May, 1911, was located about three miles south of the Madras Port Trust. Later, after working for the Port Trust for a few months, he moved to his grandmother's house on Saiva Muthiah Mudali Street, which was much closer to his place of work. At this time, Ramanujan's mother, Komalatammal, and wife, Janaki, came to live with him.

On page 350 in his first notebook [**225**], Ramanujan wrote

    Sanitary Engineer Office
    Accountant General Office
    Postal Audit Office
    (Comptroller Post Office)

(Ramanujan may have written one or two other words beside this list, but the writing is too faint to determine this.) Evidently, when Ramanujan began to consider employment, he was advised by friends to inquire at these offices. In fact, prior to taking a position at the Madras Port Trust Office, Ramanujan had a temporary position for six weeks, probably in 1911, in the Madras Accountant General's Office at a salary of 20 rupees per month.

The initials F.A. stand for First Arts. Ramanujan took the F.A. Examination unsuccessfully in 1906, and so failed to obtain the necessary qualification for entry to the University of Madras.

CHAPTER 1. RAMANUJAN IN MADRAS

Two memos from Madras Port Trust Office

16. S. Ramanujan <u>a new man</u>, a matriculate, to be appointed to the Accounts Section in III class 4th grade.

About the man to be newly entertained No. 7., an F.A., T.A. says, possesses a good knowledge of English. <u>No. 16 is reported by M.G.R. to be a mathematical genius.</u> Mr. Middlemast says of him to be "exceptionally intelligent in Mathematics!"

Commentary. Evidently, the Madras Port Trust Office assigned numbers to their employees. Thus, Ramanujan was given the number 16, and another new employee the number 7.

Entertained is a curious word, and one might think it should be *Entered,* but the same usage occurs in another memo from the Madras Port Trust in late 1919 and in S. Narayana Aiyar's letter of 10 March 1921.

The identities of T.A. and M.G.R. are not known.

The Chief Accountant to S. Ramanujan

From
        The CHIEF ACCOUNTANT
        Madras Port Trust

To
        M.R.Ry. S. Ramanujan
        8, Summer House
        Triplicane

Sir

With reference to his application dated 9th Feby. 1912, S. Ramanujan is informed that he has been appointed by the Chairman as a Clerk in this office on 25 February, 1912. He is to report himself for duty on 1st March, 1912.

                              Initials (illegible)
                              O/C (Officer in Charge)
                              For Chief Accountant

Commentary. Some of the words in this note cannot be read with certainty. There is an illegible handwritten comment in the left-hand margin, and the writer has not given Ramanujan's correct address. The courtesy title M.R.Ry. is an abbreviation for MahaRaja Rajashry, which may be approximately translated 'respected (shry) king (Raja) of great kings (Maharaja)'. They also occur before S. Narayana Aiyar's name in the postcard addressed to him by Arthur Davies, dated 7 June 1913. It is interesting that Ramanujan was to begin his duties on a Saturday.

An office memo from the Madras Port Trust indicates that Ramanujan worked as a "bill clerk." For a brief time, he also served as a "pilotage clerk."

SIR FRANCIS SPRING, THE CHAIRMAN OF PORT TRUST MADRAS

C.L.T. Griffith to Sir Francis Spring 12 November 1912

College of Engineering
Madras

Dear Sir Francis,

You have in your office as an accountant on Rs. 25. a young man named S. Ramanujan who is a most remarkable mathematician. He may be a very poor accountant, but I hope you will see that he is kept happily employed until something can be done to make use of his extraordinary gifts. I am writing to one of the leading mathematical professors at home about him and sending copies of some of Ramanujan's papers and results. Our Math. professor here says that very few people could follow or criticize the work. It is of course far beyond my scope, but I happen to know who is at work in the same line at home, and I hope to get instructions as to what this fellow ought to do.

If there is any real genius in him he will have to be provided with money for books and with leisure, but until I hear from home, I don't feel sure that it is worth while spending much time or money on him.

Yours sincerely,
C.L.T. Griffith

Commentary. Sir Francis Joseph Edward Spring (1849–1933), an Irishman and a graduate of Trinity College, Dublin, entered the Indian Civil Service in 1870 as a railway engineer. For having developed the South Indian Railways System, he was named Knight Commander of the Order of the Indian Empire in 1911. Sir Francis Spring retired from the service in 1904, but remained in Madras and became Chairman of the Madras Port Trust. He was one of the first people in South India to own an automobile.

Charles Leopold Troyte Griffith (born 6 June 1872) was Professor of Civil Engineering at Madras College of Engineering. He was an Associate Member of the Institute of Civil Engineers and had received his scientific training at University College, London, between 1890 and 1892.

The mathematician "at home" to whom Griffith intends to write is M.J.M. Hill, Professor of Mathematics at University College, London.

Alfred Bourne to Sir Francis Spring   **14 November 1912**

THE DIRECTOR OF PUBLIC INSTRUCTION, MADRAS

The Old College
Nungambakam

My dear Spring,

I never heard of Ramanujan—Littlehailes is now away on the west coast—Middlemast is at the Presidency College and Graham at the Fort. I should certainly send him to see one or both of these.—I don't like the suggestion that access to a library would ruin any genius; it savours of the middle ages and if his genius is so elusive or mysterious that good mathematicians possessed besides of much common sense cannot recognise and appreciate it even if it carries them beyond their scope, I should doubt its existence.

Yours sincerely,
A.G. Bourne

Commentary. Sir Alfred Gibbs Bourne (1859–1940) was a graduate of University College, London, and was appointed Professor of Biology at the

Presidency College, Madras, in 1886. He served as Registrar of the University of Madras from 1891 to 1899 and became Director of Public Instruction in Madras in 1903. He was, for a time, a member of the Legislative Council of Madras and retired in 1914. He was a Fellow of the Royal Society and was knighted in January 1913.

Richard Littlehailes (1878–1950), with a B.A. from Balliol College, Oxford, entered the Indian Educational Service in 1903. He held various posts, including Professor of Mathematics at Presidency College, Madras, Inspector of Schools, meteorologist and Deputy Director of the Madras observatory, and Director of Public Instruction, Madras. He was Vice-Chancellor of the University of Madras from 1934 to 1937.

The Old College is an alternative name for Pachaiyappa's College that is infrequently used today. Ramanujan entered Pachaiyappa's College in 1906 but again did not pass the required examinations. The Old College is located in Nungambakam near the center of Madras.

J.F. Graham to C.L.T. Griffith 27 November 1912

ACCOUNTANT GENERAL, MADRAS

Fort St. George

Dear Griffith,

Ramanujan came to see me today. He seems to have done a great deal of work in one particular branch of calculus and from the way he has done it he must have considerable mathematical aptitude. He has read no mathematics at all except calculus apparently, and it is possible his brains are akin to those of the calculating boy!

I should say however that if he had had the training, mathematics of a certain kind—algebra, differential equations, hydrodynamics and calculus—i.e. pure theory and development of algebraic functions—juggling with symbols it has been called—would have become very easy to him.

Whether he has the stuff of great mathematicians or not I do not know. He gave me the impression of having brains. His original work is an interesting development of work already done—but interesting only to the purist. However I am not the best qualified to judge.

Middlemast's opinion would be of value.

Yours sincerely,
J.F. Graham

Commentary. The writer of the letter was John Frederick Graham (c. 1879–1916), and not W. Graham, as given in [**263**] and other sources about Ramanujan's life. As a boy, Graham attended Rutland College. This college no longer exists, but was located at 1 Rutland Square East, Dublin. The name Rutland Square has now become Parnell Square, and the school's records are no longer extant. He then proceeded to Trinity College, Dublin, where he graduated B.A. in 1901 in Mathematics and Mathematical Physics, and reached the high grade of Senior Moderator; he received his M.A. in 1905. He entered the Indian Civil Service in 1901, and in 1902 he assumed his first appointment as Assistant Collector and Magistrate. In 1907 he transferred to the post of Assistant Accountant–General, and was soon promoted to Deputy Accountant–General and later to Officiating Accountant–General. He served in World War I, rising to the rank of Lieutenant-Colonel in the Madras Volunteers; he was killed in action in France on 1 July 1916.

Fort St. George, constructed in 1642 by the British East India Company, was the administrative center of the British government in the Madras Presidency. It is currently the seat of government for the state of Tamil Nadu. Located on the coast of the Bay of Bengal, it is not far from the University of Madras.

C.L.T. Griffith to Sir Francis Spring   28 November 1912

College of Engineering

Dear Sir Francis,

I enclose the letter re Ramanujan, which Graham has been kind enough to send me. He says the man has brains, but may be of the calculating boy type: but like the other mathematicians here modestly disclaims the power of criticizing Ramanujan's work.

Ramanujan seems to have seen Mr Middlemast already and he, also, appears to be unable to criticize.

I think I was right in writing to Prof. Hill in London, and we must wait for his opinion which I hope to get during January.

Yours sincerely,
C.L.T. Griffith

M.J.M. Hill to C.L.T. Griffith   3 December 1912

UNIVERSITY OF LONDON, UNIVERSITY COLLEGE

Gower Street, London, W. C.

Dear Professor Griffith,

Your letter of the 12th November reached me a day or two ago, and I write now in order not to miss the mail.

I am sorry that the twenty years, which have passed since you were with me, prevent me from remembering anything about you but your name.

As soon as I can get more time I will look into Mr Ramanujan's paper about the Bernoulli's Numbers, but I cannot do this during term time.

One thing however is clear.

Mr Ramanujan has fallen into the pitfalls of the very difficult subject of Divergent Series.

Otherwise he could not have got the erroneous results you send me

$$1 + 2 + 3 + \cdots + \infty = -\frac{1}{12}$$
$$1^2 + 2^2 + 3^2 + \cdots + \infty^2 = 0$$
$$1^3 + 2^3 + 3^3 + \cdots + \infty^3 = \frac{1}{240}.$$

All 3 series have infinity for their sum.

The book which will be most useful to him is

Bromwich's Theory of Infinite Series, published by the Cambridge University Press (or Macmillan).

Next as to the publication of papers, if Mr Ramanujan will write out his ideas carefully and clearly on some one subject, and send them to the Secretary of the London Mathematical Society, 22 Albemarle St., London W., his paper will be referred to the Mathematician, who is an expert in the particular department of Mathematics, with which his paper deals. If it is found to be new and worthy of publication, it will be printed in the Proceedings of the Society at the expense of the Society, and 25 free copies will be sent to the author.

But he should be very careful with his Manuscript. It should be very clearly written, and should be free from errors; and he should not use symbols which he does not explain (e.g. in his printed paper he should have explained on page 1 what the symbols $c_1, c_2, c_3, \ldots$ mean; and §12 to which he refers on p. 5 does not apparently exist). He passes from §11 to §13.

What you say about him personally is very interesting, and I hope something may come of his work.

I will write later when I have had more time to look into his paper.

I remain,
With kind regards,
Yours sincerely,
M.J.M. Hill

Commentary. Micaiah John Muller Hill (1856–1929) was born at Berhampore, Bengal (narrowly escaping death during the Indian mutiny) and was Professor of Mathematics at University College, London, from 1884 until 1926. He had a reputation as a sympathetic and painstaking teacher [79]. He had been at Peterhouse in Cambridge as an undergraduate, and was classed as 4th Wrangler (equal) in the Mathematical Tripos in 1879. Hill did research on hydrodynamics and differential equations, but at the time of Griffith's letter was primarily interested in Euclid's axioms, a subject far removed from Ramanujan's interests. In 1907 and 1908, Hill published two papers [133], [134] giving an asymptotic formula for partial sums of certain hypergeometric series. In fact, Ramanujan had recorded in his notebooks [225] more general and considerably stronger results, which were probably discovered about the same time, but, at any rate, long before Griffith wrote Hill. It is unfortunate that no one at that time knew of this mutual interest of Hill and Ramanujan. See Berndt's book [35, pp. 33-34] for a description of these results.

As Hill's letter intimates, Griffith decided to approach Hill about Ramanujan's work because Hill was Griffith's mathematics professor while Griffith was a student at University College, London. It is uncertain how much mathematics was communicated to Hill. A copy of Ramanujan's first paper [205], [226, pp. 1-14], published in 1911 in the *Journal of the Indian Mathematical Society,* was sent. From Hill's remarks, it seems that a considerable list of unpublished results was also enclosed.

Evidently, Hill shared Abel's view that "divergent series are in general deadly," for, not surprisingly, he failed to discern the origin of the three results quoted in his letter. The three claims give the values of $\zeta(-n)$, for $n = 1, 2, 3$, respectively. It was natural for Ramanujan to write these facts in terms of the divergent series of the Riemann zeta-function at these values. In Ramanujan's theory of the 'constant' of a (convergent or divergent) series, the three values, $-\frac{1}{12}$, 0, and $\frac{1}{240}$ are, respectively, the constants for the three given series. It is likely that this theory was not communicated to Hill. Ramanujan's theory of divergent series is elucidated, albeit rather imprecisely, in Chapter 6 of his second notebook [225], and readers should consult Berndt's book [35] for an account of this theory.

The publisher of Bromwich's book [54] is Macmillan.

Hill's criticism about the failure to identify the coefficients c_n is justified. In printing Ramanujan's paper [205], the section heading §12 was omitted. In his *Collected Papers* [226], the oversight is corrected. In fact, Ramanujan's paper does contain some errors. However, the vast majority of the results quoted in the paper are correct, although proofs were not supplied for many of the claims. S.S. Wagstaff [272] has thoroughly and critically examined Ramanujan's paper on Bernoulli numbers and has provided proofs of those theorems not proved by Ramanujan. Some of Ramanujan's discoveries about Bernoulli numbers are recorded in his second notebook; for a discussion of this work, consult Berndt's account [32].

M.J.M. Hill to C.L.T. Griffith   **7 December 1912**

University of London, University College
                                   Gower Street, London, W. C.

Dear Mr. Griffith,

I have been going through Mr Ramanujan's paper today.

If you will look at formulae (14), (15), and (16) on pp. 5-6 you will see that he refers in the footnote for their proofs to §12 below.

He has omitted to mark §12, but I think it begins on p. 9 near the foot with the words

"Results (14) to (17), (20) and (21) can be got from the following result."

This result he numbers (28). [It is not proved anywhere in the paper, and therefore a reference should have been given to the place from which it has been taken.]

It does not however give him what he wants, for he says at the top of p. 10, If we can prove the L.H.S. of (28) can be expanded in ascending powers of $\frac{1}{x}$ with integral coefficients, then (20) and (21) are at once deduced as follows:—

But he gives no proof that he can effect the desired expansion. The same kind of thing occurs with regard to formula (16) in the first three lines of p. 11 and on the last four lines of the section which ends just above the commencement of §13 on p. 11, "from the above theorem ... of (16) is at once got."

He has in fact observed certain properties of the earlier Bernoulli numbers and assumed them to be true of them all without proof.

For these reasons I feel sure that the London Mathematical Society would not have accepted the paper for their Proceedings.

Mr. Ramanujan is evidently a man with a taste for Mathematics, and with some ability, but he has got on to wrong lines. He does not understand the precautions which have to be taken in dealing with divergent series, otherwise he could not have obtained the erroneous results you send me, viz:—

$$1 + 2 + 3 + \cdots + \infty = -\frac{1}{12}$$
$$1^2 + 2^2 + 3^2 + \cdots + \infty^2 = 0$$
$$1^3 + 2^3 + 3^3 + \cdots + \infty^3 = \frac{1}{240}.$$

The sums of n terms of these series are

$$\tfrac{1}{2}n(n+1), \quad \tfrac{1}{3}n(n+\tfrac{1}{2})(n+1), \quad [\tfrac{1}{2}n(n+1)]^2$$

and they all tend to ∞ as n tends to ∞.

I do not think you can do better for him than get him a copy of the book I recommended, Bromwich's Theory of Infinite Series, published by Macmillan and Co., who have branches in Calcutta and Bombay. Price 15/- net.

Chapter XI of this book, on Non-Convergent and Asymptotic Series, will show him the difficulties to be studied in handling divergent series; and in other parts of the book he will learn how to handle integrals with infinite limits in such a manner that they do not lead to paradoxical results.

When I was a student in Cambridge 1876-9, these things were not properly understood, and the modern theory has only recently been established on a firm basis.

Many illustrious mathematicians of earlier days stumbled over these difficulties, so it is not surprising that Mr. Ramanujan, working by himself, has obtained erroneous results. I hope he will not be discouraged.

If this book could be obtained for him, and he would work at it <u>from the beginning</u>, it would be much better for him than for those who are interested in him to spend money on printing his papers at the present time.

At the same time if he will write out what he has written on the exact values of the Elliptic Functions, to which you refer in your letter, and which he mentioned when you showed him the British Association Report for 1911, I will go into the matter.

And I hope that after Mr. Ramanujan has read Mr. Bromwich's book he will be able to write something which will stand the test of modern criticism.

With kind regards, I remain,

Yours sincerely, M.J.M. Hill

Commentary. In this second letter, Hill repeats the comments made in his previous letter in slightly more detail. Clearly, he meant to be encouraging, but it is doubtful that his remarks would have had such an effect on Ramanujan. Hill's conjecture about the beginning of §12 is correct.

C.L.T. Griffith to Sir Francis Spring   **5 January 1913**

Camp. Pallaverain

Dear Sir Francis,

I enclose Prof. Hill's last letter re Ramanujan. Will you read it and hand it on to him. I have written to Mr. Ramachandra Rao, Collector of Nellore, and asked him to buy the book mentioned, and give it to Ramanujan.

The fact that Prof. Hill is prepared to consider Ramanujan's Elliptic Functions, even though the proofs may not be of that logical completeness that is demanded now-a-days shows that it is possible that the "intuitive results" may be of interest. All the same Ramanujan ought to do his utmost to make his proofs complete.

Yours sincerely,
C.L.T. Griffith

Commentary. Although in several areas of mathematics, e.g., analytic number theory, many of Ramanujan's proofs might not have been rigorous, it is likely that almost all of his proofs in the theory of elliptic functions were indeed rigorous. Since Ramanujan had undoubtedly written to G.H. Hardy before Hill's second letter reached him via C.L.T. Griffith and Sir Francis Spring, it is doubtful that Ramanujan ever communicated to Hill any of his results in the theory of elliptic functions. Ramanujan published very little on elliptic functions, despite the fact that his notebooks contain an enormous number of original results. See Berndt's book [37] for an account of most of this remarkable work.

Camp Pallavaram, as it is currently known, is located at the south edge of Madras, south of the airport.

Chapter 2
Ramanujan's First Two Letters to Hardy and Hardy's Response

S. Ramanujan to G.H. Hardy					16 January 1913

Madras

Dear Sir,

I beg to introduce myself to you as a clerk in the Accounts Department of the Port Trust Office at Madras on a salary of only £20 per annum. I am now about 23 years of age. I have had no University education but I have undergone the ordinary school course. After leaving school I have been employing the spare time at my disposal to work at Mathematics. I have not trodden through the conventional regular course which is followed in a University course, but I am striking out a new path for myself. I have made a special investigation of divergent series in general and the results I get are termed by the local mathematicians as "startling".

Just as in elementary mathematics you give a meaning to a^n when n is negative and fractional to conform to the law which holds when n is a positive integer, similarly the whole of my investigations proceed on giving a meaning to Eulerian Second Integral for all values of n. My friends who have gone through the regular course of University education tell me that $\int_0^\infty x^{n-1} e^{-x} dx = \Gamma(n)$ is true only when n is positive. They say that this integral relation is not true when n is negative. Supposing this is true only for positive values of n and also supposing the definition $n\Gamma(n) = \Gamma(n+1)$ to be universally true, I have given meanings to these integrals and under the conditions I state the integral is true for all values of n negative and fractional. My whole investigations are based upon this and I have been developing this to a remarkable extent so much so that the local mathematicians are not able to understand me in my higher flights.

Very recently I came across a tract published by you styled *Orders of Infinity* in page 36 of which I find a statement that no definite expression has been as yet found for the number of prime numbers less than any given number. I have found an expression which very nearly approximates to the real result, the error being negligible. I would request you to go through the enclosed papers. Being poor, if you are convinced that there is anything of value I would like to have my theorems published. I have not given the actual investigations nor the expressions that I get but I have indicated the lines on which I proceed. Being inexperienced I would very highly value any advice you give me. Requesting to be excused for the trouble I give you.

I remain, Dear Sir, Yours truly,

S. Ramanujan

P.S. My address is S. Ramanujan, Clerk Accounts Department, Port Trust, Madras, India.

(1)

In page 36 it is stated that "the number of prime numbers less than

$$x = \int_2^x \frac{dt}{\log t} + \rho(x)$$

where the precise order of $\rho(x)$ has not been determined."

The precise order itself is not sufficient to find the value of $\rho(x)$. Even if it is known that $\frac{\rho(x)}{\phi(x)} = 1$ when x becomes infinite, $\phi(x)$ being a known function of x, $\rho(x)$ cannot be supposed to have been found with sufficient accuracy; for example

$$\frac{\left(x + \frac{x}{\log \log x}\right)}{x} = 1$$

when x becomes infinite, yet the difference between $x + \frac{x}{\log \log x}$ and x is very great.

From the forms of $\rho(x)$ given in page 53, viz.

$$O\left\{\frac{x}{\log x}\Delta\right\}, \quad O\left(xe^{-a\sqrt{\log x}}\right), \quad O(\sqrt{x}),$$

etc. it appears that from particular numerical values the forms have been guessed.

Even in regular functions it is difficult to have an idea of the form from the numerical values. In such a complicated function as $\rho(x)$ it is difficult to have an idea even for large values of $\log x$; for example even if we give billion for x, $\rho(x)$ is very difficult to be found.

I have observed that $\rho(e^{2\pi x})$ is of such a nature that its value is very small when x lies between 0 and 3 (its value is less than a few hundreds when $x = 3$) and rapidly increases when x is greater than 3.

I. I have found a function which exactly represents the number of prime numbers less than x, 'exactly' in the sense that the difference between the function and the actual number of primes is generally 0 or some small finite value even when x becomes infinite.

(2)

I have got the function in the form of infinite series and have expressed it in two ways.

(1) In terms of Bernoullian numbers. From this we can easily calculate the number of prime numbers up to 100 millions, with generally no error and in some cases with an error of 1 or 2.

(2) As a definite integral from which we can calculate for all values.

II. I have also got expressions to find the actual number of prime numbers of the form $An + B$, which are less than any given number however large.

The difference between the number of prime numbers of the form $4n - 1$ and which are less than x and those of the form $4n + 1$ less than x is infinite when x becomes infinite.

Theorems connected with the calculation of the difference when x is a given number and similar calculations have also been got.

III. I have found out expressions for finding not only irregularly increasing functions but also irregular functions without increase (e.g. the number of divisors of natural numbers), not merely the order but the exact form. The following are a few examples from my theorems:

(1) The numbers of the form $2^p 3^q$ less than n

$$= \frac{1}{2} \frac{\log(2n) \log(3n)}{\log 2 \log 3}$$

where p and q may have any positive integral value including 0.

(3)

(2) Let us take all numbers containing an odd number of dissimilar prime divisors, viz.

2, 3, 5, 7, 11, 13, 17, 19, 23, 29, 30, 31, 37, 41, 42, 43, 47, ...

(a)   The number of such numbers less than $n = \dfrac{3n}{\pi^2}$.

(b) $\dfrac{1}{2^2} + \dfrac{1}{3^2} + \dfrac{1}{5^2} + \dfrac{1}{7^2} + \cdots + \dfrac{1}{30^2} + \dfrac{1}{31^2} + \cdots = \dfrac{9}{2\pi^2}$.

(c) $\dfrac{1}{2^4} + \dfrac{1}{3^4} + \dfrac{1}{5^4} + \dfrac{1}{7^4} + \cdots = \dfrac{15}{2\pi^4}$.

(3) Let us take the number of divisors of natural numbers, viz. 1, 2, 2, 3, 2, 4, 2, 4, 3, 4, 2, ... (1 having 1 divisor, 2 having 2, 3 having 2, 4 having 3, 5 having 2, ...). The sum of such numbers to n terms

$$= n(2\gamma - 1 + \log n) + \frac{1}{2} \text{ of the number of divisors of } n$$

where $\gamma = .5772156649\ldots$, the Eulerian Constant.

(4) 1, 2, 4, 5, 8, 9, 10, 13, 16, 17, 18, ... are numbers which are either themselves squares or which can be expressed as the sum of two squares.

The number of such numbers greater than A and less than B

$$= K \int_A^B \frac{dx}{\sqrt{\log x}} + \theta(x)$$

where $K = .764\ldots$ and $\theta(x)$ is very small when compared with the previous integral. K and $\theta(x)$ have been exactly found though complicated.

IV. Theorems on integrals. The following are a few examples.

(1)
$$\int_0^\infty \frac{1 + \left(\frac{x}{b+1}\right)^2}{1 + \left(\frac{x}{a}\right)^2} \frac{1 + \left(\frac{x}{b+2}\right)^2}{1 + \left(\frac{x}{a+1}\right)^2} \frac{1 + \left(\frac{x}{b+3}\right)^2}{1 + \left(\frac{x}{a+2}\right)^2} \cdots dx$$
$$= \frac{\sqrt{\pi}}{2} \frac{\Gamma(a + \frac{1}{2})}{\Gamma(a)} \frac{\Gamma(b+1)}{\Gamma(b + \frac{1}{2})} \frac{\Gamma(b - a + \frac{1}{2})}{\Gamma(b - a + 1)}.$$

(4)

(2)
$$\int_0^\infty \frac{1}{\left\{1 + \left(\frac{x}{a}\right)^2\right\}\left\{1 + \left(\frac{x}{a+1}\right)^2\right\}\left\{1 + \left(\frac{x}{a+2}\right)^2\right\}\cdots} \frac{dx}{\left\{1 + \left(\frac{x}{b}\right)^2\right\}\left\{1 + \left(\frac{x}{b+1}\right)^2\right\}\cdots}$$
$$= \frac{\sqrt{\pi}}{2} \frac{\Gamma(a + \frac{1}{2})}{\Gamma(a)} \frac{\Gamma(b + \frac{1}{2})}{\Gamma(b)} \frac{\Gamma(a + b)}{\Gamma(a + b + \frac{1}{2})}.$$

(3) If
$$\int_0^\infty \frac{\cos nx}{e^{2\pi\sqrt{x}} - 1} dx = \phi(n),$$

then
$$\int_0^\infty \frac{\sin nx}{e^{2\pi\sqrt{x}} - 1} dx = \phi(n) - \frac{1}{2n} + \phi\left(\frac{\pi^2}{n}\right)\sqrt{\frac{2\pi^3}{n^3}}.$$

$\phi(n)$ is a complicated function. The following are certain special values.

$$\phi(0) = \frac{1}{12}; \quad \phi\left(\frac{\pi}{2}\right) = \frac{1}{4\pi}; \quad \phi(\pi) = \frac{2 - \sqrt{2}}{8}; \quad \phi(2\pi) = \frac{1}{16};$$

$$\phi\left(\frac{2\pi}{5}\right) = \frac{8 - 3\sqrt{5}}{16}; \quad \phi\left(\frac{\pi}{5}\right) = \frac{6 + \sqrt{5}}{4} - \frac{5\sqrt{10}}{8}; \quad \phi(\infty) = 0;$$

$$\phi\left(\frac{2\pi}{3}\right) = \frac{1}{3} - \sqrt{3}\left(\frac{3}{16} - \frac{1}{8\pi}\right).$$

(4)
$$\int_0^\infty \frac{dx}{(1+x^2)(1+r^2x^2)(1+r^4x^2)(1+r^6x^2)\cdots} = \frac{\pi}{2(1+r+r^3+r^6+r^{10}+\cdots)}$$
where 1, 3, 6, 10, ... are sums of natural numbers.

(5)
$$\int_0^\infty \frac{\sin 2nx}{x(\cosh \pi x + \cos \pi x)} dx$$
$$= \frac{\pi}{4} - 2\left(\frac{e^{-n}\cos n}{\cosh \frac{\pi}{2}} - \frac{e^{-3n}\cos 3n}{3\cosh \frac{3\pi}{2}} + \frac{e^{-5n}\cos 5n}{5\cosh \frac{5\pi}{2}} - \cdots\right).$$

(6)
$$\int_0^\infty \tan^{-1}\frac{2nz}{n^2+x^2-z^2}\frac{dz}{e^{2\pi z}-1}$$
can be exactly found if $2n$ is any integer and x any quantity.

(5)

V. Theorems on summation of series; e.g.

(1)
$$\frac{1}{1^3}\cdot\frac{1}{2^1} + \frac{1}{2^3}\cdot\frac{1}{2^2} + \frac{1}{3^3}\cdot\frac{1}{2^3} + \frac{1}{4^3}\cdot\frac{1}{2^4} + \cdots$$
$$= \frac{1}{6}(\log 2)^3 - \frac{\pi^2}{12}\log 2 + \left(\frac{1}{1^3} + \frac{1}{3^3} + \frac{1}{5^3} + \cdots\right).$$

(2) $\quad 1 + 9\cdot\left(\frac{1}{4}\right)^4 + 17\cdot\left(\frac{1\cdot 5}{4\cdot 8}\right)^4 + 25\cdot\left(\frac{1\cdot 5\cdot 9}{4\cdot 8\cdot 12}\right)^4 + \cdots = \dfrac{2\sqrt{2}}{\sqrt{\pi}\{\Gamma(\frac{3}{4})\}^2}.$

(3) $\quad 1 - 5\cdot\left(\frac{1}{2}\right)^3 + 9\cdot\left(\frac{1\cdot 3}{2\cdot 4}\right)^3 - \cdots = \dfrac{2}{\pi}.$

(4) $\quad \dfrac{1^{13}}{e^{2\pi}-1} + \dfrac{2^{13}}{e^{4\pi}-1} + \dfrac{3^{13}}{e^{6\pi}-1} + \cdots = \dfrac{1}{24}.$

(5) $\quad \dfrac{\coth \pi}{1^7} + \dfrac{\coth 2\pi}{2^7} + \dfrac{\coth 3\pi}{3^7} + \cdots = \dfrac{19\pi^7}{56700}.$

(6) $\quad \dfrac{1}{1^5 \cosh \frac{\pi}{2}} - \dfrac{1}{3^5 \cosh \frac{3\pi}{2}} + \dfrac{1}{5^5 \cosh \frac{5\pi}{2}} - \cdots = \dfrac{\pi^5}{768}.$

(7)
$$\frac{1}{(1^2+2^2)(\sinh 3\pi - \sinh \pi)} + \frac{1}{(2^2+3^2)(\sinh 5\pi - \sinh \pi)}$$
$$+ \frac{1}{(3^2+4^2)(\sinh 7\pi - \sinh \pi)} + \cdots = \frac{1}{2\sinh \pi}\left(\frac{1}{\pi} + \coth \pi - \frac{\pi}{2}\tanh^2 \frac{\pi}{2}\right).$$

(8) $$\frac{1}{\left(25+\frac{1^4}{100}\right)(e^\pi+1)} + \frac{3}{\left(25+\frac{3^4}{100}\right)(e^{3\pi}+1)} + \frac{5}{\left(25+\frac{5^4}{100}\right)(e^{5\pi}+1)} + \cdots$$
$$= \frac{\pi}{8}\coth^2\frac{5\pi}{2} - \frac{4689}{11890}.$$

(9) $$\frac{1}{1^7\cosh\frac{1}{2}\pi\sqrt{3}} - \frac{1}{3^7\cosh\frac{3\pi}{2}\sqrt{3}} + \cdots = \frac{\pi^7}{23040}.$$

(10) $$\left\{1+\left(\frac{n}{1}\right)^3\right\}\left\{1+\left(\frac{n}{2}\right)^3\right\}\left\{1+\left(\frac{n}{3}\right)^3\right\}\cdots$$

can always be exactly found if n is any integer positive or negative.

(11) $$\frac{2}{3}\int_0^1 \frac{\tan^{-1}x}{x}dx - \int_0^{2-\sqrt{3}}\frac{\tan^{-1}x}{x}dx = \frac{\pi}{12}\log(2+\sqrt{3}).$$

(6)

VI. Theorems on transformation of series and integrals, e.g.

(1) $$\pi\left(\frac{1}{2} - \frac{1}{\sqrt{1}+\sqrt{3}} + \frac{1}{\sqrt{3}+\sqrt{5}} - \frac{1}{\sqrt{5}+\sqrt{7}} + \cdots\right)$$
$$= \frac{1}{1\sqrt{1}} - \frac{1}{3\sqrt{3}} + \frac{1}{5\sqrt{5}} - \cdots.$$

(2) $$\frac{\log 1}{\sqrt{1}} - \frac{\log 3}{\sqrt{3}} + \frac{\log 5}{\sqrt{5}} - \frac{\log 7}{\sqrt{7}} + \cdots$$
$$= \left(\frac{1}{4}\pi - \frac{1}{2}\gamma - \frac{1}{2}\log 2\pi\right)\left(\frac{1}{\sqrt{1}} - \frac{1}{\sqrt{3}} + \frac{1}{\sqrt{5}} - \cdots\right),$$

where $\gamma = .5772\ldots$, the Eulerian constant.

(3) $$1 - \frac{x^2 3!}{(1!2!)^3} + \frac{x^4 6!}{(2!4!)^3} - \frac{x^6 9!}{(3!6!)^3} + \cdots$$
$$= \left\{1 + \frac{x}{(1!)^3} + \frac{x^2}{(2!)^3} + \cdots\right\}\left\{1 - \frac{x}{(1!)^3} + \frac{x^2}{(2!)^3} - \cdots\right\}.$$

(4) If
$$\int_0^a \phi(p,x)\cos nx\, dx = \psi(p,n),$$

then
$$\frac{\pi}{2}\int_0^a \phi(p,x)\phi(q,nx)\, dx = \int_0^\infty \psi(q,x)\psi(p,nx)\, dx.$$

(5) If $\alpha\beta = \pi$, then

$$\sqrt{\alpha}\int_0^\infty \frac{e^{-x^2}}{\cosh \alpha x}dx = \sqrt{\beta}\int_0^\infty \frac{e^{-x^2}}{\cosh \beta x}dx.$$

(6) If $\alpha\beta = \pi^2$, then

$$\frac{1}{\sqrt[4]{\alpha}}\left\{1+4\alpha\int_0^\infty \frac{xe^{-\alpha x^2}}{e^{2\pi x}-1}dx\right\} = \frac{1}{\sqrt[4]{\beta}}\left\{1+4\beta\int_0^\infty \frac{xe^{-\beta x^2}}{e^{2\pi x}-1}dx\right\}.$$

(7)
$$n\left(e^{-n^2} - \frac{e^{-\frac{1}{3}n^2}}{3\sqrt{3}} + \frac{e^{-\frac{1}{5}n^2}}{5\sqrt{5}} - \cdots\right)$$
$$= \sqrt{\pi}(e^{-n\sqrt{\pi}}\sin n\sqrt{\pi} - e^{-n\sqrt{3\pi}}\sin n\sqrt{3\pi} + \cdots).$$

(8) If n is any positive integer excluding 0,

$$\frac{1^{4n}}{(e^\pi - e^{-\pi})^2} + \frac{2^{4n}}{(e^{2\pi} - e^{-2\pi})^2} + \frac{3^{4n}}{(e^{3\pi} - e^{-3\pi})^2} + \cdots$$
$$= \frac{n}{\pi}\left\{\frac{B_{4n}}{8n} + \frac{1^{4n-1}}{e^{2\pi}-1} + \frac{2^{4n-1}}{e^{4\pi}-1} + \frac{3^{4n-1}}{e^{6\pi}-1} + \cdots\right\}$$

where $B_2 = \frac{1}{6}$, $B_4 = \frac{1}{30}$, ...

(7)

VII. Theorems on approximate integration and summation of series.

(1) $\quad 1^2\log 1 + 2^2\log 2 + 3^2\log 3 + \cdots + x^2\log x$
$$= \frac{1}{6}x(x+1)(2x+1)\log x - \frac{1}{9}x^3 + \frac{1}{4\pi^2}\left(\frac{1}{1^3} + \frac{1}{2^3} + \frac{1}{3^3} + \cdots\right)$$
$$+ \frac{x}{12} - \frac{1}{360x} + \cdots.$$

(2) $\quad 1 + \frac{x}{1!} + \frac{x^2}{2!} + \frac{x^3}{3!} + \cdots + \frac{x^x}{x!}\theta = \frac{e^x}{2}$

where $\theta = \frac{1}{3} + \frac{4}{135(x+k)}$ where k lies between $\frac{8}{45}$ and $\frac{2}{21}$.

(3) $\quad 1 + \left(\frac{x}{1!}\right)^5 + \left(\frac{x^2}{2!}\right)^5 + \left(\frac{x^3}{3!}\right)^5 + \cdots = \frac{\sqrt{5}}{4\pi^2}\cdot\frac{e^{5x}}{5x^2 - x + \theta}$

where θ vanishes when $x = \infty$.

(4) $\dfrac{1^2}{e^x-1}+\dfrac{2^2}{e^{2x}-1}+\dfrac{3^2}{e^{3x}-1}+\dfrac{4^2}{e^{4x}-1}+\cdots$

$=\dfrac{2}{x^3}\left(\dfrac{1}{1^3}+\dfrac{1}{2^3}+\dfrac{1}{3^3}+\cdots\right)$

$-\dfrac{1}{12x}+\dfrac{x}{1440}+\dfrac{x^3}{181440}+\dfrac{x^5}{7257600}+\dfrac{x^7}{159667200}+\cdots$

when x is small. (*Note:* x may be given values from 0 to 2.)

(5) $\dfrac{1}{1001}+\dfrac{1}{1002^2}+\dfrac{3}{1003^3}+\dfrac{4^2}{1004^4}+\dfrac{5^3}{1005^5}+\cdots$

$=\dfrac{1}{1000}-10^{-440}\times 1.0125$ nearly.

(6) $\displaystyle\int_0^a e^{-x^2}\,dx=\dfrac{\sqrt{\pi}}{2}-\dfrac{e^{-a^2}}{2a}+\dfrac{1}{a}+\dfrac{2}{2a}+\dfrac{3}{a}+\dfrac{4}{2a}+\cdots$

(7) The coefficient of x^n in

$$\dfrac{1}{1-2x+2x^4-2x^9+2x^{16}-\cdots}$$

$=$ the nearest integer to $\dfrac{1}{4n}\left\{\cosh(\pi\sqrt{n})-\dfrac{\sinh(\pi\sqrt{n})}{\pi\sqrt{n}}\right\}$.

(9)

IX. Theorems on continued fractions, a few examples are:—

(1) $\dfrac{4}{x}+\dfrac{1^2}{2x}+\dfrac{3^2}{2x}+\dfrac{5^2}{2x}+\dfrac{7^2}{2x}+\cdots=\left\{\dfrac{\Gamma\left(\dfrac{x+1}{4}\right)}{\Gamma\left(\dfrac{x+3}{4}\right)}\right\}^2$.

(2) If

$$P=\dfrac{\Gamma\{\tfrac{1}{4}(x+m+n+1)\}\Gamma\{\tfrac{1}{4}(x+m-n+1)\}\Gamma\{\tfrac{1}{4}(x-m+n+3)\}\Gamma\{\tfrac{1}{4}(x-m-n+3)\}}{\Gamma\{\tfrac{1}{4}(x-m+n+1)\}\Gamma\{\tfrac{1}{4}(x-m-n+1)\}\Gamma\{\tfrac{1}{4}(x+m+n+3)\}\Gamma\{\tfrac{1}{4}(x+m-n+3)\}},$$

then

$$\dfrac{1-P}{1+P}=\dfrac{m}{x}+\dfrac{1^1-n^2}{x}+\dfrac{2^2-m^2}{x}+\dfrac{3^2-n^2}{x}+\dfrac{4^2-m^2}{x}+\cdots.$$

(3) If
$$z = 1 + \left(\frac{1}{2}\right)^2 x + \left(\frac{1\cdot 3}{2\cdot 4}\right)^2 x^2 + \cdots,$$
and
$$y = \frac{\pi}{2} \frac{1 + \left(\frac{1}{2}\right)^2 (1-x) + \left(\frac{1\cdot 3}{2\cdot 4}\right)^2 (1-x)^2 + \cdots}{1 + \left(\frac{1}{2}\right)^2 x + \left(\frac{1\cdot 3}{2\cdot 4}\right)^2 x^2 + \cdots},$$
then
$$\frac{1}{(1+a^2)\cosh y} + \frac{1}{(1+9a^2)\cosh 3y} + \frac{1}{(1+25a^2)\cosh 5y} + \cdots$$
$$= \frac{1}{2}\frac{z\sqrt{x}}{1} + \frac{(az)^2}{1} + \frac{(2az)^2 x}{1} + \frac{(3az)^2}{1} + \frac{(4az)^2 x}{1} + \cdots,$$
a being any quantity.

(4) If
$$u = \frac{x}{1} + \frac{x^5}{1} + \frac{x^{10}}{1} + \frac{x^{15}}{1} + \frac{x^{20}}{1} + \cdots$$
and
$$v = \frac{\sqrt[5]{x}}{1} + \frac{x}{1} + \frac{x^2}{1} + \frac{x^3}{1} + \cdots,$$
then
$$v^5 = u \cdot \frac{1 - 2u + 4u^2 - 3u^3 + u^4}{1 + 3u + 4u^2 + 2u^3 + u^4}.$$

(5) $$\frac{1}{1} + \frac{e^{-2\pi}}{1} + \frac{e^{-4\pi}}{1} + \frac{e^{-6\pi}}{1} + \cdots = \left(\sqrt{\frac{5+\sqrt{5}}{2}} - \frac{\sqrt{5}+1}{2}\right)\sqrt[5]{e^{2\pi}}.$$

(6) $$\frac{1}{1} - \frac{e^{-\pi}}{1} + \frac{e^{-2\pi}}{1} - \frac{e^{-3\pi}}{1} + \cdots = \left(\sqrt{\frac{5-\sqrt{5}}{2}} - \frac{\sqrt{5}-1}{2}\right)\sqrt[5]{e^{\pi}}.$$

(7) $$\frac{1}{1} + \frac{e^{-\pi\sqrt{n}}}{1} + \frac{e^{-2\pi\sqrt{n}}}{1} + \frac{e^{-3\pi\sqrt{n}}}{1} + \cdots$$
can be exactly found if n be any positive rational quantity.

(11)

XI. I have got theorems on divergent series, theorems to calculate the convergent values corresponding to the divergent series, viz.
$$1 - 2 + 3 - 4 + \cdots = \frac{1}{4},$$
$$1 - 1! + 2! - 3! + \cdots = .596\ldots,$$

$$1 + 2 + 3 + 4 + \cdots = -\frac{1}{12},$$

$$1^3 + 2^3 + 3^3 + 4^3 + \cdots = \frac{1}{120}.$$

Theorems to calculate such values for any given series (say $1 - 1^1 + 2^2 - 3^3 + 4^4 - 5^5 + \cdots$), and the meaning of such values.

I have also dealt with such questions 'When to use, where to use, and how to use such values, where do they fail and where do they not?'

I have also given meanings to the fractional and negative number of terms in a series as well as in a product and I have got theorems to calculate such values exactly and approximately. Many wonderful results have been got from such theorems; e.g.

$$\frac{1}{n} + \left(\frac{1}{2}\right)^2 \frac{1}{n+1} + \left(\frac{1\cdot 3}{2\cdot 4}\right)^2 \frac{1}{n+2} + \left(\frac{1\cdot 3\cdot 5}{2\cdot 4\cdot 6}\right)^2 \frac{1}{n+3} + \cdots$$

$$= \left\{\frac{\Gamma(n)}{\Gamma(n+\frac{1}{2})}\right\}^2 \left\{1 + \left(\frac{1}{2}\right)^2 + \left(\frac{1\cdot 3}{2\cdot 4}\right)^2 + \cdots \text{ to } n \text{ terms}\right\}.$$

It is even possible to find the true value in the cases in which the use of divergent series fails by finding the diff[eren]ce between the true and apparent values.

Commentary. Godfrey Harold Hardy (1877–1947) was born in Cranleigh, Kent, and died in Cambridge. A full account of Hardy's life and contributions to mathematics is given by E.C. Titchmarsh on pp. 1–11 of [105], and further interesting personal information can be found in C.P. Snow's introduction to *A Mathematician's Apology* [104]. For a more recent assessment of his personality, see Kanigel's biography of Ramanujan [142]. In the following brief sketch of Hardy's life, we have, in essence, reproduced J.C. Burkill's entry in [57]; for clarity, in a few places, we have added, subtracted, or altered a few words.

Hardy was the elder of two children of Isaac Hardy, a master at Cranleigh School, and Sophia Hall. The parents were intelligent and mathematically minded, but lack of money had precluded them from a university education. They provided an enlightened upbringing for Hardy and his sister.

The freedom to ask questions and to probe led Hardy to an early established disbelief in religious doctrine. (As a fellow of New College, Oxford, he refused to enter the chapel to take part in electing a warden.) Neither Hardy nor his sister married, and he owed much to her devoted care throughout his life, particularly in his later years.

As a boy, Hardy showed all-round ability with a precocious interest in numbers. At the age of thirteen he moved from Cranleigh School with a scholarship to Winchester College, to this day a famous nursery of mathematicians. He went on to Trinity College, Cambridge, in 1896, was fourth

G. H. Hardy

Wrangler in the mathematical tripos in 1898, was elected a Fellow of Trinity in 1900, and won (with J.H. Jeans) a Smith's Prize in 1901. Success in the tripos depended on efficient drilling in solving problems quickly. Hardy, resenting the routine of the famous "coach" R.R. Webb, had the good fortune to be transferred to A.E.H. Love. No description of Hardy's development into a mathematician can be so vivid as his own:

"My eyes were first opened by Professor Love, who taught me for a few terms and gave me my first serious conception of analysis. But the great debt which I owe to him was his advice to read Jordan's famous *Cours d'analyse*; and I shall never forget the astonishment with which I read that remarkable work, the first inspiration for so many mathematicians of my generation, and learnt for the first time as I read it what mathematics really meant ([**104**, sec. 29])."

Hardy flung himself eagerly into research and between 1900 and 1911 wrote many papers on the convergence of series and integrals and allied topics. He was elected a Fellow of the Royal Society in 1910. Although this work established his reputation as an analyst, his greatest service to mathematics in this early period was *A Course of Pure Mathematics* [**89**]. This work was the first rigorous English exposition of number, function, limit, and so on adapted to the undergraduate, and thus it transformed university teaching.

The quotation from the *Apology* continues, "The real crises of my life came ten or twelve years later, in 1911, when I began my long collaboration with Littlewood, and in 1913, when I discovered Ramanujan."

J.E. Littlewood, eight years younger than Hardy, proved in 1910 the Abel–Tauber theorem that, if na_n is bounded and $\sum a_n x^n \to s$, as $x \to 1-$, then $\sum a_n = s$. The two then entered into a collaboration which was to last thirty-five years. They wrote nearly a hundred joint papers.

This partnership of Hardy and Littlewood has no parallel, and it is remarkable that, at its greatest intensity (1920–1931), Hardy lived in Oxford and Littlewood in Cambridge. They set up a body of axioms expressing the freedom of their collaboration, for example, "When one received a letter from the other he was under no obligation to read it, let alone to answer it." The final writing of the papers was done by Hardy.

Hardy called his discovery of Srinivasa Ramanujan the one romantic incident of his life. In his three years of health and activity, Ramanujan and Hardy had arrived at spectacular solutions of problems about the partition of numbers which called forth the full power of the Indian's natural insight and the Englishman's mastery of the theory of functions. They collaborated in seven papers.

Hardy was a lecturer at Trinity College until 1919, when he became Savilian Professor of Geometry at Oxford; there he founded a flourishing school of research. For the year 1928–1929 he went to Princeton, exchanging places with Oswald Veblen. He returned to Cambridge in 1931, succeeding E.W. Hobson as Sadleirian Professor of Pure Mathematics; he held this chair until his retirement in 1942.

Besides Littlewood and Ramanujan, Hardy collaborated with many other mathematicians, including E.C. Titchmarsh, A.E. Ingham, E. Landau, G. Pólya, E.M. Wright, W.W. Rogosinski, and M. Riesz. He had an exceptional gift for working with others, as he had for leading young mathematicians in their early days of research.

Hardy had one ruling passion—mathematics. Apart from that his main interest was in ball games, particularly cricket, of which he was a stylish player and an expert critic. Some of his interests and antipathies are revealed by this list of six New Year wishes which he sent on a postcard to a friend in the 1920's: (1) prove the Riemann hypothesis; (2) make 211 not out in the fourth innings of the last test match at the Oval; (3) find an argument for the nonexistence of God which shall convince the general public; (4) be the first man at the top of Mt. Everest; (5) be proclaimed the first president of the U.S.S.R., Great Britain, and Germany; and (6) murder Mussolini.

Hardy was generally recognized as the leading English pure mathematician of his time. His writings attest both his technical power and his mastery of English prose. The photographs in the *Collected Papers* show his finely cut features and something of his physical grace. His liveliness and enthusiasm are vivid in the memory of all who knew him. He received awards from many universities and academies, being elected in 1947 *associé étranger* of the Paris Academy of Sciences—of whom there are only ten from all nations in all subjects.

In the letter, Ramanujan incorrectly gave his age as 23, instead of 25. It has been suggested that Ramanujan thought that Hardy might more favorably consider his discoveries if he were a bit younger than 25. However, in a society that does not require its citizens frequently to record their birthdates, in contrast to most Western countries, it seems more likely that Ramanujan simply forgot or miswrote his correct age.

It is ironic that Ramanujan unknowingly began the description of his accomplishments with some of his most unimportant results. His theory of divergent series is chiefly founded upon the Euler–Maclaurin summation formula and is based on tenuous and imprecise principles and ideas. Ramanujan associates a "constant" with each convergent or divergent series. If the series converges, this constant is the sum of the series, but if the series diverges, the constant arises from the constant term in the Euler–Maclaurin summation formula. This theory is vaguely described by Ramanujan in Chapter 6 of his second notebook [**225**]. See Hardy's book [**103**] and Berndt's [**32**, Chapter 6] account for more details about Ramanujan's ideas on divergent series.

It might be surmised that Ramanujan referred to the analytic continuation of the gamma function in the second paragraph of his letter. Ramanujan had no knowledge of this well-known principle of complex analysis, but his *definition* of $\Gamma(n)$ for negative values of n is precisely what one obtains by ordinary analytic continuation. His definition extends to other integrals as well and can be found in Entry 1 of Chapter 13 of his second notebook [**225**]. See Berndt's book [**35**, pp. 186-187] for a discussion of these ideas.

Ramanujan's theorems were written on eleven pages, and we have reproduced his numbering in our transcription of his letter. Note that pages (8) and (10) are missing.

As we shall see, Ramanujan's formulas or approximations for $\pi(x)$, the number of primes less than or equal to x, are much less precise than he thought. Ramanujan's first observation that $\rho(e^{2\pi x})$ is less than a few hundred when x is less than 3 is not correct. For example, $e^{6\pi} = 153552935.2...$, but $\rho(10^8) = 755$.

Ramanujan discovered two infinite series that closely approximate $\pi(x)$; one is originally due to J.P. Gram [**83**] and the other was first established by B. Riemann [**244**], [**245**]. The one involving Bernoulli numbers is that of Gram. The integral analogue mentioned in (2) of page 2 is, in fact, original with Ramanujan and was first rigorously proved by Hardy [**102**], [**106**, pp. 234–238]. These and further approximate formulas are found on pages 317–318 of Ramanujan's second notebook. Ramanujan's claim that his formulas approximated $\pi(x)$ within a bounded amount is false. Hardy [**110**, Chapter 2] and Berndt [**40**, Chapter 24] have discussed in detail the faulty proofs that led to Ramanujan's erroneously optimistic approximations. Ignorance of the theory of functions of a complex variable, interchanging limiting operations, false analogies between series and integrals, "asymptotic expansions" about several singularities, and an inadequate understanding of Riemann–Stieltjes integrals are some of the primary reasons for Ramanujan's wavering predictions. It is interesting that in 1913 S.N. Aiyar [**2**] communicated some of Ramanujan's discoveries on prime numbers to the *Journal of the Indian Mathematical Society*.

Ramanujan never recorded his corresponding formulas for the number of primes in an arithmetic progression. Possibly, he found analogues of some of his formulas for $\pi(x)$.

Let $\pi_{4,j}(x)$ denote the number of primes in the arithmetic progression $4n + j$ ($j = 1, 3$). Then Ramanujan claims that

$$\Delta(x) := \pi_{4,3}(x) - \pi_{4,1}(x) \to \infty$$

as $x \to \infty$. Although, in 1853, P.L. Chebychev [**65**], [**66**] first offered heuristic reasoning for the possible approximation

$$\Delta(x) \approx \frac{1}{2}\pi(x^{\frac{1}{2}}),$$

Ramanujan's claim is false. In fact, Hardy and Littlewood [**116**] (Hardy [**106**, pp. 20–97]) showed in 1918 that $\Delta(x)$ changes sign infinitely often. For a further discussion of Ramanujan's work on this problem and for several references to the literature, see Berndt's book [**40**, Chapter 24].

The general idea employed by Ramanujan for the results announced in paragraph III is probably the following imprecise theorem found on the initial pages of Ramanujan's third notebook [**225**, p. 362], which we now quote exactly.

THEOREM. *If A, B, C are quantities so taken that*

$$\frac{1}{A^k} + \frac{1}{B^k} + \frac{1}{C^k} + \cdots = \frac{a}{(k-\alpha)^r}$$

when $k = \alpha$ *(the only pole) then the no. of such quantities less than z is*

$$\int \frac{a(\log z)^{r-1}}{z^{1-\alpha}\Gamma(r)} dz.$$

CHAPTER 2. FIRST TWO LETTERS TO HARDY

Uncharacteristically, Ramanujan offers a remark indicating the main idea of his proof, and from this we have been able to reconstruct his full argument. The primary step involves unjustifiably equating two integrands as k tends to α. Ramanujan's "proof" also assumes that r is a positive integer, although in many applications r is not an integer. Thus, without additional hypotheses, Ramanujan's Theorem is false, and counterexamples can be given. For a more complete discussion of Ramanujan's claim, see Berndt's book [40, pp. 62-66].

Ramanujan's Theorem (1) under III on page 2 is found in a more general form on page 309 in his second notebook [225], wherein 2 and 3 are replaced by a and b, respectively. This result is one of only a small handful for which Ramanujan offers proofs in the notebooks. The argument is given in complete detail in Berndt's book [40, pp. 66-69] and has some of the same defects described in previous paragraphs. In particular, Ramanujan, without justification, equates integrands of two Riemann-Stieltjes integrals as a certain parameter approaches a limit.

It is easy to see that Ramanujan's problem of finding the number of numbers of the form $a^p b^q$ that are less than n is equivalent to finding the number of lattice points in the triangle

$$u \geq 0, \quad v \geq 0, \quad (\log a)u + (\log b)v \leq \log n.$$

Thus, the problem can be considered as one in geometry or in diophantine approximation.

As is to be expected, Ramanujan considered his approximation for this number of lattice points to be more accurate than really is the case. However, asymptotically, as n tends to ∞, Ramanujan's claim is correct. A.M. Ostrowski [179] and Hardy and Littlewood [120], [121] (Hardy [105, pp. 136-157, 159-196]) first attacked the problem rigorously in 1922. The magnitude of the error terms for the main approximation depends upon the rationality or irrationality of $\frac{\log a}{\log b}$. In his book [110, Chapter 5], Hardy offers a lengthy discussion of this lattice point problem with complete proofs. The exposition in Berndt's book [40, Chapter 23] emphasizes Ramanujan's ideas. In closing, we remark that M. Hausman and H.N. Shapiro [128] have found explicit sequences of n tending to ∞ for which Ramanujan's claim is true, in that the error remains bounded for such sequences.

We next examine Ramanujan's claims under (2) at the top of page 3. Part (a) is a well-known result normally expressed in the following equivalent form. Let $Q(x)$ denote the number of squarefree integers $\leq x$. Then as x tends to ∞,

$$Q(x) \sim \frac{6x}{\pi^2}.$$

For example, see Hardy and Wright's text [126, pp. 269-270] for a proof. Parts (b) and (c) are proved in Ramanujan's fourth published paper [209], [226, pp. 20-21] and can also be found in Ramanujan's second notebook

[225] (Berndt [32, Chapter 5, Section 30]).

The problem of estimating the summatory function for $d(n)$, the number of positive divisors of the positive integer n, is one of the most famous unsolved problems in the analytic theory of numbers. As with other arithmetical problems, Ramanujan's approximation is not as accurate as he claimed, although asymptotically it is correct. Define the "error term" $\Delta(x)$ by

$$\sum_{n \leq x} d(n) = x \log x + (2\gamma - 1)x + \Delta(x),$$

where γ denotes Euler's constant. The exact order of $\Delta(x)$ is not known. In 1916, Hardy [92], [106, pp. 268-292] proved that

(*) $$\Delta(x) \neq O(x^{\frac{1}{4}}),$$

which, of course, disproves Ramanujan's claim. The best result in this direction is currently due to J.L. Hafner [87], who has shown that (*) still holds if $x^{\frac{1}{4}}$ is multiplied by certain logarithmic factors. On the other hand, it is elementary that $\Delta(x) = O(x^{\frac{1}{2}})$; see, e.g., Hardy and Wright's book [126, pp. 264-265]. Improving on methods of H. Iwaniec and C.J. Mozzochi [139], M.N. Huxley [137] currently owns the best result

$$\Delta(x) = O(x^{\frac{23}{73}+\epsilon}).$$

It is conjectured that

$$\Delta(x) = O(x^{\frac{1}{4}+\varepsilon}),$$

for every positive ε. For very complete historical discussions of the 'divisor' problem, see the books of A. Ivić [138] and E. Krätzel [147].

Ramanujan offers his result (3) on page 3 at two places in his second notebook [225], Section 2 of Chapter 15, and page 304 in the unorganized portions. See also Berndt's books [35, p. 304] and [40, p. 57].

In (4) on page 3, $\theta(x)$ should be replaced by $\theta(B)$ in each instance. Ramanujan's claim was first established by Landau [149], [151, pp. 59-66], [150, pp. 644, 649-669] in 1908 and is elegantly sketched by Hardy in his book [110, pp. 61-63]. At the end of his discussion, Hardy [110, p. 63] remarks " ... and it would be very interesting to know just how Ramanujan came to this conclusion." G.N. Watson [288] likewise expressed wonder about Ramanujan's discovery when he wrote "The most amazing thing about this formula is that it was discovered, apparently independently, by Ramanujan in his early days in India, and it appears in its appropriate place in his manuscript note-books." Indeed, it appears on page 307 of the second notebook. More surprisingly, Ramanujan sketches a 'proof' of his claim in the third notebook [225, p. 363]. More space (the entire page) is devoted to this heuristic argument than to any other argument or proof in the notebooks. The third notebook might not have been available to Hardy and Watson; Watson's personal handwritten copy of the notebooks on file at Trinity College, Cambridge, does not contain the third notebook. Ramanujan's argument hinges upon the Theorem above, and so he claimed a much smaller error term $\theta(x)$ than was justified. For complete discussions of Ramanujan's arguments and further results in the literature, see Berndt's account [40, pp. 62–66].

As Hardy remarked in the introduction to Ramanujan's *Collected Papers* [226, p. xxiv], "There are regions of mathematics in which the precepts of modern rigour may be disregarded with comparative safety, but the Analytic Theory of Numbers is not one of them, ..." That Ramanujan conceived these problems, sometimes before anyone else had done so, with no contact with the European mathematical community, and that he correctly obtained the dominant terms in asymptotic formulas are astounding achievements that should not be denigrated because of his unrigorous, but clever, arguments.

Next, we discuss the six results on integrals listed under the heading IV.

The first two results were proved by Ramanujan in his paper [212], [226, pp. 53–58]. Further proofs of (2) have been given by Preece [182].

The function ϕ was extensively studied by Ramanujan in [214], [226, pp. 59–67], but it should be noted that his definition of ϕ in [214] is slightly different from the one given in the letter to Hardy. The theorem stated here is equivalent to $(10')$ in [214], where a proof may be found. Most of the quoted values for ϕ are also given in this paper. This result is also found in Ramanujan's second notebook [225]. See Berndt's book [40, pp. 296–303] for further discussion.

Formula (4) was first posed by Ramanujan as a problem [206], [226, p. 326], and was later stated by him in his paper [212], [226, pp. 53–58]. In Section 6 of his paper [212], Ramanujan confesses that "My own proofs of the above results make use of a general formula, the truth of which depends on conditions which I have not yet investigated completely. A direct proof depending on Cauchy's theorem will be found in Mr Hardy's note which follows this paper." (The paper of Hardy to which Ramanujan refers is [91], [108, pp. 594–597].) The general formula to which Ramanujan alludes is an integral interpolation formula, which is akin to the Lagrange inversion formula. A rigorous proof of this general formula can be found in Hardy's book [110, pp. 188–190]. Ramanujan states this formula in his first quarterly report, and (4) is one of the several examples that he gives to illustrate his theorem. In fact, Ramanujan [212], [226, p. 56] offers a more general formula

$$(4') \quad \int_0^\infty x^{s-1} \frac{(1+aqx)(1+aq^2x)\cdots}{(1+x)(1+qx)(1+q^2x)\cdots} dx = \frac{\pi}{\sin s\pi} \prod_{n=1}^\infty \frac{(1-q^{n-s})(1-aq^n)}{(1-q^n)(1-aq^{n-s})},$$

which can be regarded as a q-analogue of the classical beta function. For excellent discussions of this beautiful and important result, see R. Askey's papers [14], [15]. Ramanujan also recorded $(4')$ in Entry 14 of Chapter 16 of his second notebook [225]; see Berndt's book [37, p. 29] for further discussion and references.

The series in the denominator on the right side of (4) is a theta-function, and so, by a simple application of the Jacobi triple product identity, the right side of (4) has the alternative representation

$$\frac{\pi}{2} \prod_{n=1}^{\infty} \frac{(1-r^{2n-1})}{(1-r^{2n})}.$$

Formula (5) is valid only when $\text{Re}(n) > -\frac{1}{2}\pi$. It also must be assumed that if $\text{Re}(n) > 0$, then $|\text{Im}(n)| < \frac{1}{2}\pi$, and if $-\frac{1}{2}\pi < \text{Re}(n) < 0$, then $|\text{Im}(n)| < \frac{1}{2}\pi + \text{Re}(n)$. Preece [182] established (5) by contour integration.

By using Binet's representation for $\log \Gamma(t)$, Preece [182] found a short, elegant proof of (6).

All of the results under V were either discussed or proved by Watson [275].

Formula (1) is a special case of one of the functional equations for the trilogarithm and is found in Section 7 of Chapter 9 in Ramanujan's second notebook [225]. See Berndt's book [32, p. 249] or Watson's paper [275] for more details.

The elegant series summation (2) is a special instance of Dougall's theorem, and both can be found in Chapter 10 of Ramanujan's second notebook [225]; see Berndt's book [35, pp. 11, 24]. The evaluation (3) is a special instance of a limiting case of Dougall's theorem, namely Corollary 5 in Section 7 of Chapter 10 in the second notebook (Berndt [35, pp. 16, 23–24]). Both of these observations were first made by Hardy [99], [100, pp. 517–518]. In fact, (3) was first established in 1859 by Bauer [27].

The series evaluation (4) is a special case of the more general formula

$$(4') \qquad \sum_{k=1}^{\infty} \frac{k^{4n+1}}{e^{2\pi k} - 1} = \frac{B_{4n+2}}{8n+4},$$

where n is a positive integer and B_j, $0 \le j < \infty$, denotes the jth Bernoulli number. There exist many proofs of (4') in the literature, but the first proof appears to be by J.W.L. Glaisher [81] in 1889. For references to many more proofs of (4) and (4') and for further generalizations of (4), consult Berndt's books [35, pp. 261–262] and [41]. Formula (5) is also a special instance of a more general result, namely,

$$(5') \qquad \sum_{k=1}^{\infty} \frac{\coth(\pi k)}{k^{2n+1}} = 2^{2n} \pi^{2n+1} \sum_{k=0}^{n+1} (-1)^{k+1} \frac{B_{2k}}{(2k)!} \frac{B_{2n+2-2k}}{(2n+2-2k)!},$$

where n is an odd positive integer and B_j denotes the jth Bernoulli number. Cauchy [60, pp. 320, 361] devised a general method for evaluating the series on the left side of (5') but did not give the general formula (5'). Apparently, (5') is originally due to Lerch [160] in 1901. Readers should see Berndt's book [35, p. 293] for further references to the many proofs of (5') or of special cases that can be found in the literature.

Like the two previous formulas, (6) can be found in Chapter 14 of Ramanujan's second notebook [225] and is a particular instance of a more general formula, viz.,

$$(6') \quad \sum_{k=0}^{\infty} \frac{(-1)^k}{(2k+1)^{4n+1} \cosh(\frac{1}{2}(2k+1)\pi)} = \frac{1}{4}\left(\frac{\pi}{2}\right)^{4n+1} \sum_{k=0}^{2n} (-1)^k \frac{E_{2k}}{(2k)!} \frac{E_{4n-2k}}{(4n-2k)!},$$

where n is *any* integer and E_j, $0 \le j < \infty$, denotes the jth Euler number. In fact, Ramanujan recorded an even more general result than $(6')$ in Entry 21(ii) of Chapter 14 in his second notebook [225]. It is curious that the first proof of $(6')$ as well as the aforementioned more general theorem is by S.L. Malurkar [165] in 1925 in the *Journal of the Indian Mathematical Society*. The notebooks were not available to the mathematical public at that time, and so Malurkar did not realize that his theorem had been previously recorded by Ramanujan in his notebooks. For references to further proofs, see Berndt's book [35, pp. 276-277, 295].

Both (7) and (8) contain misprints, and these same errors also occur in their appearances as Entries 25 (xi), (xii) in Chapter 14 of the second notebook [225]. In (7), $(\sinh\{(2k+1)\pi\} - \sinh \pi)$ should be replaced by $(\cosh\{(2k+1)\pi\} - \cosh \pi)$, $k \ge 0$. In (8), the right-hand side should be multiplied by (-1). Watson's [275] proof of (7) uses contour integration and has a few errors. Berndt [35, p. 297] offers another proof by contour integration. However, the best proof is by R. Sitaramachandrarao [261] who shows that (7) follows from a previous entry in Chapter 14; see also Berndt's account [35, pp. 297-298]. Watson's [275] proof of (8) also employs contour integration. A simpler proof, using another entry from Chapter 14, has been found by Berndt [35, pp. 298-299].

The series evaluation (9) is a special case of the formula

$$(9') \quad \sum_{k=0}^{\infty} \frac{(-1)^k}{(2k+1)^{6n+1} \cosh\left\{\frac{(2k+1)\pi\sqrt{3}}{2}\right\}} = \frac{1}{2}(-1)^{n+1} \pi^{6n+1} \sum_{k=0}^{3n} \frac{E_{2k+1}(0)}{(2k+1)!} \frac{B_{6n-2k}}{(6n-2k)!} \cos\left\{\frac{(2k+1)\pi}{3}\right\},$$

where n is any nonzero integer, $E_j(x)$, $j \ge 0$, denotes the jth Euler polynomial, and B_j, $j \ge 0$, denotes the jth Bernoulli number. This formula is essentially due to Cauchy [60, p. 316], although he does not explicitly state it. The first explicit statement and proof of $(9')$ is due to Berndt [29, Theorem 7.5], where further references to the literature may be found. Formula (9) can also be found in Section 10 of Chapter 18 in Ramanujan's second notebook [225], [37, pp. 161-162].

Watson [275] has given a short proof of (10).

The equality (11) is proved in Ramanujan's paper [213], [226, pp. 40-43].

All of the results listed under VI were either proved or discussed by Preece in his paper [183].

Equation (1) is simply a special case of the functional equation of the Dirichlet L-function

$$L(s) := \sum_{k=0}^{\infty} (-1)^k (2k+1)^{-s}, \qquad \text{Re}(s) > 0.$$

Ramanujan, in fact, discovered the functional equation of $L(s)$ and recorded it as Entry 18 of Chapter 7 in his second notebook [225] (Berndt [32, pp. 171–172]).

As Preece [183] observed, (2) follows from differentiating the functional equation of $L(s)$ and then setting $s = \frac{1}{2}$.

Formula (3) is the special instance $m = n = 0$ of the more general formula

$$_0F_2(m+1, n+1; x)\,_0F_2(m+1, n+1; -x)$$
$$= \sum_{k=0}^{\infty} \frac{(-1)^k (m+n+2k+1)_k x^{2k}}{(m+1)_k (n+1)_k (m+1)_{2k} (n+1)_{2k} k!},$$

found as Entry 16 of Chapter 11 in Ramanujan's second notebook [225] (Berndt [35, pp. 59–60, 62–63]), and later is explicitly stated as Example 2 of Section 20 [225], [35, pp. 62–63]. Here $_0F_2$ is a generalized hypergeometric function. Ramanujan proved many theorems for products of hypergeometric functions and stimulated much subsequent research by W.N. Bailey and others on this topic.

The next result (4) was recorded by Ramanujan as Entry 21 of Chapter 13 in his second notebook [225] and follows from Parseval's theorem for Fourier transforms. See Berndt's book [35, p. 224] for conditions of validity. Ramanujan gave a formal proof in his paper [212], [226, pp. 53–58].

Ramanujan [208], [226, pp. 324–325], offered (5) as a problem to the *Journal of the Indian Mathematical Society*. See [226, p. 325] or [212], [226, pp. 53–58] for his solution. The result also appears in his notebooks [225] (Berndt [35, p. 225]).

Formula (6) was also established by Ramanujan in his paper [212], [226, pp. 53-58].

As Preece [183] has shown, (7) follows from Poisson's summation formula for Fourier sine transforms.

Preece's [183] proof of (8) is by contour integration.

Most of the results under VII were proved by Watson in his paper [276].

Watson's proof of (1) employs the Abel-Plana summation formula. However, Ramanujan undoubtedly deduced (1) from a more general result that he established in Entry 27(a) in Chapter 9 of his second notebook [225]. We shall indicate how to do this, but we shall forego the straightforward calcula-

CHAPTER 2. FIRST TWO LETTERS TO HARDY   41

tions. In the aforementioned entry, Ramanujan offers an asymptotic formula, as x tends to ∞, for

$$\sum_{k=1}^{x} k^r \log k, \qquad r > -1.$$

See Berndt's book [**32**, pp. 273–276]. Set $r = 2$ in this formula. For the calculations, the value of $C_2 = -\zeta'(-2)$ is needed, and this is found on page 276. An asymptotic formula for $\sum_{k=1}^{x} k^2$ is also required, and this is found in (27.6) on page 274. The reader should have no trouble furnishing the details.

One of the most famous problems that Ramanujan [**207**], [**226**, pp. 323–324] submitted to the *Journal of the Indian Mathematical Society* is a variant of (2). Another version of (2) can be found in Section 48 of Chapter 12 in Ramanujan's second notebook [**225**]. In the former variant, Ramanujan asked that $\frac{1}{3} < \theta < \frac{1}{2}$ be shown. Both Szegö [**264**], [**265**, pp. 143–150] and Watson [**278**] established these inequalities. Watson devoted a tremendous effort to establishing Ramanujan's claimed inequalities for k. He showed that $k > \frac{2}{21}$ for all x sufficiently large and presented considerable evidence that $\frac{2}{21} < k < \frac{8}{45}$ for all $x > 0$. To the best of our knowledge, these inequalities have never been completely proved or disproved. The problem in the *Journal of the Indian Mathematical Society* generated several analogues and generalizations. See Berndt's book [**35**, pp. 181–182, 193] for a discussion of this literature.

Watson [**276**] showed that (3) follows from a general theorem of Barnes [**26**]. F.W.J. Olver [**178**, pp. 307–309] has given a different proof of a generalization of (3), wherein each exponent 5 on the left side is replaced by n, where n is any positive number. See also Section 10 of Chapter 13 in Ramanujan's second notebook [**225**] for this same generalization (Berndt [**35**, p. 214]).

To correct a misprint in (4), we have replaced $-\frac{x}{1440}$ by $\frac{x}{1440}$. Watson's [**276**] proof of (4) employs a variation of the Abel–Plana summation formula and is the only proof known to us.

Proofs of (5) have been given by both Watson [**276**] and Szegö [**264**], [**265**, pp. 143–150].

The continued fraction formula (6) is originally due to Laplace [**152**] in 1805 and was first rigorously proved by Jacobi [**140**] in 1834. Watson's [**276**] proof of (6) is quite elegant. Ramanujan stated (6) in Section 43 of Chapter 12 of his second notebook [**225**]. See Berndt's book [**35**, p. 166] to see how (6) easily follows from other results established by Ramanujan.

Declaration (7) is of enormous interest. It is not correct as it stands, but a corrected version can be found in the famous paper of Hardy and Ramanujan [**124**] (Hardy [**105**, p. 334], Ramanujan [**226**, p. 304]), wherein their asymptotic formula for the partition function $p(n)$ is established. The claim (7) is an analogue of this theorem in that the generating function for $p(n)$, the Dedekind eta-function, is essentially replaced by the classical theta-function $\theta(\tau)$, where $x = -e^{\pi i \tau}$. Littlewood [162] (see also Andrews' book [6, pp. 68–69]) has written that in his collaboration with Hardy on $p(n)$, Ramanujan insisted that a highly accurate formula for $p(n)$ existed. This persistence especially pushed Hardy to their joint discovery of the amazingly precise asymptotic formula for $p(n)$. Thus, (7) demonstrates that the foundation for Ramanujan's confidence originated in India several years earlier.

In 1936, while preparing the lectures for his graduate course in analytic number theory, H. Rademacher [187], [189, pp. 108–121], in fact, discovered an exact formula for $p(n)$. (See Berndt's paper [38] for further historical background on Rademacher's discovery.) At about the same time, A. Selberg [255, pp. 695–706] also proved this exact formula, but he never published his result. Shortly thereafter, H.S. Zuckerman [301] found an exact formula for the coefficient in (7). Simpler formulas for the coefficients of the reciprocals of the classical theta–functions, as well as simpler proofs, have been derived by L.A. Goldberg [82].

In their paper [124], Hardy and Ramanujan introduced the famous 'circle method', which since then has been one of the most powerful tools in additive number theory. The 'circle method' was developed further in joint papers of Hardy and Littlewood. Although references in the literature occasionally attach Ramanujan's or Vinogradov's names to this method, most authors refer to the 'Hardy–Littlewood circle method.' Evidently, most mathematicians opine that Ramanujan played no role in the development of the 'circle method.' However, Ramanujan's contributions may have been underestimated. We discussed earlier the Landau–Ramanujan problem on sums of squares and Ramanujan's claim of a more accurate error term than is justified. Ramanujan's argument, as reconstructed by Berndt [40, pp. 62–66], shows that Ramanujan attempted to employ the asymptotic behaviors of the generating function about 'several' singularities. However, he did not properly implement this novel idea. The crux of the 'circle method' depends upon precisely determining the behavior of the generating function about selected singularities. It thus seems that the idea of employing knowledge about several singularities is due to Ramanujan, and therefore he should be given more credit for the invention of the 'circle method.'

For expository accounts of the 'circle method' emphasizing Ramanujan's contributions, see papers of R. Balasubramanian [23] and K. Ramachandra [196]. R. C. Vaughan's book [269] provides accounts of many diverse problems that are amenable to attacks by the 'circle method.'

As indicated above, the page containing formulas from Section VIII on elliptic functions has been lost.

The continued fraction formula (1), valid for $\text{Re}(x) > 0$, is Corollary 1 in Section 25 of Chapter 12 in Ramanujan's second notebook [225] (Berndt [35, p. 145]) and is a special case of a much more general theorem, Entry 39 in Chapter 12. In fact, (1) is due to G. Bauer [28] in 1872. See also Preece's paper [184]. Setting $x = 1$ in (1), we obtain Lord Brouncker's continued fraction for π,

$$\pi = \frac{4}{1} + \frac{1^2}{2} + \frac{3^2}{2} + \frac{5^2}{2} + \cdots .$$

For a lively historical discussion of Lord Brouncker's continued fraction, see Dutka's paper [77].

The beautiful continued fraction representation (2) was first established by Preece [185]. Perron [181, p. 34, eq. (15)] gave a different proof. L. Jacobsen [141] showed that (2) is valid provided that n is an odd integer, or m is an even integer, or that $\text{Re}(x) > 0$ with m and n arbitrary complex numbers. A completely different proof of (2), based upon generalized hypergeometric series, was found by L.-C. Zhang [299]. Ramanujan also recorded (2) as Entry 34 in Chapter 12 of his second notebook [225] (Berndt [35, p. 156]).

The next result was first proved in print by Preece [184] and is also Entry 12(ii) in Chapter 18 of the second notebook [225] (Berndt [37, pp. 163–165]). Formula (3) is valid for $0 < x < 1$ and $a > 0$.

The algebraic relation between u and v given in (4) is recorded by Ramanujan on page 289 of his second notebook. The first proof is by L. J. Rogers [247] in 1920. Eight years later, Watson [279] gave another proof. Ramanujan discovered many beautiful algebraic relations involving the Rogers–Ramanujan continued fraction. See the monograph [13] by Andrews, Berndt, Jacobsen, and Lamphere for proofs of most of these theorems.

Results (5)–(7) were first established by Watson [279]. Ramanathan [198], [199], [200], [201], [203] has established (5) and (6) and many further beautiful results of this sort. Formulas (5) and (6) are found in Section 39 of Chapter 16 of Ramanujan's second notebook [225] (Berndt [37, pp. 83–86]). In Ramanujan's *Collected Papers* [226, p. xxvii], (6) contains a misprint. For further work of Ramanujan on the Rogers–Ramanujan continued fraction and related results see [12] and [13].

It does not seem to have been previously noticed in print that a second page, listing the results under X, of Ramanujan's first letter to Hardy has been lost. Since nothing from this page is quoted by Hardy in Ramanujan's *Collected Papers* [226], this page evidently was lost before the publication of [226]. (In this connection, see Hardy's letter of 23 July 1940 to A.F. Scholfield.) In his commentary on the Rogers–Ramanujan identities on page 344 of Ramanujan's *Collected Papers* [226], Hardy claims "They were rediscovered nearly 20 years later by Mr Ramanujan, who communicated them to me in a letter from India in February 1913." However, Ramanujan's letter of 27 February 1913 to Hardy does not contain the Rogers–Ramanujan identities, unless a portion of the letter is lost, which does not seem likely. Thus, if Hardy is correct and Ramanujan did communicate the Rogers–Ramanujan identities to him, it would seem that they were probably recorded in the missing section X of the first letter. Moreover, such content in Section X would be a natural sequel to the continued fractions offered in Section IX.

The four divergent series and the values assigned to them by Ramanujan were discussed by Watson [280]. These values are the "constants" for the associated series arising in Ramanujan's theory of divergent series; see Berndt's book [32, Chapter 6, especially pp. 134–135]. The latter two evaluations give the values of the Riemann zeta-function at the arguments -1 and -3, respectively.

The last result in Ramanujan's first letter coincides with Entry 29(b) of Chapter 10 of Ramanujan's second notebook [225]. Watson [280] published the first proof, and a flurry of papers establishing similar results followed. See Bailey's book [21, pp. 92–95] and Berndt's book [35, pp. 39–40] for further references.

Bertrand Russell to Lady Ottoline Morrell   **2 February 1913**

<u>Sunday night</u>

My Darling,

No letter from you came this morning which was sad, but I expect it will come tomorrow by 1st post. I have had a very full day, Sanger and Norton in and out perpetually. I enjoyed Sanger very much indeed. But à propos of Wittgenstein I shocked him by saying I thought the present society a feeble lot. I finished reading poor Bosanquet, and will write my review of him tomorrow. Also I wrote a longish criticism of Karin's paper, which I sent her. In Hall I saw Arthur Dakyns and Halvey. Before dinner North appeared, much excited because he has decided to be an engineer, which is what he always wished, only it was thought his health wouldn't stand it; he is very happy about it. In Hall I found Hardy, and Littlewood in a state of wild excitement, because they believe they have discovered a second Newton, a Hindu clerk in Madras on £20 a year. He wrote to Hardy telling of some results he has got, which Hardy thinks quite wonderful, especially as the man has had only an ordinary school education. Hardy has written to the Indian Office and hopes to get the man here at once. It is private at present. I am quite excited hearing of it.

It is very hard to get time for Matter in term time. When I get back from London I shall have Royce, proofs, Preface to write, lectures, lunch and walk with Ward—besides Wittgenstein and the other people who drop in. The result is that I shall hardly have one spare moment till Friday evening, and then I shall be sleepy, so that brings it to Saturday. Of course Bosanquet has taken time. But practically I <u>can't</u> do <u>very</u> much except in vacations. In fact, I see I shan't get really seriously to work on it till after America. After that, I must do whatever is necessary in order to get on with it—leave here, probably. When one thinks of work, the years slip by at such a dreadful pace that one will be dead before one is really under way. It is so different from

CHAPTER 2. FIRST TWO LETTERS TO HARDY

thinking of emotions, which makes life seem enormously long. One comes to so many ends and new beginnings, whereas in work one is always on the threshold. I find that in philosophy I grow every year more open-minded, less the slave of habitual opinions. It is a comfort, especially as I have taken great pains to have it so and avoid slavery to mental habits.

I am <u>very</u> happy, full of life and energy. I am longing to hear from you Dearest. All my love is with you every moment, my Heart.

Your B.

Commentary. The letter was written by the eminent and controversial philosopher Bertrand William Arthur Russell, O.M., F.R.S., the third Earl Russell (1872–1970), to Lady Ottoline Violet Anne Morrell (1873–1938). For the inclusion of the letter, we are indebted to the Harry Ransom Humanities Center at the University of Texas in Austin, and to Nicholas Griffin, the author of a book of selected letters of Bertrand Russell [85]. The letter above, which is one of more than 1000 letters that Russell wrote to Ottoline Morrell, is not included in Griffin's book.

Ottoline Morrell was the wife of Philip Edward Morrell, a Liberal Member of Parliament, and was the centre and patroness of a distinguished literary and artistic circle in Garsington and London. Russell entered Trinity College, Cambridge, in 1890 and was classed as 7th Wrangler in 1893. The following year, he obtained First Class Honours in the Moral Sciences Tripos. He was, during his long life, the author of more than 70 books and pamphlets. His magnum opus, written in collaboration with Alfred North Whitehead (1861–1947), was the monumental three-volume treatise *Principia Mathematica* [**250**], which has been called the greatest single contribution to logic since Aristotle. It was an attempt, not entirely successful, to reduce the whole of mathematics to logic. Of the numerous books that have been written about Russell and his philosophy, we single out two: (i) A.J. Ayer's *Russell* [**19**], which gives an account, in a relatively short compass, of the major incidents of his life and an exposition of his philosophy; (ii) G.H. Hardy's *Bertrand Russell and Trinity* [**111**], which is an account of Russell's removal from his Fellowship at Trinity during World War I, a result of his conviction under the Defence of the Realm Act because of his writings against conscription and the severe sentences imposed on conscientious objectors.

The date of the letter, 2 February, is not given but is inferred from what is known of Russell's movements at the time, and shows that Ramanujan's first letter of 16 January must have reached Cambridge at the latest by 1 February. It is of interest that, when he met Russell, Hardy had already written to the India Office. Information about the other people mentioned in Russell's letter can be found in Griffin's book, which contains other letters to Ottoline Morrell. When the letter was written, Russell was working on a theory of matter, which explains the reference to the subject at the beginning of the second paragraph.

G. H. Hardy to S. Ramanujan 8 February 1913

Trinity College
Cambridge

Dear Sir,

I was exceedingly interested by your letter and by the theorems which you state. You will however understand that, before I can judge properly of the value of what you have done, it is essential that I should see proofs of some of your assertions.

Your results seem to me to fall into roughly 3 classes:

(1) there are a number of results which are already known, or are easily deducible from known theorems;

(2) there are results which, so far as I know, are new and interesting, but interesting rather from their curiosity and apparent difficulty than their importance;

(3) there are results which appear to be new and important, but in which almost everything depends on the precise rigour of the methods of proof which you have used.

As instances of these 3 classes I may mention

(1) (a) the formula

$$\frac{\log 1}{\sqrt{1}} - \frac{\log 3}{\sqrt{3}} + \cdots = \left(\frac{\pi}{4} - \frac{\gamma}{2} - \frac{1}{2}\log 2\pi\right)\left(\frac{1}{\sqrt{1}} - \frac{1}{\sqrt{3}} + \cdots\right)$$

(easily deducible from a known theorem about $1^{-s} - 3^{-s} + 5^{-s} + \ldots$, which is to be found, e.g. in Bromwich's <u>Infinite Series</u>),

(b) if $\alpha\beta = \pi$

$$\sqrt{\alpha}\int_0^\infty \frac{e^{-x^2}}{\cosh \alpha x}dx = \text{etc.}$$

(this I proved myself in a paper in the <u>Quarterly Journal of Math.</u>),

(c) some at any rate of your theorems about B_n in your printed paper,

(d) $\qquad 1^2 \log 1 + 2^2 \log 2 + \cdots + x^2 \log x = \text{etc.}$

(a special case of the 'Euler–Maclaurin formula', given e.g. in Bromwich),

(e) I think your expression of $\int_0^a e^{-x^2}dx$ as a continued fraction,

(f) I think a number of your special results on series

(e.g. $\dfrac{1^{13}}{e^{2\pi} - 1} + \cdots = \dfrac{1}{24}$.

I fancy this is not hard to prove by Cauchy's Theory of Residues).

I need not say that, if what you say about your lack of training is to be taken literally, the fact that you should have rediscovered such interesting results is all to your credit. But you must be prepared for a certain amount of disappointment of this kind.

There are also of course known theories of divergent series, fractional orders of differentiation and integration, and so on. I should be extremely interested to compare your theories with these.

In this class also come some of your theorems about numbers. Thus the formula you give for the sum of all the divisors of all numbers $< n$ is known; and so is your formula $K \int_0^x \frac{dx}{\sqrt{\log x}} +$ etc. for the number of numbers of the form $p^2 + q^2$.

But I want particularly to see your proofs of your assertions here. You will understand that, in this theory, <u>everything</u> depends on rigorous exactitude of proof.

I should add that theorems of this character are not only interesting, but very difficult and important. If you have made sound and independent proofs of them, it would be, in my opinion, a very remarkable achievement. They are much more noteworthy theorems than those on p. 3.

<u>(2)</u> Among results of class (2) I may specially instance the following (of which I should like to see proofs).

(1) If
$$\int_0^\infty \frac{\cos nx}{e^{2\pi\sqrt{x}} - 1} dx = \phi(n),$$
then
$$\int_0^\infty \frac{\sin nx}{e^{2\pi\sqrt{x}} - 1} = \phi(n) - \frac{1}{2n} + \phi\left(\frac{\pi^2}{n}\right)\sqrt{\frac{2\pi^3}{n^3}},$$

(2) $\int_0^\infty \frac{dx}{(1+x^2)(1+r^2x^2)(1+r^4x^2)\cdots} = \frac{\pi}{2(1 + r + r^3 + r^6 + r^{10} + \cdots)},$

(3) the coefficient of x^n in $\frac{1}{1 - 2x + 2x^4 - \cdots}$ is the nearest integer to
$$\frac{1}{4n}\left\{\cosh \pi\sqrt{n} - \frac{\sinh \pi\sqrt{n}}{\pi\sqrt{n}}\right\}$$

—<u>this</u> theorem, I may say, both I and my colleague Mr Littlewood, who is also much interested in your work, think <u>cannot</u> be true.

(8) $\quad \dfrac{4}{x} + \dfrac{1^2}{2x} + \dfrac{3^2}{2x} + \cdots = \left(\dfrac{\Gamma\left(\frac{x+1}{4}\right)}{\Gamma\left(\frac{x+3}{4}\right)}\right)^2,$

(9) If
$$u = \frac{x}{1} + \frac{x^5}{1} + \cdots, \qquad v = \frac{\sqrt[5]{x}}{1} + \frac{x}{1} + \cdots$$
then
$$v^5 = \frac{1 - 2u + 4u^2 - 3u^3 + u^4}{1 + 3u + 4u^2 + 2u^3 + u^4},$$

(10) the particular formula for
$$\frac{1}{1} + \frac{e^{-2\pi}}{1} + \frac{e^{-4\pi}}{1} + \cdots.$$

(3) In this class I should put (assuming the proofs to be rigorous) some of your theorems about prime numbers—e.g. the expression you say you have for the number of primes $< x$, which is nearly exact.

It is of course possible that some of the results I have classed under (2) are really important, as examples of general methods. You always state your results in such particular forms that it is difficult to be sure about this.

I hope very much that you will send me <u>as quickly as possible</u> at any rate a few of your proofs, and follow this more at your leisure by a more detailed account of your work on primes and divergent series. It seems to me quite likely that you have done a good deal of work worth publication; and, if you can produce satisfactory demonstrations, I should be very glad to do what I can to secure it.

I have said nothing about some of your results—notably those about elliptic functions. I have not got them to refer to, as I handed them to another mathematician more expert than I in this special subject.

Hoping to hear from you again as soon as possible.

I am
Yours very truly,
G.H. Hardy

Further notes suggested by Mr Littlewood

1. $p(x)$ is known to be $O\left(e^{-\alpha\sqrt{\log x}}\right)$ for a certain α, and <u>not</u> to be $O\left(x^{\frac{1}{2}-\delta}\right)$ for any positive δ. The exact order of $p(x)$ is intimately connected with a very famous problem as to the roots of a certain function: <u>if</u> these are all real, $p(x) = O\left(x^{\frac{1}{4}+\delta}\right)$ for every positive δ. The converse also holds, and an independent rigorous proof of $p(x) = O\left(x^{\frac{1}{4}+\delta}\right)$ would be of extreme importance.

Please send the <u>formula</u> for the number of primes and as much proof as possible quickly.

2. Do you mean to have strictly proved that

(no. of primes $4x - 1$) - (no. of primes $4x + 1$) $\to \infty$?

This would be more than is known. It is known the excess is <u>sometimes</u> large.
  In the formula

$$f(x) = K \int_A^B \frac{dx}{\sqrt{\log x}} + \theta(x),$$

what is the <u>order</u> of θ? The best known result is

$$f(x) \sim \frac{Kx}{\sqrt{\log x}},$$

where

$$K = \frac{1}{\sqrt{(1 - \frac{1}{3^2})(1 - \frac{1}{7^2})(1 - \frac{1}{(11)^2})\cdots}}.$$

Does this agree with your formula?

3. In your formula for $\sum$ (number of divisors of n) do you mean that the error is < the last term you retain. If so it should be important. But to get the <u>leading terms</u> is comparatively easy.

Of course in all these questions everything depends on <u>absolute</u> rigour.

G.H.H.

Commentary. Since we offered comments on each result of Ramanujan's letter of 16 January 1913 in our Commentary after that letter, we have no further comments here about the mathematics contained therein.

The paper of Hardy mentioned under (b) of the first group of results classified by Hardy is [88], [112, pp. 10-24].

As with his letter to M.J.M. Hill, Ramanujan evidently also sent Hardy an offprint of his paper on Bernoulli numbers [205].

The expert on elliptic functions was probably Arthur Berry (1862-1929) of King's College, Cambridge. He had been Senior Wrangler in 1885, and it is known that he lectured on elliptic integrals at Cambridge. His list of publications is small, but it is of interest that it includes one on differential equations written in collaboration with M.J.M. Hill. Other possibilities are G.N. Watson (1886-1965) and L.J. Mordell (1888-1971), who were both in Cambridge early in 1913, but it is doubtful that they were known to be expert in elliptic function theory at that early stage in their careers.

At the head of the letter, to the left of the address, S. Narayana Aiyar has noted: This is the reply received to our first letter from Mr Hardy. S.N.

J. E. Littlewood

John Edensor Littlewood (1885–1977) was born in Rochester in Kent, but spent the years 1892–1899 at Wynberg near Cape Town, South Africa, where his father (who had been 9th Wrangler in the Mathematical Tripos in 1882) had gone to assume the post of headmaster of a new school. The son returned to England to complete his school education at St. Paul's School, London, and then matriculated at Trinity College, Cambridge, with an Entrance Scholarship in 1903. In 1905 he was classed as Senior Wrangler (bracketed with J. Mercer). He began research in the summer of 1906 under the supervision of E.W. Barnes. His first paper on integral functions of order zero was published the following year. It is of interest that Barnes then suggested to Littlewood that he should try to prove the Riemann hypothesis; the problem's notorious difficulty and importance were not mentioned to Littlewood.

Apart from a brief period between 1907 and 1910 as a Lecturer at Manchester University and war service working on ballistics from 1914 to 1918,

Littlewood spent all his working life in Cambridge, as a Fellow of Trinity, Cayley Lecturer, and ultimately, from 1928 to 1951, as Rouse Ball Professor. In 1911, he began his long and fruitful collaboration with G.H. Hardy. During this period their fame as mathematical analysts and analytical number theorists was worldwide, and their advice on the filling of academic posts was widely sought. Littlewood was elected a Fellow of the Royal Society in 1916. In addition to mathematical books and papers, he published in 1953 a miscellany of articles and anecdotes, written in a lighter vein, which give an insight into his life and times in Cambridge [163]. For a detailed account of his important mathematical work, see Burkill's obituary notice [58].

Littlewood also collaborated with other mathematicians, such as R.E.A.C. Paley, M.L. Cartwright, and A.C. Offord, but not with Ramanujan, although he was familiar with his work and closely examined with Hardy the results stated by Ramanujan in his first letter of 16 January 1913. Moreover, as recounted by E. Shils [260, p. 416], "Professor Littlewood once told me that he had been assigned by Hardy to the task of bringing Ramanujan up to date in the more rigorous methods of European mathematics which had emerged subsequently to the state reached by Ramanujan's studies in India; he said that it was extremely difficult because every time some matter, which it was thought Ramanujan needed to know, was mentioned, Ramanujan's response was an avalanche of original ideas which made it almost impossible for Littlewood to persist in his original intention."

Gilbert T. Walker to Francis Dewsbury   **26 February 1913**

Madras

To
    The Registrar of the
        University of Madras.

Sir,

I have the honour to draw your attention to the case of S. Ramanujan, a clerk in the Accounts Department of the Madras Port Trust. I have not seen him, but was yesterday shown some of his work in the presence of Sir Francis Spring. He is, I am told, 22 years of age and the character of the work that I saw impressed me as comparable in originality with that of a Mathematics fellow in a Cambridge College; it appears to lack, however, as might be expected in the circumstances the completeness and precision necessary before the universal validity of the results could be accepted. I have not specialised in the branches of pure mathematics at which he has worked, and could not therefore form a reliable estimate of his abilities, which might be of an order to bring him a European reputation. But it was perfectly clear to me that the university would be justified in enabling S. Ramanujan for a few years at least to spend the whole of his time on mathematics without any anxiety as to his livelihood, and I would suggest that they should communicate with Mr. G.H. Hardy, Fellow of Trinity College, Cambridge with whom he is already in correspondence and assure Mr. Hardy of their interest in him.

> I have the honour to be
> Sir
> Your most obedient servant
> Gilbert T. Walker,
> Director General of Observatories

Commentary. This letter is found in Srinivasan's collection [263].

Ramanujan's age was 25, not 22, when this letter was written.

Sir Gilbert Thomas Walker (1868–1958) was classed as Senior Wrangler in Part I of the Mathematical Tripos in 1889. A Fellow of Trinity College (1891–1904), he served as Director–General of Indian Observatories from 1904 to 1924. Subsequently (1924–1935), he became Professor of Meteorology at Imperial College, London.

The Observatory was in the district of Madras called Nungambakkam and located on the south bank of the River Cooum, about $2\frac{1}{2}$ miles from the Bay of Bengal.

While he was Director–General of Indian Observatories, Walker became absorbed with forecasting the duration and intensity of the monsoon. The importance of the monsoon had long been recognized, and the ability to forecast it months in advance was of great economic value. He averaged surface temperatures and precipitation values for individual months for each of the many oceanic island stations near the equator, and investigated time series of those averages over decades. He discovered a 26-month oscillation, which has come to be known as Walker's quasi-biennial oscillation or, for brevity, Walker circulation. Walker's method of forecasting was empirical, as he took the monthly precipitation anomalies from the last 26 months and superimposed them on the mean precipitation values to make predictions for the next 26 months. This is systematically unsatisfying, because there is no cause–effect understanding of the method. However, Walker's method represented a substantial improvement over that of the time, and his predictions were valuable.

This oscillation is now also recognized in wind direction in the high troposphere (above 25,000 feet) in the tropics. These winds change direction with a period of about 26 months, a phenomenon now known as the Southern Oscillation, the lower half of the el Niño–Southern Oscillation. The el Niño and Southern Oscillation are dynamically related. Thus high pressures over the Pacific Ocean coincide with low pressures over the Indian Ocean, the presence of the latter during winter indicating a likelihood of heavy summer rains. Up until Walker's time, weather studies and forecasting were "local," with weather conditions in distant places being considered irrelevant.

Thus with his brilliant empirical work, Walker was one of the first to recognize that the winds of the hemisphere are spatially continuous and related, and so provided a paradigm for modeling the general circulation of the atmosphere. The current Director-General of Indian Observatories, Dr. R.P. Sarker [**80**, p. 191], asserts that "Sir Gilbert is the father of monsoon studies." (For a description of the Indian monsoon and its effects on the lives of Indian people, see A. Frater's delightful book [**80**].)

Francis Dewsbury (1872–1946) was educated at Sedbergh School, Cumbria, and at St John's College, Cambridge, where he was admitted in 1889. He took the Law Tripos and graduated B.A. and LL.B. in 1892; he qualified as a solicitor in 1896. Between 1898 and 1903 he was engaged in schoolteaching, latterly at Dunnheved College, Launceston, Cornwall. He then became district clerk to various Education Committees. He moved to India, and in the India Office Lists for 1909 to 1924 is listed as Registrar of the University of Madras. He retired in the latter year and returned to England. According to his successor as Registrar, William McLean, who died in Glasgow on 8 January 1994 at the age of 94, Dewsbury spent some time after his return at Reading helping that University with an appeal for funds. He then moved to a house in Sedbergh near his old school, where he died.

S. Ramanujan to G.H. Hardy 27 February 1913

(1)

Madras Port Trust Office
Accounts Department.

Dear Sir,

I am very much gratified on perusing your letter of the 8th February 1913. I was expecting a reply from you similar to the one which a Mathematics Professor at London wrote asking me to study carefully Bromwich's Infinite Series and not fall into the pitfalls of divergent series. I have found a friend in you who views my labours sympathetically. This is already some encouragement to me to proceed with my onward course. I find in many a place in your letter rigorous proofs are required and so on and you ask me to communicate the methods of proof. If I had given you my methods of proof I am sure you will follow the London Professor. But as a fact I did not give him any proof but made some assertions as the following under my new theory. I told him that the sum of an infinite number of terms of the series:- $1 + 2 + 3 + 4 + \cdots = -\frac{1}{12}$ under my theory. If I tell you this you will at once point out to me the lunatic asylum as my goal. I dilate on this simply to convince you that you will not be able to follow my methods of proof if I indicate the lines on which I proceed in a single letter. You may ask how you can accept results based upon wrong premises. What I tell you is this. Verify the results I give and if they agree with your results, got by treading on the groove in which the present day mathematicians move, you should at least grant that there may be some truths in my fundamental basis. So what I now want at this stage is for eminent professors like you to recognize that there

(2)

is some worth in me. I am already a half starving man. To preserve my brains I want food and this is now my first consideration. Any sympathetic letter from you will be helpful to me here to get a scholarship either from the University or from the Government.

With respect to the mathematics portion of your letter it is the results that you class under 1st head and which you say are already known or are easily deducible from known theorems which encourage me now to proceed onward. For my results are verified to be true even though I may take my stand upon slender basis. I may now assure myself that my results and my method of proof are as rigid as ether. Suppose I say that ether is rigid to one who does not know the ether hypothesis. He will simply laugh. The results I gave in my letter to you were only examples derived from substitution of particular values in some of the theorems I got. This time I give you some more general than the previous ones but still only particular cases of my theorems.

You may judge me hard that I am silent on the methods of proof. I have to re-iterate that I may be misunderstood if I give in a short compass the lines on which I proceed. It is not on account of my unwillingness on my part but because I fear I shall not be able to explain everything in a letter. I do not mean that the methods should be buried with me. I shall

(3)

have them published if my results are recognised by eminent men like you. You ask me to give you the expression I have got for the number of prime numbers within a given number. These are the expressions that I have obtained for the number of primes less than a give[n] number.

1. The number of prime numbers less than e^a

$$= \int_0^\infty \frac{a^x \, dx}{x S_{x+1} \Gamma(x+1)},$$

where

$$S_{x+1} = \frac{1}{1^{x+1}} + \frac{1}{2^{x+1}} + \frac{1}{3^{x+1}} + \frac{1}{4^{x+1}} + \cdots.$$

2. The number of prime numbers less than $n =$

$$\frac{2}{\pi} \left\{ \frac{2}{B_2} \left(\frac{\log n}{2\pi} \right) + \frac{4}{3 B_4} \left(\frac{\log n}{2\pi} \right)^3 + \frac{6}{5 B_6} \left(\frac{\log n}{2\pi} \right)^5 + \frac{8}{7 B_8} \left(\frac{\log n}{2\pi} \right)^7 + \cdots \right\},$$

where $B_2 = \frac{1}{6}$, $B_4 = \frac{1}{30}$, $B_6 = \frac{1}{42} \ldots$, the Bernoullian numbers.

CHAPTER 2. FIRST TWO LETTERS TO HARDY   55

3. The number of prime numbers less than $n =$

(*) $$\int_\mu^n \frac{dx}{\log x} - \frac{1}{2}\int_\mu^{\sqrt{n}} \frac{dx}{\log x} - \frac{1}{3}\int_\mu^{\sqrt[3]{n}} \frac{dx}{\log x} - \frac{1}{5}\int_\mu^{\sqrt[5]{n}} \frac{dx}{\log x} \bigg|$$
$$+ \frac{1}{6}\int_\mu^{\sqrt[6]{n}} \frac{dx}{\log x} - \frac{1}{7}\int_\mu^{\sqrt[7]{n}} \frac{dx}{\log x} \bigg| + \frac{1}{10}\int_\mu^{\sqrt[10]{n}} \frac{dx}{\log x}$$
$$- \frac{1}{11}\int_\mu^{\sqrt[11]{n}} \frac{dx}{\log x} \bigg| - \frac{1}{13}\int_\mu^{\sqrt[13]{n}} \frac{dx}{\log x} + \frac{1}{14}\int_\mu^{\sqrt[14]{n}} \frac{dx}{\log x} \bigg|$$
$$+ \frac{1}{15}\int_\mu^{\sqrt[15]{n}} \frac{dx}{\log x} - \frac{1}{17}\int_\mu^{\sqrt[17]{n}} \frac{dx}{\log x} \bigg| - \frac{1}{19}\int_\mu^{\sqrt[19]{n}} \frac{dx}{\log x} + \cdots,$$

where $\mu = 1.45136380$ nearly.

The numbers 1, 2, 3, 5, 6, 7, 10, 11, 13, 14, 15, 17, 19, ... above are numbers containing dissimilar prime divisors; hence 4, 8, 9, 12, 16, 18, 20, ... are excluded. (Plus sign for even number of prime divisors and minus sign for odd number of prime divisors.) As soon as a term becomes less than unity in practical calculation we should stop at the term before any [vertical] red line marked above and not anywhere; hence the first four terms are necessary even when n is very small.

(4)

Prime numbers begin with 2 and not with 1.

For practical calculations,

$$\int_\mu^n \frac{dx}{\log x} = n\left\{\frac{1}{\log n} + \frac{1!}{(\log n)^2} + \frac{2!}{(\log n)^3} + \cdots + \frac{(k-1)!}{(\log n)^k}\theta\right\},$$

where

$$\theta = \left(\frac{2}{3} - \delta\right) + \frac{1}{\log n}\left\{\frac{4}{135} - \frac{\delta^2(1-\delta)}{3}\right\}$$
$$+ \frac{1}{(\log n)^2}\left\{\frac{8}{2835} + \frac{2\delta(1-\delta)}{135} - \frac{\delta(1-\delta^2)(2-3\delta^2)}{45}\right\} + \cdots,$$

where $\delta = k - \log n$.

It is better to choose k to be the integer just greater than $\log n$.

The number of primes less than $50 = 15$ according to my formula 14.9,

the number of primes less than $300 = 62$ according to my formula 61.9,

the number of primes less than $1000 = 168$ according to my formula 168.2,

and so on.

I have also found expressions for the number of prime numbers of a given form (say of the form $24n + 17$) less than any given number.

Primes of the form $4n + 1 =$ primes of the form $6n + 1$,

primes of the form $4n-1$ = primes of the form $6n-1$,
primes of the form $8n+1$ = primes of the form $12n+1$.
Those of the forms $8n+3$, $8n+5$, $8n+7$, $12n+5$, $12n+7$ and $12n+11$ are all equal. But
(primes of the form $4n-1$) − (those of the form $4n+1$) $\to \infty$,
(primes of the form $6n-1$) − (those of the form $6n+1$) $\to \infty$,
(primes of the form $8n+3$) − (those of the form $8n+1$) $\to \infty$,
(primes of the form $12n+5$) − (those of the form $12n+1$) $\to \infty$.

I have not merely shown that the difference tends to infinity but found out expressions (like those for prime numbers) for the difference, within any given number.

With kind regards
Yours very truly
S. Ramanujan

(5)

(a) From the formulae given on page 3 you may infer that the best known order of $p(x)$ viz. $O\left(e^{-\alpha\sqrt{\log x}}\right)$ is wrong.

(b) The order of $\theta(x)$ which you asked in your letter is $\dfrac{\sqrt{x}}{\sqrt{\log x}}$.

(c) The sum of the number of divisors of n natural numbers $= n(2c - 1 + \log n) + \frac{1}{2}t_n + E$, where $c = .5772\ldots$, t_n, the nth term i.e. the number of divisors of n, and E, the error in the approximation.

E is of lower order than $\log n$ whatever be the value of n. In practical calculations you will find E is very small.

(d) The coefficient of x^n in $\dfrac{1}{1-2x+2x^4-\cdots}$

$$= \frac{1}{4n}\left\{\cosh \pi\sqrt{n} - \frac{\sinh \pi\sqrt{n}}{\pi\sqrt{n}}\right\} + F(\cos \pi\sqrt{n}) + f(\sin \pi\sqrt{n}).$$

I have not written here the forms of F and f as they are very irregular and complicated, and their values are very small, in most cases a very small proper fraction. In a few cases they assume some small finite values. Hence the coefficient of x^n is an integer very near to

$$\frac{1}{4n}\left\{\cosh \pi\sqrt{n} - \frac{\sinh \pi\sqrt{n}}{\pi\sqrt{n}}\right\},$$

and not always the nearest integer to it as I hastily wrote to you in my previous letter. Yet in many cases you will find the coefficient to be the nearest integer though not always. At present we may be contented with the result viz. The coefficient of x^n in the above function divided by

$$\frac{1}{4n}\left\{\cosh \pi\sqrt{n} - \frac{\sinh \pi\sqrt{n}}{\pi\sqrt{n}}\right\}$$

is very very nearly equal to unity for all values of n, from 0 to ∞ and very rapidly approaches 1 when n becomes infinity.

(e) All kinds of examples written to you in my previous letter and some more examples which will be found in the following pages are not ends in themselves but all similar questions can be solved from my theorems.

(6)

(1) If
$$F(x) = \frac{1}{1} + \frac{x}{1} + \frac{x^2}{1} + \frac{x^3}{1} + \frac{x^4}{1} + \frac{x^5}{1} + \cdots,$$
then

(1a) $\left\{\frac{\sqrt{5}+1}{2} + e^{-2\alpha/5} F(e^{-2\alpha})\right\} \left\{\frac{\sqrt{5}+1}{2} + e^{-2\beta/5} F(e^{-2\beta})\right\} = \frac{5+\sqrt{5}}{2}$

with the condition $\alpha\beta = \pi^2$.

N. B. It is always possible to find exactly the value of
$$\frac{1}{1} + \frac{e^{-\pi\sqrt{n}}}{1} + \frac{e^{-2\pi\sqrt{n}}}{1} + \frac{e^{-3\pi\sqrt{n}}}{1} + \cdots$$
and similar continued fraction if n be any rational quantity. e.g.

(1b) $\frac{1}{1} + \frac{e^{-2\pi\sqrt{5}}}{1} + \frac{e^{-4\pi\sqrt{5}}}{1} + \frac{e^{-6\pi\sqrt{5}}}{1} + \frac{e^{-8\pi\sqrt{5}}}{1} + \cdots$

$$= e^{2\pi/\sqrt{5}} \left\{ \frac{\sqrt{5}}{1 + \sqrt[5]{5^{\frac{3}{4}} \left(\frac{\sqrt{5}-1}{2}\right)^{5/2} - 1}} - \frac{\sqrt{5}+1}{2} \right\}.$$

The above theorem is a particular case of a theorem on the continued fraction

(1c) $\frac{1}{1} + \frac{ax}{1} + \frac{ax^2}{1} + \frac{ax^3}{1} + \frac{ax^4}{1} + \frac{ax^5}{1} + \cdots,$

which is a particular case of the continued fraction

(1d) $\frac{1}{1} + \frac{ax}{1+bx} + \frac{ax^2}{1+bx^2} + \frac{ax^3}{1+bx^3} + \cdots,$

which is a particular case of a general theorem on continued fractions.

2(i) $4 \int_0^\infty \frac{xe^{-x\sqrt{5}}}{\cosh x} dx = \frac{1}{1} + \frac{1^2}{1} + \frac{1^2}{1} + \frac{2^2}{1} + \frac{2^2}{1} + \frac{3^2}{1} + \frac{3^2}{1} + \cdots.$

(ii) $2 \int_0^\infty \frac{x^2 e^{-x\sqrt{3}}}{\sinh x} dx = \frac{1}{1} + \frac{1^3}{1} + \frac{1^3}{3} + \frac{2^3}{1} + \frac{2^3}{5} + \frac{3^3}{1} + \frac{3^3}{7} + \cdots.$

(3) $1 - 5 \cdot \left(\frac{1}{2}\right)^5 + 9 \cdot \left(\frac{1 \cdot 3}{2 \cdot 4}\right)^5 - 13 \cdot \left(\frac{1 \cdot 3 \cdot 5}{2 \cdot 4 \cdot 6}\right)^5 + \cdots = \frac{2}{\{\Gamma(\frac{3}{4})\}^4}.$

(4) $$\frac{1}{1^3}\left(\coth \pi x + x^2 \coth \frac{\pi}{x}\right) + \frac{1}{2^3}\left(\coth 2\pi x + x^2 \coth \frac{2\pi}{x}\right) +$$
$$\frac{1}{3^3}\left(\coth 3\pi x + x^2 \coth \frac{3\pi}{x}\right) + \cdots = \frac{\pi^3}{90x}(x^4 + 5x^2 + 1).$$

(5) $$\frac{1^5}{e^{2\pi}-1}\frac{1}{2500+1^4} + \frac{2^5}{e^{4\pi}-1}\frac{1}{2500+2^4} + \frac{3^5}{e^{6\pi}-1}\frac{1}{2500+3^4} + \cdots$$
$$= \frac{123826979}{6306456} - \frac{25\pi}{4}\coth^2 5\pi.$$

(7)

(6) If
$$v = \frac{x}{1} + \frac{x^3 + x^6}{1} + \frac{x^6 + x^{12}}{1} + \frac{x^9 + x^{18}}{1} + \cdots,$$
then

(i) $$x\left(1 + \frac{1}{v}\right) = \frac{1 + x + x^3 + x^6 + x^{10} + \cdots}{1 + x^9 + x^{27} + x^{54} + x^{90} + \cdots},$$

(ii) $$x^3\left(1 + \frac{1}{v^3}\right) = \left(\frac{1 + x + x^3 + x^6 + x^{10} + \cdots}{1 + x^3 + x^9 + x^{18} + x^{30} + \cdots}\right)^4.$$

(7) If n is any odd integer,
$$\frac{1}{\cosh\frac{\pi}{2n} + \cos\frac{\pi}{2n}} - \frac{1}{3\left(\cosh\frac{3\pi}{2n} + \cos\frac{3\pi}{2n}\right)} + \frac{1}{5\left(\cosh\frac{5\pi}{2n} + \cos\frac{5\pi}{2n}\right)}$$
$$-\frac{1}{7\left(\cosh\frac{7\pi}{2n} + \cos\frac{7\pi}{2n}\right)} + \cdots = \frac{\pi}{8}.$$

(8) $$\int_0^\infty \frac{(1+ab^2x^2)(1+ab^4x^2)(1+ab^6x^2)(1+ab^8x^2)\cdots}{(1+x^2)(1+b^2x^2)(1+b^4x^2)(1+b^6x^2)\cdots}dx$$
$$= \frac{\pi}{2(1+b+b^3+b^6+b^{10}+b^{15}+\cdots)}\frac{(1-ab^2)(1-ab^4)(1-ab^6)\cdots}{(1-ab)(1-ab^3)(1-ab^5)\cdots}.$$

(9) $$\int_0^a \frac{\sin z}{z}dz = \frac{\pi}{2} - r\cos(a - \theta),$$
where
$$r = \sqrt{\int_0^\infty \frac{e^{-z}}{z}\log\left(1 + \frac{z^2}{a^2}\right)dz}$$
and

$$\theta = \tan^{-1}\left\{\frac{\int_0^\infty \frac{ze^{-z}}{a^2+z^2}dz}{\int_0^\infty \frac{ae^{-z}}{a^2+z^2}dz}\right\}.$$

(10) If

$$F(\alpha,\beta,\gamma,\delta,\epsilon) = 1 + \frac{\alpha}{1!}\cdot\frac{\beta}{\delta}\cdot\frac{\gamma}{\epsilon} + \frac{\alpha(\alpha+1)}{2!}\cdot\frac{\beta(\beta+1)}{\delta(\delta+1)}\cdot\frac{\gamma(\gamma+1)}{\epsilon(\epsilon+1)} + \cdots,$$

then

$$F(\alpha,\beta,\gamma,\delta,\epsilon) = \frac{\Gamma(\delta)\Gamma(\delta-\alpha-\beta)}{\Gamma(\delta-\alpha)\Gamma(\delta-\beta)}F(\alpha,\beta,\epsilon-\gamma,\alpha+\beta-\delta+1,\epsilon)$$

$$+ \frac{\Gamma(\delta)\Gamma(\epsilon)\Gamma(\alpha+\beta-\delta)\Gamma(\delta+\epsilon-\alpha-\beta-\gamma)}{\Gamma(\alpha)\Gamma(\beta)\Gamma(\epsilon-\gamma)\Gamma(\delta+\epsilon-\alpha-\beta)}$$

$$\times F(\delta-\alpha,\delta-\beta,\delta+\epsilon-\alpha-\beta-\gamma,\delta-\alpha-\beta+1,\delta+\epsilon-\alpha-\beta).$$

(8)

(11) When x is not great,

$$xe^{\frac{1}{2}}\left\{e^{-\frac{1}{2}(1+x)^2} + e^{-\frac{1}{2}(1+2x)^2} + e^{-\frac{1}{2}(1+3x)^2} + \cdots\right\}$$

$$= \frac{1}{1+} \frac{1}{1+} \frac{2}{1+} \frac{3}{1+} \frac{4}{1+} \cdots$$

$$- \frac{x}{2} + \frac{x^2}{12} + \frac{x^4}{360} + \frac{x^6}{5040} + \frac{x^8}{60480} + \frac{x^{10}}{1710720} + \cdots.$$

(12) If $\frac{1}{2}\pi\alpha = \log\tan\left\{\frac{1}{4}\pi(1+\beta)\right\}$, then

$$\frac{1^2+\alpha^2}{1^2-\beta^2}\left(\frac{3^2-\beta^2}{3^2+\alpha^2}\right)^3\left(\frac{5^2+\alpha^2}{5^2-\beta^2}\right)^5\left(\frac{7^2-\beta^2}{7^2+\alpha^2}\right)^7\left(\frac{9^2+\alpha^2}{9^2-\beta^2}\right)^9\cdots = e^{\pi\alpha\beta/2}.$$

(13)
$$\frac{a}{1+n} + \frac{a^2}{3+n} + \frac{(2a)^2}{5+n} + \frac{(3a)^2}{7+n} + \cdots$$

$$= 2a\int_0^1 z^{n/\sqrt{1+a^2}}\frac{dz}{\left(\sqrt{1+a^2}+1\right) + z^2\left(\sqrt{1+a^2}-1\right)},$$

which is a particular case of the continued fraction

$$\frac{a}{p+n} + \frac{1\cdot p\cdot a^2}{p+n+2} + \frac{2(p+1)a^2}{p+n+3} + \cdots,$$

which is a particular case of a corollary to a theorem on transformation of integrals and continued fractions.

(14) If

$$F(\alpha,\beta) = \alpha + \frac{(1+\beta)^2+k}{2\alpha} + \frac{(3+\beta)^2+k}{2\alpha} + \frac{(5+\beta)^2+k}{2\alpha} + \cdots,$$

then
$$F(\alpha, \beta) = F(\beta, \alpha).$$

(15) If
$$F(\alpha, \beta) = \frac{\alpha}{n} + \frac{\beta^2}{n} + \frac{(2\alpha)^2}{n} + \frac{(3\beta)^2}{n} + \cdots,$$
then
$$F(\alpha, \beta) + F(\beta, \alpha) = 2F\left(\frac{1}{2}(\alpha + \beta), \sqrt{\alpha\beta}\right).$$

(16) If
$$F(\alpha, \beta) = \tan^{-1}\left\{\frac{\alpha}{x} + \frac{\beta^2 + k^2}{x} + \frac{\alpha^2 + (2k)^2}{x} + \frac{\beta^2 + (3k)^2}{x} + \cdots\right\},$$
then
$$F(\alpha, \beta) + F(\beta, \alpha) = 2F\left(\frac{1}{2}(\alpha + \beta), \frac{1}{2}(\alpha + \beta)\right).$$

(9)

(17) If
$$\frac{1 + \left(\frac{1}{2}\right)^2 (1-k) + \left(\frac{1\cdot 3}{2\cdot 4}\right)^2 (1-k)^2 + \cdots}{1 + \left(\frac{1}{2}\right)^2 k + \left(\frac{1\cdot 3}{2\cdot 4}\right)^2 k^2 + \cdots} = \sqrt{210},$$
then
$$k = (\sqrt{2} - 1)^4 (2 - \sqrt{3})^2 (\sqrt{7} - \sqrt{6})^4 (8 - 3\sqrt{7})^2$$
$$\times (\sqrt{10} - 3)^4 (4 - \sqrt{15})^4 (\sqrt{15} - \sqrt{14})^2 (6 - \sqrt{35})^2.$$

(18) If
$$\frac{1 + \frac{1\cdot 2}{3^2}(1-\alpha) + \frac{1\cdot 2\cdot 4\cdot 5}{3^2\cdot 6^2}(1-\alpha)^2 + \cdots}{1 + \frac{1\cdot 2}{3^2}\alpha + \frac{1\cdot 2\cdot 4\cdot 5}{3^2\cdot 6^2}\alpha^2 + \cdots} = 5\frac{1 + \frac{1\cdot 2}{3^2}(1-\beta) + \cdots}{1 + \frac{1\cdot 2}{3^2}\beta + \cdots},$$
then
$$(\alpha\beta)^{\frac{1}{3}} + \{(1-\alpha)(1-\beta)\}^{\frac{1}{3}} + 3\{\alpha\beta(1-\alpha)(1-\beta)\}^{\frac{1}{6}} = 1.$$

(19) If
$$\frac{1 + \frac{1\cdot 3}{4^2}(1-\alpha) + \frac{1\cdot 3\cdot 5\cdot 7}{4^2\cdot 8^2}(1-\alpha)^2 + \cdots}{1 + \frac{1\cdot 3}{4^2}\alpha + \frac{1\cdot 3\cdot 5\cdot 7}{4^2\cdot 8^2}\alpha^2 + \cdots} = 5\frac{1 + \frac{1\cdot 3}{4^2}(1-\beta) + \cdots}{1 + \frac{1\cdot 3}{4^2}\beta + \cdots},$$
then
$$(\alpha\beta)^{\frac{1}{2}} + \{(1-\alpha)(1-\beta)\}^{\frac{1}{2}} + 8\{\alpha\beta(1-\alpha)(1-\beta)\}^{\frac{1}{6}}$$
$$\times [(\alpha\beta)^{\frac{1}{6}} + \{(1-\alpha)(1-\beta)\}^{\frac{1}{6}}] = 1.$$

(20) If
$$F(\alpha) = \frac{\int_0^{\pi/2} \frac{d\phi}{\sqrt{1-(1-\alpha)\sin^2\phi}}}{\int_0^{\pi/2} \frac{d\phi}{\sqrt{1-\alpha\sin^2\phi}}}$$

and
$$F(\alpha) = 3F(\beta) = 5F(\gamma) = 15F(\delta),$$

then

(i) $\quad [(\alpha\delta)^{\frac{1}{8}} + \{(1-\alpha)(1-\delta)\}^{\frac{1}{8}}][(\beta\gamma)^{\frac{1}{8}} + \{(1-\beta)(1-\gamma)\}^{\frac{1}{8}}] = 1$

(ii) $\quad (\alpha\delta)^{\frac{1}{8}} - \{(1-\alpha)(1-\delta)\}^{\frac{1}{8}} = (\beta\gamma)^{\frac{1}{8}} - \{(1-\beta)(1-\gamma)\}^{\frac{1}{8}}$

(iii) $\quad [1 + (\beta\gamma)^{\frac{1}{8}} + \{(1-\beta)(1-\gamma)\}^{\frac{1}{8}}][1 - (\alpha\delta)^{\frac{1}{8}} - \{(1-\alpha)(1-\delta)\}^{\frac{1}{8}}]$
$$= 2\{16\alpha\beta(1-\alpha)(1-\beta)(1-\gamma)(1-\delta)\}^{\frac{1}{24}}$$

(iv) $\quad \dfrac{1 + (\beta\gamma)^{\frac{1}{8}} + \{(1-\beta)(1-\gamma)\}^{\frac{1}{8}}}{1 - (\alpha\delta)^{\frac{1}{8}} - \{(1-\alpha)(1-\delta)\}^{\frac{1}{8}}} = \left\{\dfrac{\beta\gamma(1-\beta)(1-\gamma)}{\alpha\delta(1-\alpha)(1-\delta)}\right\}^{\frac{1}{8}}$

(v) $\quad (\alpha\beta\gamma\delta)^{\frac{1}{8}} + \{(1-\alpha)(1-\beta)(1-\gamma)(1-\delta)\}^{\frac{1}{8}}$
$\quad\quad + \{16\alpha\beta\gamma\delta(1-\alpha)(1-\beta)(1-\gamma)(1-\delta)\}^{\frac{1}{24}} = 1$

(vi) $\quad Q + \dfrac{1}{Q} = \left(P + \dfrac{1}{P}\right)\sqrt{2},$

where
$$Q = \left\{\frac{\alpha\delta(1-\alpha)(1-\delta)}{\beta\gamma(1-\beta)(1-\gamma)}\right\}^{\frac{1}{16}}$$

and
$$P = \{256\alpha\beta\gamma\delta(1-\alpha)(1-\beta)(1-\gamma)(1-\delta)\}^{\frac{1}{48}}.$$
(10)

(21) If $F(\alpha) = $ the same in (20) and if
$$F(\alpha) = 3F(\beta) = 13F(\gamma) = 39F(\delta)$$
or $\quad F(\alpha) = 5F(\beta) = 11F(\gamma) = 55F(\delta)$
or $\quad F(\alpha) = 7F(\beta) = 9F(\gamma) = 63F(\delta),$

then

$$\frac{\{(1-\alpha)(1-\delta)\}^{\frac{1}{8}} - (\alpha\delta)^{\frac{1}{8}}}{\{(1-\beta)(1-\gamma)\}^{\frac{1}{8}} - (\beta\gamma)^{\frac{1}{8}}} = \frac{1 + \{(1-\alpha)(1-\delta)\}^{\frac{1}{4}} + (\alpha\delta)^{\frac{1}{4}}}{1 + \{(1-\beta)(1-\gamma)\}^{\frac{1}{4}} + (\beta\gamma)^{\frac{1}{4}}}$$

$$= \frac{\left(\frac{1}{2}\left\{1 + (\alpha\delta)^{\frac{1}{2}} + [(1-\alpha)(1-\delta)]^{\frac{1}{2}}\right\}\right)^{\frac{1}{2}} - \{(1-\alpha)(1-\delta)\}^{\frac{1}{8}}}{(\beta\gamma)^{\frac{1}{8}} + \{(1-\beta)(1-\gamma)\}^{\frac{1}{8}}}$$

$$= \left\{\frac{1 + (\frac{1}{2})^2\beta + \cdots}{1 + (\frac{1}{2})^2\alpha + \cdots} \frac{1 + (\frac{1}{2})^2\gamma + \cdots}{1 + (\frac{1}{2})^2\delta + \cdots}\right\}^{\frac{1}{2}}.$$

(22) If $F(\alpha) =$ the same in (20) or (21) and if

$$F(\alpha) = 3F(\beta) = 29F(\gamma) = 87F(\delta)$$

or $\quad F(\alpha) = 5F(\beta) = 27F(\gamma) = 135F(\delta)$

or $\quad F(\alpha) = 7F(\beta) = 25F(\gamma) = 175F(\delta)$

or $\quad F(\alpha) = 9F(\beta) = 23F(\gamma) = 207F(\delta)$

or $\quad F(\alpha) = 11F(\beta) = 21F(\gamma) = 231F(\delta)$

or $\quad F(\alpha) = 13F(\beta) = 19F(\gamma) = 247F(\delta)$

or $\quad F(\alpha) = 15F(\beta) = 17F(\gamma) = 255F(\delta),$

then

$$\left(\frac{1}{2}\left\{1 + (\beta\gamma)^{\frac{1}{2}} + [(1-\beta)(1-\gamma)]^{\frac{1}{2}}\right\}\right)^{\frac{1}{2}} + (\beta\gamma)^{\frac{1}{8}}$$

$$+ \{(1-\beta)(1-\gamma)\}^{\frac{1}{8}} + \{\beta\gamma(1-\beta)(1-\gamma)\}^{\frac{1}{8}}$$

$$= (1 + (\alpha\delta)^{\frac{1}{4}} + \{(1-\alpha)(1-\delta)\}^{\frac{1}{4}})$$

$$\times \left\{\frac{1 + (\frac{1}{2})^2\alpha + \cdots}{1 + (\frac{1}{2})^2\beta + \cdots} \frac{1 + (\frac{1}{2})^2\gamma + \cdots}{1 + (\frac{1}{2})^2\delta + \cdots}\right\}^{\frac{1}{2}}.$$

(23) $\quad (1 + e^{-\pi\sqrt{1353}})(1 + e^{-3\pi\sqrt{1353}})(1 + e^{-5\pi\sqrt{1353}}) \cdots$

$$= \sqrt[4]{2}e^{-\frac{1}{24}\pi\sqrt{1353}} \times \sqrt{\sqrt{\frac{569 + 99\sqrt{33}}{8}} + \sqrt{\frac{561 + 99\sqrt{33}}{8}}}$$

$$\times \sqrt{\sqrt{\frac{25 + 3\sqrt{33}}{8}} + \sqrt{\frac{17 + 3\sqrt{33}}{8}}} \times \sqrt[4]{\frac{\sqrt{123} + 11}{\sqrt{2}}}$$

$$\times \sqrt[8]{10 + 3\sqrt{11}} \times \sqrt[8]{26 + 15\sqrt{3}} \times \sqrt[12]{\frac{6817 + 321\sqrt{451}}{\sqrt{2}}}.$$

Commentary. Ramanujan's second letter as reproduced in his *Collected Papers* [226] is not complete, for missing material is marked by ... at certain places. In our transcription above, we have indicated Ramanujan's pagination.

The example $1 + 2 + 3 + 4 + \cdots = -\frac{1}{12}$ is found in Ramanujan's first letter to Hardy, and we discussed it in our commentary above.

The results quoted in 1. and 2. of page (3) are those described less explicitly under (1) and (2) in Section I of the first letter. Ramanujan always used S_x instead of the customary notation $\zeta(x)$ for the Riemann zeta–function. Note that Ramanujan's sign convention for the Bernoulli numbers is not that most commonly used today. See Hardy's paper [102], [106, pp. 234-238] and book [110, Chapter 2] or Berndt's book [40, pp. 129, 124] for further discussions of these formulas.

The formula for the number of primes less than n given at the beginning of 3. is Riemann's series for $\pi(n)$, except that the lower limit for each integral is normally taken to be 0. More precisely,

$$\pi(x) \approx \sum_{n=1}^{\infty} \frac{\mu(n)}{n} \mathrm{Li}(x^{1/n}),$$

where μ denotes the Möbius function and

$$\mathrm{Li}(x) = \int_0^x \frac{dt}{\log t}.$$

The number μ in Ramanujan's integrals is defined to be that unique positive number such that

$$PV \int_0^{\mu} \frac{dt}{\log t} = 0.$$

Ramanujan's numerical value of μ is not quite correct. Soldner [176, p. 88] calculated the value $\mu = 1.4513692346$, but this is also not quite correct. We used MACSYMA to determine the correct value $\mu = 1.4513692349$. Vertical bars are inserted after certain terms according to the following rules. The first bar is inserted after the 4th term. Between subsequent bars an equal number of positive and negative terms appear with the bars placed to minimize the number of terms in each interval. Ramanujan does not adhere to this latter rule in the first interval because of the dominance of the leading term $\mathrm{Li}(n)$.

The calculations quoted three paragraphs later were based on the formula (*) and not on the formula immediately prior to this brief table. Using *Mathematica,* we checked Ramanujan's calculations and found that just four terms of Riemann's series (*) give the values 14.90, 61.85, and 168.22 as approximations for $\pi(x)$ when $x = 50$, 300, and 1000, respectively. Thus, indeed, Ramanujan's calculations are correct.

The next formula for $\mathrm{Li}(n)$ offered by Ramanujan is equivalent to a result of Gram [83], who expanded θ in powers of $\frac{1}{k}$ instead of $\frac{1}{\log n}$. Berndt [40, pp. 124–126] has shown that this asymptotic expansion is a special case of a considerably more general asymptotic formula.

Ramanujan did not elaborate on his "expressions for the number of prime numbers of a given form." However, many of the formulas in ordinary prime number theory have analogues in the theory of primes in arithmetic progressions.

The statements about "equal" numbers of primes are true, if equality is interpreted asymptotically, and follow from the prime number theorem for arithmetic progressions.

The next quadruple of claims is false. The first is mentioned in Ramanujan's first letter to Hardy, and we discussed it extensively in our commentary above.

As we remarked in our commentary on the first letter, the assertion in (b) about the error term for the counting function for integers that can be represented as the sum of two squares is false.

The claim about the sum of the divisor function in (c) is a refinement of one made in Ramanujan's first letter and is still not valid. As we previously indicated, Hardy proved that $E \neq O(n^{\frac{1}{4}})$.

The next result is a corrected, but imprecise, version of a claim made in the first letter. See our commentary above for several remarks on this fascinating claim.

We now offer remarks about (1)–(23).

The continued fraction $F(x)$ is the famous Rogers–Ramanujan continued fraction, which converges to the value

$$(1') \qquad \frac{(x; x^5)_\infty (x^4; x^5)_\infty}{(x^2; x^5)_\infty (x^3; x^5)_\infty},$$

where

$$(a; q)_\infty = \prod_{n=0}^{\infty} (1 - aq^n), \qquad |q| < 1.$$

This result was first discovered by L.J. Rogers [246] in 1894 and rediscovered by Ramanujan, although he does not record the value in his letters to Hardy. The Rogers–Ramanujan continued fraction is found in Entry 38(iii) of Chapter 16 in Ramanujan's second notebook [225] (Berndt [37, p. 79]).

The beautiful relation (1a) is Entry 39(i) of Chapter 16 in Ramanujan's second notebook [225] and was first proved in print by Watson [281]. Ramanujan stated a companion result in Entry 39(ii). K.G. Ramanathan [198], [200] has not only proved these two results but has established additional theorems of this type. See Berndt's book [37, p. 84] for further remarks.

The first proof of (1b) is by Watson [281]. In his first and "lost" notebooks, Ramanujan recorded further explicit evaluations of the Rogers–Ramanujan continued fraction. For proofs of most of these claims, see papers by Ramanathan [198], [199], [200], [201] and Berndt and H. H. Chan [43].

The continued fractions (1d) and (1c) are, respectively, in Entry 15 and its corollary in Chapter 16 of Ramanujan's second notebook [225]. More precisely, the continued fraction in (1d) converges to

$$(1'') \quad \frac{\sum_{n=0}^{\infty} \dfrac{a^n x^{n(n+1)}}{(-bx;x)_n (x;x)_n}}{\sum_{n=0}^{\infty} \dfrac{a^n x^{n^2}}{(-bx;x)_n (x;x)_n}}, \quad |x| < 1,$$

where

$$(a;q)_n = \prod_{k=0}^{n-1} (1 - aq^k).$$

For a proof of this result and for many references to generalizations that can be found in the literature, see Berndt's book [37, Sections 15, 38, pp. 30–31, 79–80].

To obtain the evaluation $(1')$ from $(1'')$, it is necessary to use the famous Rogers–Ramanujan identities

$$\sum_{n=0}^{\infty} \frac{x^{n^2}}{(x;x)_n} = \frac{1}{(x;x^5)_\infty (x^4;x^5)_\infty}$$

and

$$\sum_{n=0}^{\infty} \frac{x^{n(n+1)}}{(x;x)_n} = \frac{1}{(x^2;x^5)_\infty (x^3;x^5)_\infty},$$

where $|x| < 1$. See our commentary on Ramanujan's first letter about his communication of them to Hardy. The identities were first proved by Rogers [246] in 1894. They are recorded in the second notebook as Entries 38 (i), (ii) in Chapter 16. Although Ramanujan did not discover a proof of these identities until 1916, R.A. Askey has pointed out that they follow as limiting cases of Entry 7 in Chapter 16, which was found by Ramanujan earlier in India. Ramanujan must have had a proof of Entry 7, because it is too complicated to discover without having a proof. For Askey's argument, a short history of the Rogers–Ramanujan identities, and many references to the literature, see Berndt's book [37, Section 38, pp. 77–79]. In particular, Andrews' monograph [8] and paper [9] should be consulted. This last paper [9] provides a discussion of all of the many known proofs of the Rogers–Ramanujan identities.

Ramanujan does not divulge the "general theorem on continued fractions" to which he alludes. We offer two candidates for this general theorem. The first is a theorem of Andrews [5, Theorem 6], and the second is a result of M. Hirschhorn [135], [136]. For a discussion of these theorems, see papers of Andrews, Berndt, Jacobsen, and Lamphere [12], [13].

The elegant continued fraction representations in (2) have been established by Preece [185], [180]. On the left side of (ii) we corrected two misprints by replacing $\cosh x$ by $\sinh x$ and 4 by 2.

As Watson [**282**] has demonstrated, (3) follows from Whipple's transformation. See Berndt's book [**35**, p. 41] for the same proof and further discussion.

For a proof of (4), see Preece's paper [**186**].

Watson [**282**] has proved (5).

For proofs of (6) (i), (ii), see Watson's paper [**281**]. These equalities are also recorded as Entry 1 (i) in Chapter 20 of Ramanujan's second notebook [**225**]. Note that the series on the right sides of (i) and (ii) are theta-functions. H.H. Chan [**62**] has proved many beautiful theorems for the continued fraction in (6) that are analogues of Ramanujan's results for the Rogers–Ramanujan continued fraction. In particular, he has established analogues of results in Section (1) and has found several modular equations for this continued fraction.

Preece [**186**] has provided a proof of (7).

Formula (8) is a special case of the general result (4') discussed in our commentary on the formulas under IV in Ramanujan's first letter to Hardy. It is also recorded by Ramanujan in his paper [**212**, eq. (23)], [**226**, p. 58].

Watson [**282**] has established (9).

The transformation (10) was proved by Hardy in his paper [**99**, p. 498–499], [**107**, pp. 511–512], wherein several of Ramanujan's contributions to hypergeometric series are discussed. For another proof, see Bailey's tract [**21**, p. 21]. This result is also recorded as Entry 29(i) in Chapter 11 of Ramanujan's second notebook [**225**].

The series on the right side of (11) should be interpreted as an asymptotic series as $x \to 0+$. Watson [**282**] established (11) by proving a more general result found on page 276 of Ramanujan's second notebook [**225**]. See also the memoir of Andrews, Berndt, Jacobsen, and Lamphere [**13**], where the domains of validity of Ramanujan's general theorem are extended.

The intriguing formula (12) was established by Ramanujan in his paper [**213**, eq. (17)], [**226**, p. 41].

Formula (13) was established by Preece [**186**], and the generalization vaguely suggested by Ramanujan is probably the one found by Preece [**186**].

Formula (14) was proved by Preece [**184**] for $\alpha > 0$ and $\beta > 0$. Further proofs of the elegant, symmetrical formula (14) have been given by Perron [**181**, p. 37, eq. (31)] and Ramanathan [**202**]. See also a paper of Jacobsen [**141**]. This result was also recorded by Ramanujan as Entry 27 in Chapter 12 in his second notebook [**225**] (Berndt [**35**, p. 146]).

What an exquisite result (15) is! In words, the continued fraction F evaluated at the arithmetic and geometric means of α and β is equal to the arithmetic mean of $F(\alpha, \beta)$ and $F(\beta, \alpha)$. This result was first established by Preece [**186**] using the theory of elliptic functions and is valid for $n > 0$. Furthermore, (15) is found as a corollary in Section 12 of Chapter 18 in Ramanujan's second notebook [**225**]. A different proof of (15), which is likely to be closer to that found by Ramanujan, is given in Berndt's account [**37**, pp. 164–165].

The functional equation in (16) is also an enchantingly elegant result. In words, the function F evaluated, in each variable, at the arithmetic mean of α and β is equal to the arithmetic mean of $F(\alpha, \beta)$ and $F(\beta, \alpha)$. As Preece [185] has shown, (16) is a corollary of formula (2) in Section IX of Ramanujan's first letter to Hardy. This corollary also appears in Section 34 of Chapter 12 of Ramanujan's second notebook [225].

The curious result in (17) is a singular modulus and was proved by Watson [283].

Near the end of his epic paper [210], [225, pp. 23–39], Ramanujan hints at three alternative theories of elliptic functions in which the classical base (or nome) is replaced by one of three other bases. In the classical theory, the hypergeometric function $_2F_1(\frac{1}{2}, \frac{1}{2}; 1; x)$ plays a central role. In the new theories, this hypergeometric function is replaced by $_2F_1(\frac{1}{3}, \frac{2}{3}; 1; x)$, $_2F_1(\frac{1}{6}, \frac{5}{6}; 1; x)$, or $_2F_1(\frac{1}{4}, \frac{3}{4}; 1; x)$. Up to now, these theories had never been developed, but some results in these directions were established by K. Venkatachaliengar [270], J. and P. Borwein [51], and the Borweins and F. Garvan [52]. On pages 257–262 in his second notebook [225], Ramanujan offers several results in the three theories, and all of these claims have now been established by Berndt, Bhargava, and Garvan [42]. The result in (18) is, in fact, a modular equation of degree 5 in the first mentioned alternative theory and is found on page 259. The Borweins [51] state this modular equation and assert that they have proved it by the theory of modular forms; more details are supplied in a subsequent paper by the Borweins and Garvan [52]. See Berndt, Bhargava, and Garvan's paper for a still easier proof. Similarly, in (19), Ramanujan offers a modular equation of degree 5 in the last mentioned alternative theory. This modular equation is easier to prove than the former one; see [42].

In (20), Ramanujan communicates six 'mixed' modular equations, or modular equations of composite degree, in the classical theory. The degrees of α, β, γ, and δ are, respectively, 1, 3, 5, and 15. All of these results are either recorded in Entry 11 of Chapter 20 of Ramanujan's second notebook or are easy consequences of parts of this entry. Thus, (i) arises by combining Entries 11(i),(ii); (ii) coincides with Entry 11(iii); both (iii) and (iv) are combinations of Entries 11(iv),(v); (v) is the same as Entry 11(xiv); and (vi) coincides with Entry 11(xv). We indicate some misprints that occur in Ramanujan's letter and *Collected Papers* [226, pp. xxix, 353]; on the right side of (iii), replace 16 by 256 and insert $\gamma\delta$ under the radical sign; in (v), replace 16 by 256. In his brief discussion of Ramanujan's work on modular equations, Hardy [110, p. 220] recorded two of Ramanujan's modular equations from Entry 11 but made two sign errors; in the second equation offered by Hardy, replace the minus signs before the second and fourth expressions on the left side by plus signs. H. Weber [292] found a proof of (v) above and established another modular equation that arises by combining two parts of

Entry 11. For complete proofs of all the results in (20) and Entry 11, see Berndt's book [**37**, pp. 383–395].

Ramanujan offers further mixed modular equations in (21). The equality of the first, second, and fourth expressions in Ramanujan's conclusion is found in Entry 19(iii) of Chapter 20 in his second notebook [**225**], while the third equality is located in Entry 21(ii) of the same chapter. See Berndt's treatise [**37**, pp. 426–430, 435–439] for proofs.

The mixed modular equation in (22), valid for seven distinct sets of moduli, is found in Entry 24(i) of Chapter 20 in Ramanujan's second notebook [**225**], and a proof can be found in Berndt's book [**37**, pp. 449–453].

The last result (23) gives the value of the class invariant G_{1353}, and a proof has been constructed by Watson [**284**]. In Ramanujan's *Collected Papers* [**226**, p. xxix], replace 123 by $\sqrt{123}$.

Ramanujan published (without proofs) [**210**], [**226**, pp. 23–39] only a tiny fraction of his prodigious outpouring of modular equations found in his second notebook [**225**]. Proofs of all these modular equations can be found in Berndt's books [**37**], [**40**]. For introductory accounts of Ramanujan's work on modular equations, see Hardy's book [**110**, Chapter XII], and papers by Berndt [**33**], [**34**] and Ramanathan [**204**].

J.E. Littlewood to G.H. Hardy

Dear Hardy,

The stuff about primes is wrong. (3) is the usual formula save for the terms coming from the zeros of $\zeta(s)$. It's known that

$$\pi(x) - \mathrm{Li}(x) + \frac{1}{2}\mathrm{Li}(\sqrt{x}) \neq O(\sqrt{x}/\log x).$$

From his formula it would follow that the left side is $O(\mathrm{Li}(\sqrt[3]{x}))$. $\left[\sum_{n=2}^{\infty} \frac{\mu(n)}{n} \text{ is convergent } : \int^{\sqrt[3]{x}} \frac{dx}{\log x} \text{ is a convergence factor}\right]$. As for (1) it is substantially equivalent to (3). Write $\sum_{n=3}^{\infty} \frac{\mu(n)}{n^{1+x}}$ for S_{1+x} and integrate term by term. (2) I think comes by contour integration from (1), though I haven't quite managed to get it. S_{2n} is substantially something like $\zeta(\overline{-2n-1})$.

I have a vague theory as to how his mistakes have come about. I imagine that he is satisfied if he can convince himself that his results are correct and he has probably staked on certain operations on divergent series involving primes being legitimate. His results are just what one would get if $\zeta(s)$ had <u>no</u> zeros in $\sigma > 0$. This he might have inferred from the product form $\prod(1 - p^{-s})$, but I still can't believe he knows any function theory. I imagine, however, that he has supposed, on the grounds that $\zeta(s) \neq 0$ for $0 < \sigma < 1$, that the series $\sum \mu(n)/n^\sigma$ can be treated as if it were convergent for $\sigma > 0$, and that this might lead to his formula (3). After all, it is only in this subject that work with divergent series leads one badly wrong, and it is not surprising that he would have been caught, unsuspicious as he presumably is of the diabolical malice inherent in primes.

(b) is doubtless correct, and (c) may be for all I know [it is a very interesting question — E is known to be $O(\sqrt[3]{x})$]. But of course all this is hopelessly incorrect.

I've looked at some of the other results.

(d) is still wrong, of course, rather a howler.

(6)(i) is correct up to x^{33}. (ii) I've not tried.

(8). Here is a mild verification.

Equate coefficients of a.

$$\frac{b^2}{1-b^2} \int_0^\infty \frac{x^2 dx}{\Pi} = I(b - b^2 + b^3 - \cdots)$$

where

$$\Pi = \prod_0^\infty (1 + x^2 b^{2n}), \quad I = \frac{\pi}{2(1 + b + b^3 + \cdots)}$$

Hence

$$\int_0^\infty \frac{x^2 dx}{\Pi} = \frac{1-b}{b} I,$$

and so

$$\int_0^\infty \frac{dx}{(1 + b^2 x^2)\cdots} = \int \frac{(1+x^2)dx}{\Pi} = \frac{1}{b} I,$$

which agrees if we write $bx = y$. (Might perhaps be rather trivial.)

(9) All terms in

$$\left\{\int_0^\infty \frac{ze^{-z}}{z^2 + a^2} dz\right\}^2 + \left\{\int_0^\infty \frac{ae^{-z}}{a^2 + z^2} dz\right\}^2 = \int_0^\infty \frac{e^{-z}}{z} \log\left(1 + \frac{z^2}{a^2}\right) dz,$$

which, if we expand each side in descending powers of a, appears to be correct, though I haven't verified all the algebra. Not very exciting of course, though I should say rather a tour de force without Cauchy's theorem.

(11) I imagine he's omitted $e^{-1/x}$'s. This should be easily verifiable. Alas! — an unprecedented disaster — I've come down without T.&M.

My hopes now are that he has made important discoveries about continued fractions and elliptic functions.

(1) seems to suggest that $F(x)$ is connected with ϑ-functions or perhaps some new class of functions with analogous transformation formulae, that he has discovered. Confer 18, 19, where the left sides are queer modifications of K'/K. 17 is an expression for $\vartheta(0, \sqrt{210})$ isn't it? Surely this and 20–23 are new, and if so, very exciting? I can believe that he's at least a Jacobi.

How maddening his letter is in the circumstances. I rather suspect he's afraid that you'll steal his work.

By the way I saw (at the London Mathematical Society) his name as an author in some Indian Journal of Mathematics: there was nothing by him in that particular part though.

Ever yours,
J.E.L.

Commentary. This letter is undated but was likely written in the latter half of March 1913 from Treen in Cornwall, where Littlewood often spent the Easter vacation. This letter has been reproduced in the collection of material containing the 'lost notebook' [**227**, pp. 380–383].

Littlewood's comment on (d) probably refers to the statement (d) on page 5 of Ramanujan's letter. If this is the case, the comment is slightly uncharitable, since, as we have previously noted, Ramanujan's statement is 'very close' to being correct.

The concluding step in Littlewood's partial proof of (8) (on page 7 of Ramanujan's letter) depends on the truth of (8) for $a = 0$. In fact, this special case is formula (4) on page 4 of Ramanujan's *first* letter to Hardy.

Littlewood's comment on (11) is not correct, since the formula is true as it stands. T.&M. denotes the four-volume treatise on elliptic functions by Tannery and Molk [**266**]

The "Indian Journal of Mathematics" is obviously the *Journal of the Indian Mathematical Society*.

Francis Dewsbury to B. Hanumantha Rao   10 March 1913

I have the honour, by direction of the Syndicate, to refer to your Board the accompanying letter from the Director–General of Observatories bringing to the notice of the University the original character of the work in mathematics by one S. Ramanujan, a clerk in the Port Trust Office, and stating that the University would be justified in enabling him for a few years to spend his whole time on mathematics without any anxiety as to his livelihood, and to request the favour of the Board's recommendation thereon.

Commentary. This and the following two letters are transcripts of original letters found in the Tamil Nadu Archives. No salutations or closings are given.

Bhimasena Hanumantha Rao was Chairman of the Board of Studies of Mathematics at the University of Madras and Deputy Accountant-General, Madras.

Arthur Davies to B. Hanumantha Rao   10 March 1913

I am directed by my Committee to forward to you the enclosed copy of a letter dated 3rd February, 1913, from the Secretary for Indian Students in

London, about a gentleman named S. Ramanujan as it is understood that his case is being considered by the Board of Studies.

Mr. Ramanujan has definitely declined going to England.

Secretary for Indian Students to
Secretary to the Advisory Committee for Indian Students, Madras
3 February 1913

I have an enquiry from one of the Mathematical Tutors at Trinity College, Cambridge, about a young Indian called S. Ramanujan, who is a Clerk in the Accounts Department of the Port Trust at Madras. It appears that Mr. Ramanujan has sent to the Tutor of Trinity a series of mathematical studies and theorems, which are so remarkable and which show such an extraordinary aptitude for Mathematics, that, if they are his own work, as I assume they are, the Tutor in question thinks it possible that he may prove to be a Mathematician of the very highest class. If this be so, it would seem worth while enquiring whether some means cannot be found of getting him a Cambridge education.

Would you very kindly make full enquiries and report to me about this young man? If the work sent to Cambridge is really his own work, evolved by a man without any University training, we ought to see if anything can be done for him.

Commentary. The "Tutor" is clearly G.H. Hardy.

Chapter 3
Preparing to Go to England

B. Hanumantha Rao to S.N. Aiyar 13 March 1913

Tiruva Aeswaranpet

My dear Mr. Narayana Aiyar,

I am calling a meeting of the Board of Studies in Mathematics in the Senate House at 5.30 p.m. on Wednesday the 19th instant for the purpose of considering what we can do for S. Ramanujan. Will you kindly come over and assist us? You have seen some of his results and can help us to understand them better than the author himself.

Yours sincerely,
B. Hanumantha Rao

Commentary. Tiruva Aeswaranpet is part of Triplicane. Adjacent to Tiruva Aeswaranpet lies Hindu High School, which many of India's most illustrious citizens have attended.

S. Narayana Aiyar was born on 15 December 1874 at Cumbum, near Madurai in South India. His father and forefathers were the traditional astrologers in the Temple of Goddess Meenakshi at Madurai. His family was so poor that once during his early school days, at the time of the annual school Inspection, he had to share his only shirt with his elder brother. Having struggled in his early life, he had an affectionate concern for the welfare of everyone with whom he worked.

S.N. Aiyar held an M.A. in mathematics from St. Joseph's College in Trichinopoly, where he remained to become a lecturer in mathematics. He joined the Public Works Department in Madras in 1900. Sir Francis Spring asked him to join the Madras Port Trust as Office Manager. Later Narayana Aiyar was appointed Chief Accountant and thus became the highest ranking Indian at the Port Trust, which he served until 1934. Often, he worked on mathematics with Ramanujan through the late hours of the night. Perhaps more than any other Indian mathematician, he could provide the desired assessment of Ramanujan's talents.

S. Narayana Aiyar, Manager of Port Trust, Madras,
a loyal friend and supporter of Ramanujan

In 1907 S.N. Aiyar became a founding member of the Indian Mathematical Club, which soon became the Indian Mathematical Society. He served as the Club's first Assistant Secretary from 1907 to 1910 and as Treasurer of the Society from 1914 to 1928.

Narayana Aiyar was known for his simplicity bordering on austerity. He was a religious person without being demonstratively orthodox and was pro-

CHAPTER 3. PREPARING TO GO TO ENGLAND

foundly influenced by the Sage of Arunachala, Ramana Maharishi of Tiruvannamalai. S.N. Aiyar was a strong, but quiet, advocate of Indian independence. He died on 17 January 1937. S.N. Aiyar's son, N. Subbanarayanan, has written a poignant description of his father and his relationship with Ramanujan [**263**, pp. 112–115]. For further information, see [**239**].

It is well known [**226**] that in 1910 Ramanujan asked the founder of the Indian Mathematical Club and Society, V.R. Aiyar, for a position in his office. Although no known correspondence with V.R. Aiyar about Ramanujan survives, it is perhaps fitting that a short account of V.R. Aiyar's life be given here. For more details, see [**169**].

V. Ramaswami Aiyar was born on 4 August 1871 at Satyamangalam in the Coimbatore District in South India. His father died six months after his birth, and one of his uncles helped raise him. V.R. Aiyar received B.A. and M.A. degrees from Presidency College, Madras. While at Presidency College, he made several contributions to mathematical journals and the *Educational Times*. The editor of the *Times*, thinking that his contributions must be those of a college professor, addressed him as Professor Ramaswami, and the title stuck with him ever afterward, even though he never became a professor. From 1893 to 1898, V.R. Aiyar taught at colleges in Mysore. In 1898, he joined the Madras Civil Service and in 1901 became Deputy Collector, retiring in 1926.

V.R. Aiyar officially formed the Indian Mathematical Club (Society) on 4 April 1907 with 20 original members. He acted as its first Secretary until 1910, and from 1926 to 1930 he served as President. He also founded the *Journal of the Indian Mathematical Society* in 1909. The *Journal* remains today as perhaps the leading Indian mathematical journal. He died on 22 January 1936.

Memo from Sir Francis Spring

Dr. Gilbert Walker's visit

On February 25th 1913 Dr. Gilbert Walker, F.R.S., himself a fine mathematician and Senior Wrangler, now head of the Indian Meteorological Department, paid a visit to the harbour in connection with our tidal observatory. I took the opportunity of bringing the work of Mr. S. Ramanujan to his notice through Mr. S. Narayana Aiyar, himself a mathematician, and Honorary Secretary of the Indian Mathematical Society.

Dr. Gilbert Walker disclaims ability himself to judge of some of Mr. Ramanujan's work and said that Mr. Hardy, of Trinity College Cambridge, was in his opinion best competent to arrive at a judgment of the true value of the work.

Mr. Ramanujan had already been in correspondence with Mr. Hardy, a letter of whose dated 8th February, 1913 — just 17 days before Dr. Walker's visit here, is in the file.

B. Hanumantha Rao to Francis Dewsbury 25 March 1913

In returning herewith the letters from Dr. Walker and Mr. Arthur Davies, I have the honour to inform you that the Board of Studies in Mathematics has examined the work of S. Ramanujan and is of opinion that he is a person of very remarkable intuition in the domain of Pure Mathematics, especially in the Theory of Numbers and Series. His results appear to be wonderful; but he is not, now, able to present any intelligible proof of some of them.

S. Ramanujan, who is now 23 years old, is an undergraduate of this University, having gone through the old F.A. Course. He has sufficient knowledge of English and is not too old to learn the modern methods from books. The Board does not consider that he will profit by the course of studies, usually given in our affiliated Colleges.

If Ramanujan were placed in such a position as to devote his whole time to the study of Mathematics and have access to the mathematical books of the University Library, he might turn out highly original work. For this purpose it will be necessary to keep him and his family above want and to free him from all anxiety about his livelihood. The Board, therefore, recommends that he be awarded a scholarship of Rs75/- per mensem.

If the Syndicate accept this recommendation, Ramanujan may be required to furnish the Board of Studies, once in three months, with a statement of the work done by him during the previous quarter. The Scholarship may, in the first instance, be granted for a period of two years, with the proviso that it may be discontinued at any time on the report of the Board of Studies.

Commentary. The letter above is a copy of the original letter found in the Tamil Nadu Archives.

The information in the first sentence of the second paragraph is not correct.

G.H. Hardy to S. Ramanujan 26 March 1913

Trinity College
Cambridge

Dear Mr Ramanujan,

Since I wrote to you last I have heard from Mr Littlewood to whom I sent your last letter to me, and I have considered further some of your results.

Your formula (3) is wrong (that is to say the formula

$$\int_\mu^n \frac{dx}{\log x} - \frac{1}{2} \int_\mu^{\sqrt{n}} \frac{dx}{\log x} - \frac{1}{3} \ldots \ldots \right).$$

There are (as I think I suggested in my last letter, and as I can say more definitely now) other terms to be considered, which arise from the complex roots of the Riemann Zeta function $1^{-s} + 2^{-s} + 3^{-s} \ldots$. Your formula would be correct if there were no such roots; but it is certain that there are. From your formula it would follow that

$$\pi(n) - \int^n \frac{dx}{\log x} + \frac{1}{2} \int^{\sqrt{n}} \frac{dx}{\log x} = O \frac{\sqrt[3]{n}}{\log n}$$

and it has been definitely proved that this is untrue.

Mr Littlewood says that your first formula (1)—the integral—is substantially equivalent to (3). He also says that

$$\theta(n) = O \sqrt{\frac{n}{\log n}}$$

and your formula for the sum of the divisors of $1, 2, \ldots, n$, with error of order less than that of $\log n$, are very likely correct; but of course everything depends on the proofs.

We are still doubtful about your formula (even as altered) for the coefficient of x^n in $\frac{1}{1-2x+2x^4-\cdots}$.

The theory of divergent series leads, as a rule, to correct results. The important exceptions to this general rule lie in the theory of prime numbers, the most difficult of <u>all</u> branches of pure mathematics. If, as seems to me probable, it has led you here with mistakes, my view that what you have done is very remarkable would not in any way be shaken; the results you gave concerning continued fractions and elliptic functions are enough in themselves to show it (assuming, as I am quite willing to do, that they are correct).

Mr Littlewood suggested to me also that your unwillingness to give proofs was probably due to apprehensions as to the use I might make of your results. Let me put the matter quite plainly to you. You have in your possession now 3 long letters of mine, in which I speak quite plainly about what you have proved or claim to be able to prove. I have shown your letters to Mr Littlewood, Dr Barnes, Mr Berry, and other mathematicians. Surely it is obvious that, if I were to attempt to make any illegitimate use of your results, nothing would be easier for you than to expose me. You will, I am sure, excuse my stating the case with such bluntness: I should not do so if I were not genuinely anxious to see what can be done to give you a better chance of making the best use of your obvious mathematical gifts.

What I should like above all is a definite proof of some of your results concerning continued fractions of the type

$$\frac{x}{1} + \frac{x^2}{1} + \frac{x^3}{1} + \cdots ;$$

and I am quite sure that the wisest thing you can do, in your own interests, is to let me have one as soon as possible.

> I am
> Yours very sincerely
> G.H. Hardy.

Commentary. Excerpts from the letter above were published by Srinivasan [**263**, p. 56]. We are grateful to him for supplying us with a complete copy of the letter.

Francis Dewsbury to the Educational Department   5 April 1913

From
> Francis Dewsbury Esq., B.A., LL.B.
> *Registrar, University of Madras*

To
> The Secretary to Government
> Educational Department.

Sir,

I have the honour, by the direction of the Syndicate, to forward herewith the enclosed copy correspondence relating to S. Ramanujan, a clerk in the Accounts Department of the Port Trust Office, Madras, who appears to be a mathematician of exceptional promise. The syndicate referred his case to the Board of Studies in Mathematics for investigation, with a request for a recommendation from the Board thereon. It will be seen from the reply received from the Chairman of the Board that it is of opinion that S. Ramanujan should receive special assistance from the University in order to enable him to devote all his time to the study of Mathematics. The Syndicate has accepted the recommendation of the Board that a scholarship of Rs75/- per mensem for two years be granted to him on the condition laid down by the Board, and has directed me to approach Government with a request for sanction to this payment. The Regulations of the University do not at present provide for such a special scholarship. But the Syndicate assumes that Section XV of the Act of Incorporation and Section 3 of the Indian Universities Act, 1904, allow of the grant of such a scholarship, subject to the express consent of the Governor of Fort St George in Council.

> I have the honour to be
> Sir
> Your most obedient servant
> Francis Dewsbury
> Registrar.

CHAPTER 3. PREPARING TO GO TO ENGLAND

Memo from files of the Education Department   **5 April 1913**

<u>Subject</u>

Grant of a special scholarship of
Rs.75 <u>per mensem</u> for a period
of two years to S. Ramanujan
a clerk in the Accounts Department of the Port Trust Office,
Madras.

———-

Please see the regulations governing research studentships printed on page 3 of G.O.No.625, Edl., dated 25th November 1909. These regulations apply only to <u>graduates</u>. In the present case the student is only an under–graduate (F.A.) and any scholarship that is given to him should be outside these regulations. The amount and the tenure of the scholarship proposed are the same as those provided in the regulations. There seems to be no objection to sanction the special scholarship recommended by the syndicate. The general provisions of section 3 of the Indian Universities Act, 1904, and section SV of the Act of Incorporation appear to cover this expenditure. A draft order is submitted below for approval. After the issue of the order, the file may perhaps be sent to the Marine Department for perusal.

ORDER:

The Government are pleased to sanction the proposal of the Syndicate to grant S. Ramanujan a special scholarship of Rs.75 <u>per mensem</u> tenable for a period of two years.

Commentary. There is an added handwritten note indicating that it is unnecessary to send the file to the Marine Department.

Memo from the Madras Port Trust Office   **11 April 1913**

Now that the Madras University has been pleased to grant S. Ramanujan a monthly scholarship of Rs.75 I beg to propose that S. Ramanujan be granted two years' leave on loss of pay from the 1st May 1913 with the option of returning to us whenever the University ceases to pay him the scholarship. C.P.T. has power to sanction this without going to the Board.

A separate proposal will be submitted later for carrying on his work during his absence.

C.A.C.

Commentary. The initials C.A.C. stand for S. Narayana Aiyar, the Manager of the Madras Port Trust under its chairman (C.P.T.) Sir Francis Spring. They possibly denote Chairman of the Accounts Committee, or Chief Accounts Clerk, since Narayana Aiyar was Chief Accountant of the Trust. The initials C.P.T. denote Chairman Port Trust. The initials C.A.C. and C.P.T. were used in memoranda circulated between S.N. Aiyar and Sir Francis Spring and were initialed S.N. and F.J.E.S., respectively.

S. Ramanujan to Francis Dewsbury   **12 April 1913**

> Accounts Department
> Port Trust Office, Madras,

To
> The Registrar,
>> University of Madras.

Through
> The Chairman,
>> Madras Port Trust.

Sir,

I have the honour to acknowledge with thanks the receipt of your letter No. 1631 dated the 9th April 1913 offering me a scholarship of Rs. 75 per mensem tenable for two years. In reply I beg to inform you that I am willing to take up the scholarship from the 1st May and that I shall abide by the conditions detailed in your above letter. The Chairman of the Port Trust will relieve me from my present work on the 1st proximo.

> I beg to remain
> Sir
> Your Obedient Servant,
> S. Ramanujan

Commentary. The conditions set forth in the letter from the University of Madras offering Ramanujan a scholarship were that Ramanujan write quarterly reports on his research that were to be submitted to the Board of Studies in Mathematics. Ramanujan wrote three of these reports, dated 5 August 1913, 7 November 1913, and 9 March 1914, before departing for England. Unfortunately, the University of Madras has evidently lost the original quarterly reports, but fortunately a handwritten copy of them was made in 1925 by T.A. Satagopan. This copy was sent to Hardy along with a handwritten copy of Ramanujan's notebooks also made by Satagopan. G.N. Watson subsequently made a second handwritten copy of the reports. Both of these copies of the quarterly reports are now on file in the library at Trinity College, Cambridge. All of the mathematical results related in these reports have been described in detail in Berndt's book [32]. A notable feature of the reports is that, in contrast to his notebooks and letters to Hardy, Ramanujan provides arguments, some of which are not rigorous, for his claims.

Commentary. Ramanujan's third letter to Hardy, dated 17 April 1913, is only partially reproduced in his *Collected Papers* [**226**, pp. xxix-xxx]. Furthermore, in the library at Trinity College, Cambridge, there are two fragments which were reproduced in the publication of the lost notebook [**227**, pp. 376–377] and which are probably parts of, respectively, this third letter and Ramanujan's letter of 22 January 1914 to Hardy. The first fragment follows Ramanujan's signature below.

S. Ramanujan to G.H. Hardy   **17 April 1913**

Madras Port Trust
Accounts Department

Dear Sir,

I am in receipt of your letter of the 26th ultimo. I am a little pained to see what you have written at the suggestion of Mr. Littlewood. I am not in the least apprehensive of my method being utilised by others. On the contrary my method has been in my possession for the last eight years and I have not found anyone to appreciate the method. As I wrote in my last letter I have found a sympathetic friend in you and I am willing to place unreservedly in your possession what little I have. It was on account of the novelty of the method I have used that I am a little diffident even now to communicate my own way of arriving at the expressions I have already given. But still in this letter I have attempted to give a demonstration which would be acceptable to you all.

You speak of having written to me three long letters. But I have received only two. Your first communication to me and the letter of the 26th ultimo. I am anxious to know what your other communication contained. Very probably my not replying to you to what was contained in the letter not received by me made you write in the strain you have done at the suggestion of Mr. Littlewood. I was wondering why you never wrote anything about my personal question. Very probably the second communication contained something about it. I am glad to inform you that the local University has been pleased to grant me a scholarship of £60 per annum for two years and this was at the instance of Dr. Walker F.R.S. Head of the Meteorological Department in India to whom my thanks are due. The scholarship will help me a great deal for two years.

My knowledge of English being poor I find it difficult to collect my thoughts and put them in a form presentable to you. I have tried this time to give you a proof in connection with the expression I have for the distribution of primes. I do not agree with you when you say that my formula

$$\int^n \frac{dx}{\log x} - \frac{1}{2} \int^{\sqrt{n}} \frac{dx}{\log x} - \cdots$$

is wrong and that it has been proved definitely that

$$\pi(n) - \int^n dx + \frac{1}{2} \int^{\sqrt{n}} \frac{dx}{\log x}$$

is not of the order

$$\frac{\sqrt[3]{n}}{\log n}.$$

My belief is that the expression I have given will be sufficient for practical purposes and if there are some terms which are omitted they will be terms of a very low order. I give as much as possible in the modern mathematical language a proof of the theorem on prime numbers and a continued fraction.

I am delighted to hear that not only yourself but also other mathematicians at the very fountain head of mathematical knowledge are interesting themselves in my humble work. I request you to convey my thanks not only to your good self but also to Mr. Littlewood, Dr. Barnes, Mr. Berry and others who take an interest in me.

I am
Yours very sincerely
S. Ramanujan

(1) $x \log 1 + x^2 \log 2 + x^3 \log 3 + x^4 \log 4 + x^5 \log 5 + \cdots$
$= \log 2 \left(x^2 + 2x^4 + x^6 + 3x^8 + x^{10} + 2x^{12} + x^{14} + 4x^{16} + \cdots \right)$
$+ \log 3 \left(x^3 + x^6 + 2x^9 + x^{12} + x^{15} + 2x^{18} + x^{21} + x^{24} + \cdots \right)$
$+ \log 5 \left(x^5 + x^{10} + x^{15} + x^{20} + 2x^{25} + x^{30} + x^{35} + \cdots \right)$
$+ \log 7 \left(x^7 + x^{14} + x^{21} + x^{28} + x^{35} + x^{42} + 2x^{49} + \cdots \right)$
$+ \cdots \quad + \cdots \quad + \cdots \quad + \cdots$

where 2, 3, 5, 7, 11 ... are prime numbers.

(2) $= \log 2 \left(\dfrac{x^2}{1-x^2} + \dfrac{x^4}{1-x^4} + \dfrac{x^8}{1-x^8} + \dfrac{x^{16}}{1-x^{16}} + \cdots \right)$
$+ \log 3 \left(\dfrac{x^3}{1-x^3} + \dfrac{x^9}{1-x^9} + \dfrac{x^{27}}{1-x^{27}} + \dfrac{x^{81}}{1-x^{81}} + \cdots \right)$
$+ \log 5 \left(\dfrac{x^5}{1-x^5} + \dfrac{x^{25}}{1-x^{25}} + \dfrac{x^{125}}{1-x^{125}} + \cdots \right)$
$+ \log 7 \left(\dfrac{x^7}{1-x^7} + \dfrac{x^{49}}{1-x^{49}} + \dfrac{x^{343}}{1-x^{343}} + \cdots \right)$
$+ \cdots\cdots\cdots\cdots\cdots$

changing x to $-e^{-x}$ in (1) and (2) we have

$$\log 3 \left(\frac{1}{e^{3x}+1} + \frac{1}{e^{9x}+1} + \frac{1}{e^{27x}+1} + \cdots \right)$$
$$+ \log 5 \left(\frac{1}{e^{5x}+1} + \frac{1}{e^{25x}+1} + \frac{1}{e^{125x}+1} + \cdots \right)$$
$$+ \log 7 \left(\frac{1}{e^{7x}+1} + \frac{1}{e^{49x}+1} + \frac{1}{e^{363x}+1} + \cdots \right)$$
$$+ \cdots \cdots$$

3, 5, 7, ... being prime numbers

(3)
$$= \log 2 \left(\frac{1}{e^{2x}-1} + \frac{1}{e^{4x}-1} + \frac{1}{e^{8x}-1} + \frac{1}{e^{16x}-1} + \cdots \right)$$
$$+ e^{-x} \log 1 - e^{-2x} \log 2 + e^{-3x} \log 3 - \cdots .$$

Let us suppose,

(4)
$$\phi(x) + \log 2 \left(2e^{-2x} + 4e^{-4x} + 8e^{-8x} + 16e^{-16x} + \cdots \right)$$
$$= \log 2 \left(e^{-2x} + e^{-4x} + e^{-8x} + e^{-16x} + \cdots \right)$$
$$+ \log 3 \left(e^{-3x} + e^{-9x} + e^{-27x} + e^{-81x} + \cdots \right)$$
$$+ \log 5 \left(e^{-5x} + e^{-25x} + e^{-125x} + e^{-625x} + \cdots \right)$$
$$+ \cdots \cdots$$

2, 3, 5, ... being prime numbers.
Then we see that,

(5)
$$\phi(x) - \phi(2x) + \phi(3x) - \phi(4x) + \phi(5x) - \cdots$$
$$= \log 2 \left(\frac{1}{e^{2x}+1} + \frac{1}{e^{4x}+1} + \frac{1}{e^{8x}+1} + \cdots \right)$$
$$+ \log 3 \left(\frac{1}{e^{3x}+1} + \frac{1}{e^{9x}+1} + \frac{1}{e^{27x}+1} + \cdots \right)$$
$$+ \log 5 \left(\frac{1}{e^{5x}+1} + \frac{1}{e^{25x}+1} + \frac{1}{e^{125x}+1} + \cdots \right)$$
$$+ \cdots \cdots$$
$$- \log 2 \left(\frac{2}{e^{2x}+1} + \frac{4}{e^{4x}+1} + \frac{8}{e^{8x}+1} + \cdots \right).$$

But we know that

$$\frac{1}{1-e^{-2x}} = (1+e^{-2x})(1+e^{-4x})(1+e^{-8x})(1+e^{-16x}) \cdots .$$

Taking logarithmic differentiation, we have

(6) $$\frac{2}{e^{2x}-1} = \frac{2}{e^{2x}+1} + \frac{4}{e^{4x}+1} + \frac{8}{e^{8x}+1} + \cdots.$$

Also we know that

$$\frac{1}{e^{2x}+1} = \frac{1}{e^{2x}-1} - \frac{2}{e^{4x}-1}.$$

Hence,

(7) $$\frac{1}{e^{2x}+1} + \frac{1}{e^{4x}+1} + \frac{1}{e^{8x}+1} + \cdots$$
$$= \left(\frac{1}{e^{2x}-1} - \frac{2}{e^{4x}-1}\right) + \left(\frac{1}{e^{4x}-1} - \frac{2}{e^{8x}-1}\right) + \cdots$$
$$= \frac{1}{e^{2x}-1} - \frac{1}{e^{4x}-1} - \frac{1}{e^{8x}-1} - \frac{1}{e^{16x}-1} - \cdots.$$

Therefore, by subtracting (6) and (7) in (5) we see

$$\phi(x) - \phi(2x) + \phi(3x) - \phi(4x) + \cdots$$
$$= \log 2 \left(\frac{1}{e^{2x}-1} - \frac{1}{e^{4x}-1} - \frac{1}{e^{8x}-1} - \cdots\right)$$
$$+ \log 3 \left(\frac{1}{e^{3x}+1} + \frac{1}{e^{9x}+1} + \frac{1}{e^{27x}+1} + \cdots\right)$$
$$+ \log 5 \left(\frac{1}{e^{5x}+1} + \frac{1}{e^{25x}+1} + \frac{1}{e^{125x}+1} + \cdots\right)$$
$$+ \cdots\cdots$$
$$- \frac{2\log 2}{e^{2x}-1}$$
$$= e^{-x}\log 1 - e^{-2x}\log 2 + e^{-3x}\log 3 - \cdots. \text{ by (3) .--}$$
$$= \text{a finite quantity when } x \text{ approaches } 0.$$

Hence

(8) $\phi(x)$ is finite when x approaches 0.

Applying (8) in (4) we see that

(9)
$$\log 2 \left(e^{-2x} + e^{-4x} + e^{-8x} + \cdots \right)$$
$$+ \log 3 \left(e^{-3x} + e^{-9x} + e^{-27x} + \cdots \right)$$
$$+ \log 5 \left(e^{-5x} + e^{-25x} + e^{-125x} + \cdots \right)$$
$$+ \log 7 \left(e^{-7x} + e^{-49x} + e^{-343x} + \cdots \right)$$
$$+ \cdots \cdots \cdots$$
$$- \log 2 \left(2e^{-2x} + 4e^{-4x} + 8e^{-8x} + 16e^{-16x} + \cdots \right)$$

is finite when x approaches 0.

Now let us suppose that z is a function of n such that there are n prime numbers within z. Then from (9) we see that

$$\int^{\infty} \log z \left(e^{-zx} + e^{-z^2 x} + e^{-z^3 x} + e^{-z^4 x} + \cdots \right) dn$$
$$- \log 2 \int^{\infty} 2^y e^{-2^y x} dy$$

is finite when x becomes 0, that is

(10)
$$\int^{\infty} \log z \left(e^{-zx} + e^{-z^2 x} + e^{-z^3 x} + \cdots \right) dn - \frac{1}{x}$$

is finite when x becomes 0.

Again supposing $\frac{dn}{dz} = \frac{\psi(z)}{\log z}$ and substituting this in (10) we see that $\psi(z)$ is such a function that,

(11)
$$\int^{\infty} \left(e^{-zx} + e^{-z^2 x} + e^{-z^3 x} + \cdots \right) \psi(z) dz - \frac{1}{x}$$

is finite when x becomes 0.

Hence from (11) we see that

$$\psi(z) = 1 - \frac{1}{2\sqrt{z}} - \frac{1}{3z^{\frac{2}{3}}} - \frac{1}{5z^{\frac{4}{5}}} + \cdots .$$

Therefore

(12)
$$\frac{dn}{dz} = \frac{z - \frac{\sqrt{z}}{2} - \frac{\sqrt[3]{z}}{3} - \cdots}{z \log z}$$

Hence the number of prime numbers within z

(13)
$$= \int^z \frac{u - \frac{\sqrt{u}}{2} - \frac{\sqrt[3]{u}}{3} - \frac{\sqrt[5]{u}}{5} + \cdots}{u \log u} du$$
$$= \int^z \frac{du}{\log u} - \frac{1}{2} \int^{\sqrt{z}} \frac{du}{\log u} - \frac{1}{3} \int^{\sqrt[3]{z}} \frac{du}{\log u} - \frac{1}{5} \int^{\sqrt[5]{z}} \frac{du}{\log u} + \cdots$$

Commentary. Ramanujan's argument up to the equality immediately preceding (8) is quite straightforward. The limit indicated just prior to (8) is, in fact, equal to $\log 2 - \frac{1}{2}\log(2\pi)$. This is easy to prove but also follows from Example (i) in Section 2 of Chapter 15 of Ramanujan's second notebook [**225**]. A complete argument may be found in Berndt's book [**40**, pp. 119–120]. However, the conclusion (8) is false, for Hardy [**110**, p. 38] showed that $x\phi(x)$ oscillates as x tends to 0. This renders Ramanujan's claims (9) and (10) false as well. The lower limits missing on the integrals can be set equal to 0. Observe that Ramanujan uses n, or $n(x)$, instead of $\pi(x)$ to denote the number of primes $\leq x$. The assumption that there exists a function ψ such that

$$\frac{dn}{dz} = \frac{\psi(z)}{\log z}$$

is not justified. As Ramanujan explains in his fourth letter to Hardy, $\frac{dn}{dz}$ is to be thought of as the derivative of a 'smooth' function approximating $n(z)$. Assuming this 'equality,' we easily deduce (11) from (10). The deduction from (11) of the proffered formula for ψ is unjustified, for the solution to (11) is clearly not unique. However, if we substitute Ramanujan's formula for ψ into (11), we do obtain a finite number as claimed; in fact, the number is 0. The remainder of the formal argument is straightforward. (See [**40**, pp. 119–123].)

In conclusion, Ramanujan communicated to Hardy what he considered to be a complete proof of not only Riemann's formula for $\pi(x)$ but also the prime number theorem. In his undated response given below, Hardy points out the many pitfalls in Ramanujan's argument. Readers should consult Hardy's fascinating account [**110**, Chapter 2] and Berndt's further exegesis [**40**, Chapter 24, esp. Section 10] for more details of the arguments presented in this fragment, for their deficiencies, and for further insights into Ramanujan's theory of prime numbers.

In his letter of 26 March 1913 Hardy stated that he had written all together three letters to Ramanujan. However, according to Ramanujan, only two letters were received. We have been unable to find any information about the date and contents of the missing letter.

Arthur Davies to S.N. Aiyar   7 June 1913

I have to acknowledge with thanks your letter about the scholarship that has been awarded to Mr. S. Ramanujan. —I am conveying the news to Mr. Walker at the Meteorological Office, and he will tell the gentlemen at Cambridge, who will watch his future career with the greatest interest.

                                                            Arthur Davies

Commentary. This is the message of the postcard addressed to S.N. Aiyar, with the courtesy title M.R.Ry, to which we referred on page 11. Arthur

CHAPTER 3. PREPARING TO GO TO ENGLAND

Davies was Secretary of the Madras Students Advisory Committee and was Principal and Senior Professor at the Madras Law College.

S.N. Aiyar to Sir Francis Spring   **31 October 1913**

Ramanujan and myself went to Mr. Littlehailes last Sunday. I explained to him the new method of procedure in attacking problems as used by Ramanujan. Mr. Littlehailes was pleased to see many of Ramanujan's results and promised that from the 1st of December, when he expects to be free from the meteorological work, he would regularly study Ramanujan's results. Being the Professor of Mathematics in the premier college of this presidency I am sure Mr. Littlehailes will bring into prominence before the mathematical world Ramanujan's researches.

S.N.

Commentary. S.N. Aiyar's report is initialled by Spring with the comment: I am glad to learn all this.

G.H. Hardy to S. Ramanujan

will state my conclusions at once:
(i) Your outline of the proof is too incomplete for me to pronounce a confident opinion, but it has all the appearance of being right, and it looks to me a very remarkable piece of work.

If you will send me your proof <u>written out carefully</u> (so that it is easy to follow), I will (assuming that I agree with it—of which I have very little doubt) try to get it published for you in England. Write it in the form of a paper 'On the continued fraction

$$\frac{1}{1} + \frac{1}{x} + \frac{1}{x^2} + \cdots,$$

giving a full proof of the principal and most remarkable theorem, viz. that the fraction can be expressed in finite terms when $x = e^{-\pi\sqrt{n}}$, where $\underline{n}$ is rational.
(ii) As regards your work about primes, the result is certainly wrong. Of this there is no doubt whatsoever. Results are known which definitely contradict yours. Let $\pi(x)=$ the number of primes $< x$, and

$$\operatorname{li} x = \int^x \frac{dx}{\log x}$$

You say

$$\pi(x) = \operatorname{li} x - \frac{1}{2}\operatorname{li} \sqrt{x} - \frac{1}{3}\operatorname{li} \sqrt[3]{x} \cdots.$$

This is untrue. That the first approximation is $\operatorname{li} x$ was proved in 1896 by Hadamard and de la Valleé-Poussin. So far this agrees. But it has also been proved that

$$\pi(x) - \operatorname{li} x - \frac{1}{2}\operatorname{li} \sqrt{x}$$

is sometimes (infinitely often, as x tends to ∞) of order not less than (and indeed greater than)

$$\operatorname{li}\sqrt{x} \qquad \text{or} \qquad \frac{\sqrt{x}}{\log x}$$

and <u>not</u>, as your result makes out, of order $\frac{\sqrt[3]{x}}{\log x}$. So your work <u>must</u> be wrong. And in fact your argument is unsound in several places.

You prove that, if

$$\phi(x) + \log 2 \left(2e^{-2x} + 4e^{-4x} + \cdots \right)$$
$$= \sum_{p=2,3,5,7,11,\ldots} \log p \left(e^{-px} + e^{-p^2 x} + e^{-p^3 x} + \cdots \right)$$

then

$$\phi(x) - \phi(2x) + \phi(3x) - \cdots$$
$$= e^{-x} \log 1 - e^{-2x} \log 2 + \cdots.$$

All this is correct. You infer (correctly) that $\phi(x) - \phi(2x) + \phi(3x) - \cdots$ remains finite as $x \to 0$. You give no proof: but the result is true. You then infer that $\phi(x)$ remains finite as $x \to 0$. There is no theorem I know of which warrants such a conclusion: and I do not believe it to be a correct inference (though I cannot offhand construct an example to the contrary).

However, let us suppose that $\phi(x)$ does remain finite. You have then that

$$\sum_{p=2,3,5,7,11,\ldots} \log p \left(e^{-px} + e^{-p^2 x} + \cdots \right) - \log 2 \left(2e^{-2x} + 4e^{-4x} + \cdots \right)$$

remains finite.

You then put $z(n) =$ a number such that there are n primes $\leq z$. You say that

$$\int^{\infty} \log z \left(e^{-zx} + e^{-z^2 x} + \cdots \right) dn - \log 2 \int^{\infty} 2^y e^{-2^y x} dy$$

remains finite. Here you assume that

$$2e^{-2x} + 4e^{-4x} + \cdots - \int^{\infty} 2^y e^{-2^y x} dy$$

remains finite. Perhaps you have proved this: anyhow I am prepared to believe it.

You then put

$$\frac{dn}{dz} = \frac{\psi(z)}{\log z}!$$

But n and z are discontinuous functions the one of the other: e.g.

$$n = 1 (1 \leq z < 2) \qquad n = 2(2 \leq z < 3)$$
$$n = 3(3 \leq z < 5)\ldots.$$

so that $\dfrac{dn}{dz}$ is in general zero, except for certain special values of z where it is not defined. So I can make nothing of this step.
However, you infer that

$$\int^\infty \left(e^{-zx} + e^{-z^2 x} + \cdots \right) \psi(z) dz - \frac{1}{x}$$

remains finite as $x \to 0$. Hence you deduce

$$\psi(z) = 1 - \frac{1}{2\sqrt{z}} - \frac{1}{3z^{\frac{2}{3}}} \cdots$$

How do you get this?
A simpler inference of the same sort would be from

$$\int^\infty e^{-zx} f(z) dz = \frac{1}{x} + \text{something of lower order}$$

to deduce
$$f(z) = 1 + \text{something which tends to } 0.$$

Mr Littlewood and I <u>have</u> proved <u>this</u> (under the assumptions that $f(z)$ is ≥ 0)—but even here the proof is exceedingly difficult. I can see how your result is <u>suggested</u>: but to get a rigid proof is quite a different matter.

You will see that, with all these gaps in the proof, it is no wonder that the result is wrong.
The truth is that the theory of primes is full of pitfalls, to surmount which requires the fullest of trainings in modern rigorous methods. This you are naturally without. I hope you will not be discouraged by my criticisms. I think your argument a very remarkable and ingenious one. To have proved what you claimed to have proved would have been about the most remarkable mathematical feat in the whole history of mathematics.
As regards your work on continued fractions and elliptic functions—here the difficulties to be surmounted are of an entirely different kind, and I have no reason at all to suppose that your results are not perfectly correct. I hope that you will adopt the suggestions I made at the beginning.
Try to make the acquaintance of Mr E.H. Neville, who is now in Madras lecturing. He comes from my College and you might find his advice as to reading and study invaluable.
Well, you will see from the length of this letter that answering yours is not an entirely trifling business and that I have some excuse if I have delayed.
Believe me.

Yours very sincerely,
G. H. Hardy

Commentary. Excerpts from this letter originally appeared in Srinivasan's collection [**263**, pp. 57–58], and we are grateful to him for a copy of the fuller letter transcribed above.

The first page of this letter is apparently permanently lost. It seems likely that this letter is the one written on 24 December 1913 to which Ramanujan refers in the letter below. Although Neville was not yet in Madras when Hardy wrote the letter, he probably was in Madras by the time Ramanujan received it.

The paper on continued fractions envisioned in the letter was not published. At the end of his stay in England, Ramanujan did publish his evaluation of this continued fraction, now called the Rogers–Ramanujan continued fraction [**223**]. For much of Ramanujan's work on continued fractions of the Rogers–Ramanujan type, see Berndt's book [**37**, Chapter 16] and the memoir [**13**].

S. Ramanujan to G.H. Hardy   **22 January 1914**

Madras

Dear Sir,

Received your kind letter dated 24th December 1913. I carefully went through your valuable remarks and I have written at the end of this letter some corrections I have made, for your kind consideration.

Now I learn from your letter and Mr. Neville that you are anxious to get me to Cambridge. If you had written to me previously I would have expressed my thoughts plainly to you. In February 1913 when I was in the Port Trust, the Secretary to the Students Advisory Committee of Madras wrote to me that he had been asked by Mr. Mallet of the India Office to see me and therefore I might go to him the next noon. The Chairman of the Port Trust told my superior officer to go with me and answer his questions. Accordingly the next day we went to him and he asked us whether I was prepared to go to England. While I was hesitating to reply him as the question appeared vague to me and I naturally was thinking whether I had to appear for any examination with my very poor educational qualification as I used to see students from here going to England only for appearing for some examination, my superior officer, a very orthodox Brahman having scruples to go to foreign land replied at once that I could not go to England and the matter was dropped.

After that I had no correspondence from you. Then I wrote a reminder to you which, I think, you have not received. Another thing I have to say to you is that all letters written to you, except this one and the remainder, did not contain my language. Those were written by the superior officer mentioned before, though the mathematical results and handwriting were my own. I am writing all these things plain to you so that you may judge properly my knowledge of English and power of expression of thought as they are.

CHAPTER 3. PREPARING TO GO TO ENGLAND

I went to Mr. Neville of your college who very kindly spoke to me and cleared my doubts that I need not care for my expenses, that my English will do, that I am not asked to go to England to appear for any examination and that I can remain a vegetarian there. He also pointed out to me the benefits I derive in coming in contact with modern mathematicians and modern ways of thinking. Then when I expressed my willingness to go there he said that the best time for me to go there is summer and it will be difficult for me to go in winter.

So I request that you and Mr. Littlewood will be good enough to take the trouble of getting me there within a very few months. I think I may prepare the article with your help after going to England. You may write to me to the Port Trust as usual. I request you to convey my thanks to Mr. Littlewood and beg to hear as soon as possible from you.

Yours very sincerely
S. Ramanujan

1. For all values of x I have proved that

$$\log 2 \left(e^{-x} + 2e^{-2x} + 4e^{-4x} + 8e^{-8x} + 16e^{-16x} + \cdots \right.$$
$$\left. + 1 - \frac{x}{3 \cdot 1!} + \frac{x^2}{7 \cdot 2!} - \frac{x^3}{15 \cdot 3!} + \frac{x^4}{31 \cdot 4!} - \frac{x^5}{63 \cdot 5!} + \cdots \right) = \frac{1}{x} + F(x),$$

where $xF(x)$ correct to 10 places of decimals

$$= .0000098844 \cos \left(\frac{2\pi \log x}{\log 2} + .872811 \right),$$

so my assertion in the previous letter that

$$\text{``} \log 2 \left(2e^{-2x} + 4e^{-4x} + 8e^{-8x} + \cdots \right) - \int_0^\infty 2^y e^{-x2^y} dy$$

is finite when $x = 0$" is wrong.

2. The above $F(x)$ has also this property viz.

$$F(x) - F(2x) + F(3x) - F(4x) + \cdots = 0.$$

Hence the assertion that, "If

$$\phi(x) - \phi(2x) + \phi(3x) - \phi(4x) + \cdots$$

be finite when $x = 0$, then $\phi(x)$ is necessarily finite when $x = 0$" is also wrong as you remarked. But one thing is certain; that $\phi(x)$ when $x = 0$ cannot be <u>definitely infinite</u> but will either be finite or indeterminate, i.e. infinitely often going to infinity as well as vanishing like the function

$$\frac{a\cos(b+c\log x)}{x^n}.$$

As a matter of fact $\phi(x)$ is not finite when $x=0$ in the present case.

3. I think I am correct in using $\frac{dn}{dz}$ which is not the differential coefficient of a discontinuous function but the differential coefficient of an average continuous function passing fairly (though not exactly) through the isolated points. I have used $\frac{dn}{dz}$ in finding the number of numbers of the form $2^p \cdot 3^q$, $2^p + 3^q$ &c less than z and have got correct results.

4.

(4a) $$\int_0^\infty \left(e^{-zx} + e^{-z^2 x} + e^{-z^3 x} + \cdots \right) \log z \, dn = \frac{1}{x} + f(x)$$

when x approaches 0, where $f(x)$ is of lower order than $\frac{1}{x}$ and indeterminate when $x=0$ which can be found if we know the value of

(4b) $$\sum \log p \left(e^{-px} + e^{-p^2 x} + e^{-p^3 x} + \cdots \right) - \frac{1}{x}$$

when x is very small, where $p = 2, 3, 5, 7, 11, \ldots$.

5. Even though

$$\text{Li}(n) - \frac{1}{2}\text{Li}(\sqrt{n}) - \frac{1}{3}\text{Li}(\sqrt[3]{n}) - \frac{1}{5}\text{Li}(\sqrt[5]{n}) + \cdots$$

does not actually coincide with the number of primes less than n, yet it is the average value of the number of primes less than n and so we can use it in finding the sum of n prime numbers and in other similar questions.

Thus in the case of number of divisors of n which I have found to be of order $e^{c \log n / \log \log n}$, where c is some definite and finite constant, the average value is only $2\gamma + \log n$ where γ is the Eulerian constant.

Also we have, if $a_1, a_2, a_3 \ldots a_n$ are n irregular numbers then,

$$a_1 + a_2 + a_3 + \cdots + a_n$$
$$= \int (\text{average value of } a_n) dn + \frac{1}{2} a_n + \theta(n),$$

where $\theta(n)$ is of lower order than the difference between the actual order of a_n and the average order of a_n. Thus we see that

6.

$2 + 3 + 5 + 7 + \cdots + z$ (where z is nth prime number)
$$= \left\{ \text{Li}(z^2) - \frac{1}{2}\text{Li}(z^{\frac{3}{2}}) - \frac{1}{3}\text{Li}(z^{\frac{4}{3}}) - \frac{1}{5}\text{Li}(z^{\frac{6}{5}}) + \cdots + \frac{z}{2} + \theta(z) \right\}$$

where $\theta(z)$ is of lower order than $z(\log z)^{-t}$ for any value of t, for its actual order is less than ($z - $ average value of z).

N.B. The left side i.e. the sum of the first n prime numbers is not equal to the right side as it stands but will be equal to the right side transformed to a function of n assuming that

$$n = \text{Li}(z) - \frac{1}{2}\text{Li}(\sqrt{z}) - \frac{1}{3}\text{Li}(\sqrt[3]{z}) - \cdots .$$

The above is not clearly written and I shall clearly explain to you afterwards what I mean.

Yours very sincerely
S. Ramanujan

Commentary. Except for the 'mathematical' portion of the letter previously published in the 'lost notebook' [**227**], this letter has not previously been published.

The "superior officer" who helped Ramanujan compose the previous letters to Hardy is not identified, but it undoubtedly was S.N. Aiyar. However, S.N. Aiyar's letters that follow give no indication of any regret of Ramanujan's travelling to England. (It appears that Ramanujan was somewhat embarrassed by his long silence.) Among some orthodox Brahmins, it is believed that crossing the seas makes one an outcaste and therefore unfit to attend Brahmin weddings, funerals, and other special occasions. This, however, was and still is a minority opinion. See Kanigel's book [**142**, pp. 184–185] to learn more on how Ramanujan changed his mind and overcame caste beliefs and prejudices.

Ramanujan clearly was fearful about taking any future examinations. His failures at passing examinations at the Government College in Kumbakonam and at Pachaiyappa's College in Madras had been a heavy burden to bear.

As we remarked in our commentary on Ramanujan's letter of 17 April 1913 to Hardy, Ramanujan's claims about "an average continuous function" are not very convincing. To learn how Ramanujan employed this unrigorous idea in determining the number of numbers of the form $2^p 3^q$ that are less than x, say, see Berndt's account [**40**, pp. 66–69]. Ramanujan apparently left no record of his study of numbers of the form $2^p + 3^q$.

Ramanujan's claim in §4 is correct, for the integral in (4a) is easily seen to equal the double sum in (4b).

The sum at the beginning of §5 is Riemann's series for $\pi(n)$. Ramanujan seems to be claiming that Riemann's series, in some sense, is the average value of $\pi(n)$.

Ramanujan's assertion about $d(n)$, the number of positive divisors of n, is correct, for in 1906, Wigert [**296**] proved that, for all sufficiently large positive integers n,

$$d(n) < 2\frac{\log n}{\log\log n}(1+\varepsilon),$$

for each $\varepsilon > 0$. This result was improved by Ramanujan in his famous paper [217], [226, pp. 78–128]. The claim about the average order of $d(n)$ was discussed in the commentary on Ramanujan's first letter to Hardy.

Ramanujan does not define an "irregular number," but the loose principle described afterward is valid only under very restrictive hypotheses. Evidently, this formal idea was used by Ramanujan heuristically to determine information about the average order of arithmetical functions.

E.H. Neville to Francis Dewsbury   28 January 1914

Madras

Dear Mr. Dewsbury,

The discovery of the genius of S. Ramanujan of Madras promises to be the most interesting event of our time in the mathematical world. From the first results which he communicated, the mathematicians of Cambridge at once believed that he had uncommon ability, and the effect of personal acquaintance with the man and conversation as to his methods has been in my own case to replace that belief by certainty. At the same time the importance of securing to Ramanujan a training in the refinements of modern methods and a contact with men who know what range of ideas have been explored and what have not cannot be over estimated.

Unassisted by knowledge of contemporary achievements in Europe, Ramanujan has among other things developed two of the most fruitful and subversive theories which have been studied there during the last ten or fifteen years, theories which still are to be found only in contributions to the various scientific journals and are not admitted to current text books and treatises. Who can say, had his power not been employed in the invention of these tools, what other machinery he might by now have built or what uses unnoticed by the others he might have observed for the processes themselves? Inspiration is not confined to the making of a single discovery, and it is always a loss to science when two men do the same work.

On the other hand, we have learnt in Europe what Ramanujan has not yet discovered, that the more powerful a method may be the more carefully it must be used. It is often thought that mathematical genius includes an instinct for the avoidance of fallacies, but it is not true to say more than that genius includes a potential faculty of detecting danger. A trained mathematician is often aware intuitively when special care is necessary, but that this intuition, [to] which no English analyst would trust his reputation, may be developed in a man of genius, the fact that Ramanujan himself has sent to Cambridge a number of demonstrably false results proves conclusively. At present all his results must necessarily be regarded with some suspicion till they have been independently obtained, a state of affairs which must not continue.

E. H. Neville

I see no reason to doubt that Ramanujan himself will respond fully to the stimulus which contact with Western mathematicians of the highest class will afford him. In that case his name will become one of the greatest in the history of mathematics, and the University and City of Madras will be proud to have assisted in his passage from obscurity to fame.

Yours sincerely,
E.H. Neville

Commentary. Eric Harold Neville (1889–1961) entered Trinity College, Cambridge, as an undergraduate in 1907 and was classed as Second Wrangler in the Mathematical Tripos in 1909. He was elected to a College Fellowship two years later. His Fellowship was not renewed when it terminated a few years later, and Neville suspected, rightly or wrongly, that this was because of his pacifist views, which at that time were unpopular with those in authority. In 1919 he was appointed to the Chair of Mathematics in the University College of Reading and remained there until he retired in 1954. He was a strong supporter of the Mathematical Association, serving as its President in 1934 and, for many years, as its Librarian. He was not a prolific mathematician, and is, perhaps, remembered most for his book on the Jacobian elliptic functions [172], where he modified the traditional symbolism and introduced, sadly too late for general acceptance, a more logical notation.

We owe a great debt to Neville for obtaining the confidence of Ramanujan in Madras, and to him and his wife for opening their home in Cambridge to Ramanujan during his first few weeks in Cambridge, when he must have felt his new surroundings to be very strange.

In reminiscences of S. Narayanan about M.T. Narayaniengar, Narayanan recounts that "About the year 1907 or 1908, I met him in the company of the late Mr. V. Ramaswami Iyer, the founder of our society—when all the members of the Society met in the Presidency College to welcome Mr. E.H. Neville of Cambridge—when the late Mr. S. Ramanujam read his maiden-paper on 'Elliptic Functions and Modular transformations' which elicited the admiration of Mr. Neville" [170]. The year "1907 or 1908" is clearly incorrect.

R. Littlehailes to F. Dewsbury   **29 January 1914**

> The Observatory,
> Nungambakam,
> Madras,

My dear Dewsbury,

I venture to request that you will be so good as to place before the Syndicate of this University a request that I have to make regarding Ramanujan, at present research student in Mathematics.

It is that he be granted by this University a scholarship of about £250 (sterling) together with a grant of about £100 in order to enable him to proceed to Cambridge.

Ramanujan is a man of most remarkable mathematical ability, amounting I might say to genius, whose light is metaphorically hidden under a bushel in Madras.

He has, I understand, passed the Matriculation Examination from the High School, Kumbakonam, but after beginning his Intermediate College course, he fell ill and did nothing for a couple of years. Subsequently he obtained an appointment as a clerk in the Office of the Madras Port Trust and was last year granted a Research Studentship in Mathematics.

He is now 26 years of age and has for the last 10 years made the study of Mathematics his special hobby, though he has had access to no memoirs of modern mathematical treatises until comparatively recently when he came to Madras. He has, nevertheless, developed his subject to such a remarkable degree as to leave no trace of doubt that he has most exceptional mathematical genius and for this reason it is considered most desirable that he should be granted the financial means which will enable him to proceed to Cambridge where he will have ready access to modern mathematical literature, advice and criticism which is not to be obtained in this country and the privilege of absorbing and in turn reacting upon the mathematical atmosphere for which Cambridge is so renowned. He has already formulated theorems of a most comprehensive character dealing with some of these patterns of mathematics that are on the boundaries of our present day development and treatment of the subject and it is greatly to be desired that he be given the opportunity of discussing his methods and results with some of the leading mathematicians of Europe and of having them published in the leading mathematical periodicals. I have little or indeed no hesitation in stating my opinion that if he continued in the future to develop as he has done in the past and subjects his work and methods to the test of modern mathematical criticism, then it will be only a matter of a delay of two or three years before he will be called upon to fill a chair of Mathematics at some university.

It was formally [formerly?] suggested that Ramanujan should go to Cambridge but he was unwilling to leave India, and it is only within the last fortnight that he has been persuaded to agree to leave India. He has been in correspondence with certain Cambridge Mathematicians of late, and Mr. Hardy, one of the finest of the modern Cambridge school has only recently expressed his keen disappointment at Ramanujan's refusal to leave India for Cambridge. But now that he has been persuaded to put aside his prejudices against travel, the University of Madras is afforded a unique opportunity for the promotion of research by granting—if they can and will—such financial aid as will enable Ramanujan to proceed to Cambridge.

It is probably necessary to mention that he is not overburdened with this world's goods being but passing rich on Rs.50 per mensem.

Mr. Neville tells me that he has already forwarded to you a memo dealing with Ramanujan; so it is unnecessary to state more than I have done. But in this case [if] there is any question likely to arise regarding his treatment at Cambridge, I should say that we have it on Mr. Neville's authority that Trinity College will do everything in their power to aid Ramanujan and it is even possible—though we cannot at present reckon on it—that they will subsequently give him some financial assistance.

I therefore recommend that this University grant Ramanujan a scholarship of the value of £250 per annum for one year, to be possibly extended in whole or in part for a second year on receipt of a report from the authorities of Trinity College, Cambridge, and in addition give him a sum of £100 which will be necessary for his passage to England and initial outfit.

Mr. Neville, who leaves India on March 14th, would be willing to take him to Cambridge provided a passage can be obtained in the same boat as that in which he is travelling; but if that is not possible then he would like Ramanujan to leave at the earliest possible date so that he may reach England before the end of the May term.

I trust that the Syndicate will be able to grant this scholarship which promises to bring credit on this University as being one of the foremost in the promotion of research and advancement of knowledge.

I am
Yours sincerely,
Littlehailes

Commentary. Although Ramanujan, indeed, did become ill shortly after leaving the Government College at Kumbakonam, there is no evidence that illness was a cause of his failure to pass examinations upon the completion of his first year. Neville left India at the end of February, while Ramanujan departed on March 17, three days after Neville originally was to sail. Since Littlehailes's letter indicates that reservations may have had to be made well in advance, Neville possibly gave Ramanujan his original reservation and secured his own passage about two weeks earlier so that he could meet Ramanujan on his arrival in England.

Sir Francis Spring to C.B. Cottrell   **5 February 1914**

My dear Cotterell [sic],

If I understand right, His Excellency has the Educational portfolio. So I am anxious to interest him in a matter which I presume will come before him within the next few days—a matter which under the circumstances is, I believe, very urgent. It relates to the affairs of a clerk of my office named S. Ramanujan who, as I think His Excellency has already heard from me, is pronounced by very high Mathematical authorities to be a Mathematician of a new and high, if not transcendental, order of genius.

A few months ago the Madras University gave S. Ramanujan a scholarship to enable him to fill certain gaps in his education which operated to prevent his conveying his conceptions to the outside world. Meanwhile during the last 8 or 9 months various Mathematicians in the first rank in Cambridge,

Simla and Madras have had before them selections from his work and have pronounced upon them in terms of the very highest eulogy.

Just now, as probably His Excellency is aware, a Mr. Neville, who, I think, is a Senior Wrangler and a Fellow of Trinity, Cambridge, has been in Madras giving a series of lectures on certain phases of the Higher Mathematics to Honours students and others interested. Under a mandate from Cambridge he has interested himself greatly in Ramanujan and there is every reason to hope that he may be persuaded to go to Cambridge for a year or two so that under expert guidance not only may the fruits of his genius be given to the world but also, we may hope, his own fame, future usefulness and personal prosperity may be secured—matters probably quite impossible if he remained in a backwater like Madras for the rest of his life.

I now come to the point where His Excellency may perhaps be able to interfere with advantage. Last evening I learnt from Mr. Littlehailes and others that the University Syndicate had decided, subject to sanction of Government, to set aside a sum of Rs.10,000 in order to secure Ramanujan's visit to England for a couple of years. Messrs. Littlehailes and Neville begged me to intercede with His Excellency with a view to the speedy confirmation of this action of the University Syndicate. But I wish to make it quite clear that I write under no mandate from the Syndicate but merely as a private individual interested in my own employee, Ramanujan, as well as in Mathematics.

Mr. Arthur Davies will doubtless arrange for the voyage to England and that Ramanujan's orthodoxy may be maintained unimpaired. Mr. Neville assures me that he will meet him on his arrival in London and conduct him personally to Cambridge, and that when there he will interest himself personally in his welfare, generally and in all matters of Brahmin orthodoxy, so that he may return to India without any loss of the esteem of his caste men.

I myself am very far from being Mathematician enough to express adequately what has been said to me by several who are fully qualified to express an opinion on the subject of the potential value to Science of the new line of thought in which Ramanujan's investigations lie. I am assured however by those who ought to know what I am talking about that they may conceivably be epoch–making and as such well worthy of financial support at the hands of the Madras University.

Needless to say Professor Hardy and other high Mathematicians may be trusted to give Ramanujan the fullest credit in the scientific world for his work. My reason for saying anything so obvious as this is that I am told that certain of his Indian friends have been suggesting to him that all that the Scientists of England desire is to steal his ideas—obviously an utterly impossible suggestion with men of their class.

Yours sincerely,
Francis J. E. Spring

Commentary. Cecil Barnard Cottrell, a graduate of Balliol College, Oxford, was appointed Private Secretary to His Excellency, Lord Pentland, the Governor of Madras, in 1912.

John Sinclair (1860–1925), 1st Lord Pentland of Lyth, Caithness, entered the army in 1879 and later served as Liberal Member of Parliament for Dumbartonshire and Forfarshire. He was for a time Secretary of State for Scotland under Prime Minister David Lloyd George. He was appointed Governor of Madras in 1912, and demitted office in 1919. Ramanujan owed much to his sympathetic and practical help.

We are unable to determine the identities of the mathematicians in Simla.

Neville was a Second Wrangler, not a Senior Wrangler as claimed in the letter.

C.B. Cottrell to Sir Francis Spring   **5 February 1914**

                              GOVERNMENT HOUSE
                                      MADRAS

Dear Sir Francis,
   Your letter of 5th February.
   His Excellency cordially sympathises with your desire that the University should provide Ramanujan with the means of continuing his researches, at Cambridge, and will be glad to do what he can to assist in the proposal.

                                        Yours sincerely,
                                        C.B. Cottrell

Commentary. The following document is found in the Tamil Nadu Archives.

Document No. 182 of the Educational Department   **12 February 1914**
               GOVERNMENT OF MADRAS.
               EDUCATIONAL DEPARTMENT.

READ—the following paper:—
   Letter—from F. DEWSBURY, Esq., B.A., LL.B., Registrar, University of Madras.
   To—the Secretary to Government, Educational Department.
   Dated—Senate House, the 5th February 1914.
   No.—523.

I am directed by the Syndicate to request that the Government will be pleased to accord its sanction to the appropriation for a special scholarship to S. Ramanujam of the sum of Rs. 10,000 from the balance remaining to the University from the grant of Rs. 50,000 sanctioned in G.O. No. 622, Educational, dated 1st September 1908, towards the cost of establishing University vacation lectures.

2. S. Ramanujam at present is in receipt of a special University scholarship of Rs. 75 per mensem which is granted to him by the Syndicate with the sanction of Government under G.O. No. 291, Educational, dated 9th April 1913. By the aid of this scholarship he is now devoting his whole time to the study of Mathematics.

3. The Syndicate has been advised that Ramanujam's Mathematical ability is of such extraordinary nature as to amount to genius, and that under proper training and in a suitable environment he may be expected to make most valuable contributions to the Science of Mathematics. I enclose, for the information of Government on this point, a copy of a letter, dated 29th January 1914, from Mr. R. Littlehailes, M.A., Professor of Mathematical Physics in the Presidency College, Madras, and also a copy of a memorandum from Mr. E.H. Neville, Fellow of Trinity College, Cambridge, now delivering a course of Special University Lectures in Mathematics in Madras.

4. In view of these expressions of opinion, the Syndicate considers that, if possible, Ramanujam should be sent to Cambridge to continue his Mathematical researches there, and that he should be granted a scholarship for the purpose of £250 a year for at least one year, to be extended for a second year if the reports of his work received prove favourable, and that he should be provided with the cost of his passage and a reasonable sum for outfit, involving a further outlay of at least £100. It is to meet this proposed expenditure that the Syndicate requests permission for the appropriation of Rs. 10,000 from the University Lectures in Mathematics in Madras.

5. On the 31st March 1913, the Vacation Lectures Fund amounted in cash and Government securities to Rs. 50,577-11-8. The Syndicate does not at present propose to institute any courses of vacation lectures in the near future, and suggests that an appropriation from this Fund which is not being utilised for any immediate purpose would be the most satisfactory means of meeting the cost of the proposed scholarship. There are no other funds from which such appropriation could conveniently be made, and the income from the Fee Fund is at present hardly sufficient to meet the ordinary expenditure of the University.

6. It will be seen from Mr. Littlehailes' letter that S. Ramanujam is willing to accept the scholarship if it is offered to him, and that Mr. Neville is prepared to accompany him to Cambridge and is anxious that he should reach England before the end of the Cambridge May term. In these circumstances, I have to request that orders may issue on these proposals at the earliest possible moment.

Commentary. Ramanujan and Ramanujam are two transliterations of the same Sanskrit name RAMANUJAHA, which means younger brother of Rama. In Ramanujan's native language Tamil, there are also two distinct spellings. At the Town High School and at the Government College in Kum-

RAMANUJAM HALL, TOWN HIGH SCHOOL

bakonam, rooms are named in honor of their most illustrious student, and, in each case, the spelling Ramanujam is used.

The letters from Littlehailes and Neville are given six and eight pages earlier, respectively.

Order–No. 182, Educational   **12 February 1914**

In the circumstances stated in the Registrar's letter read above, the Government sanction the appropriation of a sum not exceeding Rs. 10,000 from the University Vacation Lectures Fund for the grant to S. Ramanujam of a scholarship of £250 a year, tenable in England for a period of two years, free passage and a reasonable sum to outfit.

W. Francis,
Ag. Secretary to Government.

Commentary. We have extracted below portions of some official notes of the Educational Department.

Notes Connected with G.O. No. 182, Educational   **12 February 1914**

What caste is he? Treat as urgent.

W. Francis

S. Ramanujam appears to be a Brahman, Ayyangar. He is already in receipt of a scholarship of Rs. 75 per mensem from the University of Madras, tenable for a period of two years from April 1913.–*vide* G.O. No. 291, Educational, dated 7th April 1913. The principle of granting special scholarships from the University's funds having already been recognised by Government in the Government Order quoted above the question for consideration in connection with the present proposal of the Syndicate is whether the University vacation lectures fund can be diverted for the purpose.

... They were intended to bring the teachers of high schools into touch with the newest methods of teaching certain subjects which were prescribed by the University Regulations and it was thought that the arrangement would be needed for a few years only as the new teachers who had studied under the revised regulations would be more competent to teach the subjects in question without any special instructions. Accordingly the lectures of 1909–10 were the last, and the interest on the investment is being added to the principal. ... The Syndicate does not contemplate instituting any course of vacation lectures in the near future. The fund is not thus needed for the purpose for which it was originally intended. ...

As, however, no other object has been proposed for expenditure from the vacation lecture fund, which should apparently be diverted for other purposes, there seems to be no objection to meet the proposed scholarship from it. ...

C. Ramanujachari

This Ramanujam is a budding Newton on whose account rules and precedents may well be stretched. The praise given here to his work by Mr. Neville (a Fellow of Trinity, Cambridge, who has come out specially to give lectures in higher mathematics) could scarcely be higher.

W. Francis

It is a very remarkable proposal. I suppose it is strictly irregular, but if Mr. Neville is to be believed, some irregularity would be justified.

W.O. Horne

Commentary. Walter Francis, a graduate of Corpus Christi College, Oxford, was appointed to the Indian Civil Service in 1888, and retired in December 1916. William Ogilvie Horne, a graduate of Trinity College, Oxford, was appointed to the I.C.S. in 1880, and retired in October 1914.

C. Mallet to G.H. Hardy 11 February 1914

INDIA OFFICE,
WHITEHALL,
LONDON,
S.W.

Dear Mr. Hardy,

I quite understand your letter, and sympathise with it, but money is the vital point. I know of no means from which sufficient money is forthcoming to bring Mr. Ramanujan home to Cambridge, even for two years. If he came, I should have thought it might be worth while for him to take a Mathematical degree. I am quite clear that no money for this purpose can be got from the India Office, and I should have thought it doubtful—though on this point you know better than I—whether Trinity College or Cambridge University was likely to find any money for the purpose. What Ramanujan's friends in Madras, or the Government or the University there, may be able to do for him in regard to money, I cannot say; but as it has only been possible at present to raise for him a little scholarship of Rs. 75 a month (that is £60 a year), it seems to me very unlikely that it would be possible to raise in Madras the £200 or £300 a year which he ought to have for Cambridge. Mr. Neville's letter rather alarmed me, because it seemed to me that he was encouraging Ramanujan to come to England without any real prospect of providing for him when he got here; and there seems to me a danger that a student in India might count on Mr. Neville's kindly assurance, and assume that a Cambridge Don would find it easy to raise all the money required. We have known so many cases of Indian students, brought over here in the vague hope that somehow or other money would be found to keep them in England, with the inevitable result of disappointment and misery for them, that I am bound to urge rather seriously, the danger of tempting Indian students to come here without a very definite provision for them. I think it would be well if you would kindly write out to Mr. Neville by this mail, and urge this point upon him as plainly as you can. I would tell him that the India Office can find no money for such a purpose, and I think I would remind him, if you agree, that Cambridge is very unlikely to do so.

Yours sincerely,
C. Mallet
Secretary for Indian Students

Commentary. Charles Edward Mallet (1862–1947), barrister and historian, served as Secretary for Indian Students at the India Office in London from 1912 to 1916, and was knighted in 1917. He had earlier been a Member of Parliament and was the author of several biographical memoirs and historical books.

G.H. Hardy to E.H. Neville 12 February 1914

TRINITY COLLEGE,
CAMBRIDGE.

Dear Neville,

I'm writing in a hurry to catch the mail and warn you to be a little careful. I've been in correspondence with the India office again—I enclose the last letter. In order that Ramanujan should come—and of that I'm as anxious as ever—there must be an absolute certainty of about £250 a year. When I wrote to them before they left me under the impression that that could certainly be found (not from the India office here—there is not and never was any question of that). The sources contemplated were, I presume, Madras University or the central Government of India. According to their then version, Ramanujan's reluctance was the only obstacle. Now this man Mallet seems disposed to sing a different tune. So be very cautious. I think J.E.L. and I between us could contribute £50 for 2 years (don't tell Ramanujan so)—but that's only a very little way.

Of course what Mallet says about 'as it has only been possible etc. ... ' is bilge—£60 a year is heaps for Madras.

As regards the College I don't know at all: I'll consult Barnes without delay. He did seem to think it not inconceivable before [...]

Please keep me well informed as to the progress of the attempts to raise money in India: but don't pledge yourself to anything uncertain.

Kind regards to Mrs. Neville.

Ever yours
G.H. Hardy

Commentary. When Hardy originally offered Ramanujan the invitation to come to Cambridge, he apparently thought little about how Ramanujan's stay would be financed. A misunderstanding between Hardy and the India Office evidently led Hardy to believe that money to support Ramanujan in Cambridge could easily be found. Upon receiving Mallet's letter, Hardy began to worry and get "cold feet." It was not until 12 February 1914, just before or about the same time as Hardy's letter, that the award of a scholarship of £250 a year for two years from Madras University was finally approved.

J.E.L. is, of course, Hardy's colleague J.E. Littlewood.

The last sentence in the antepenultimate paragraph was not completed by Hardy.

The exact date of the letter is unknown, but the date 12 February 1914 must be very close if not exact.

Extract from the newspaper, the Madras Mail   15 May 1914

Mr. S. Ramanujan, of Madras, whose work in the higher mathematics has excited the wonder of Cambridge, is now in residence at Trinity. He will read mainly with three fellows of the College—Mr. Hardy, Mr. Littlewood and Mr. Neville. They are going through masses of work he has already done, and are making some surprising discoveries in it. Mr. Hardy, Fellow of Trinity, says: "The first I knew of him was about fifteen months ago. He wrote to me explaining who he was, and sent a large number of mathematical theorems which he had proved. There were a great many very remarkable results. His theorems were all in pure mathematics, particularly in the theory of numbers and the theory of elliptic functions. While many of them were quite new, others had been anticipated by writers of whom he had never heard, and of whose work he was quite innocent. That is the wonderful thing; he discovered for himself a great number of things which the leading mathematicians of the last hundred years, such as Cauchy and Jacobi, had added to the knowledge of schoolmen."

Commentary. The letters below from Ramanujan to his friend R. Krishna Rao were secured by P.K. Srinivasan [263]. R. Krishna Rao was a nephew of Dewan Bahadur Raghunatha Ramachandra Rao and grandson of Sir T. Madhava Rao. This nephew is probably the one mentioned by R. Ramachandra Rao in his biography of Ramanujan [226, p. xiii], coauthored with P.V. Seshu Aiyar.

The title Dewan denotes the financial minister of an Indian state, or, more generally, any official with considerable financial responsibilities, such as R. Ramachandra Rao, who was Collector at Nellore, a local district headquarters lying about 100 miles north of Madras in a fertile agricultural region. Bahadur (*hero* in Hindustani) is a title of respect often added to the names of officers and officials. R. Ramachandra Rao was one of the original members of the Indian Mathematical Club and helped to start its library with gifts of both books and a sum of Rs.100/-. He was President of the Indian Mathematical Society from 1915 to 1917. For further information, see [238].

In 1841 Tanjore Madhava Rao (1828–1891) entered the High School, Madras, which later became the Presidency College, and distinguished himself in mathematics, receiving in 1846 the degree of Proficient in the First Class. He served as Dewan in several Indian states, notably in Baroda from 1875 to 1882. He had a reputation as an able administrator and great statesman. He was awarded the title Raja by Queen Victoria in 1877 and was listed K.C.S.I. (Knight Commander of the Star of India). He was an ancestor also of Ramanujan's friend K. Ananda Rau, being grandfather of the latter's mother.

For further information, see [55] and [258].

S. Ramanujan to R. Krishna Rao   **30 March 1914**

Suez

Dear Mr. Krishna Row,

Reached Suez this evening. The steamer arrived on the 15th at Madras and I was very busy the next day (sending my people to Kumbakonam after packing up things, going to Wrenn Bennett to buy things, to the College, town, etc.) and so I couldn't go to you; but I told your brother to inform you of my starting on the 17th morning and thought I could see you at the Harbour.

For the first three days I was very uncomfortable and took very little food and after that I have been alright. The sea is very smooth and there is no fear of sea sickness. I do not know whether I have to go to Cambridge directly or stay at London and then go. I shall write to you after I reach England and everything is definitely settled. My best compliments to your brother and respects and warmest thanks to your uncle.

I am
Yours very sincerely
S. Ramanujan

Commentary. Wrenn Bannett (note correct spelling) was a British-owned, inexpensive department store located on General Patters Road near Triplicane. After Independence, the firm was nationalized and since then has sold only furniture.

Chapter 4
Ramanujan at Cambridge

S. Ramanujan to R. Krishna Rao
11 June 1914
Trinity College

My dear Krishna Row,

Please excuse me for the long delay in writing to you. Now I am somewhat accustomed to the living here. Till now I did not feel comfortable and I would often think why I had come here. It is due to the difficulty of getting proper food. Had it not been for the good milk and fruits here I would have suffered more. Now I have determined to cook one or two things myself and have written to my native place to send some necessary things for it.

After enjoying the pleasant voyage except for two or three days when I was seasick, I reached London on the 14th April when Mr. Neville and his brother were kindly waiting at the docks and took me to Cromwell road where I remained a few days. I came to Cambridge on the 18th evening and remained for some days in Mr. Neville's house.

Now I am living in the college and going to stay here for the future also even though it is more costly than lodging houses, as it will be inconvenient for the professors and myself if I stay outside the college.

Mr. Hardy, Mr. Neville and others here are very unassuming, kind and obliging. As soon as I came here, Mr. Hardy paid £20 to the college for my entrance and other fees and made arrangements to give me a scholarship of £40 a year. The remaining £20 may be given in due course or may be taken for the fees of the tutors.

I am attending lectures and have written two articles till now. Mr. Hardy is going to London today to read a paper on one of my results before the London Math. Society.

I hope you have passed your examination. Is your brother coming here? My respects to your uncle and compliments to your brother.

Yours sincerely
S. Ramanujan
c/o G.H. Hardy Esq.,
Trinity College

21 Cromwell Road

Commentary. Neville's younger brother Raymond recalls that when Ramanujan arrived in London aboard the S.S. Nevasa, he was met at the docks by Neville and his older brother, and driven in the latter's Jowett car to the Department of the Educational Adviser to Indian Students at 21 Cromwell Road.

The exact date when Ramanujan moved from Neville's house at 113 Chesterton Road to rooms in Trinity College is not known. Neville was clearly in error when he stated in his article on Ramanujan [**171**] that it was July 1914.

S. Ramanujan to R. Krishna Rao   **7 August 1914**

Trinity College

My dear Krishna Row,

Received your kind letter of 7th July. Very glad to hear that you have passed your apprentice examination. Glad to hear also that Mr. Ananda Row is coming here, and I am ready to help him in any sort of way I can be of use to him.

I came here at the end of the year for the climatic conditions as you know. The college was closed in the middle of June and it will be reopened in the middle of October. There is nobody here except Prof. Hardy as the examinations are all over and all have gone outside. I can write to you something interesting to you after the vacation is over. That is why I have

E. H. Neville's home at 113 Chesterton Road

nothing to write to you at present and you will excuse me for that.

I have written three papers till now. The proof sheets have come. I am writing three more papers. All will be published at the end of the vacation, i.e. in October.

It will be difficult for Mr. Ananda Row to reach London through the Channel and the Thames owing to the present war, and so it is better for him to get down at Plymouth or some other seaport.

Has your brother determined to go over here? Hoping all in your family are doing well and wishing a happy and successful career in life.

Yours sincerely
S. Ramanujan
Trinity College
Cambridge

P.S. I am living within the college premises and am cooking my food myself though it takes so much of my time. I am getting things from a company at London selling Indian things as well as from my house.

Yours sincerely
S. Ramanujan

S. Ramanujan to R. Krishna Rao   **13 November 1914**

Trinity College

My dear Krishna Rao,

Ananda Rao has joined the King's College and settled quite comfortably. He is also coming to this College to attend some lectures.

I am attending only some of the University lectures. A few students from America and Japan have come here to attend these lectures.

I am very slowly publishing my results owing to the present war. A lecturer here whom I know well and from whom I received some help to publish my results has gone to war. The other professors here whom I know have lost their interest in Mathematics owing to the present war. One of the professors here, some days back, remarked that I have come to England in the most unfortunate time.

I have changed my plan of publishing my results. I am not going to publish any of the old results in my notebook till the war is over. After coming here I have learned some of their methods. I am trying to get new results by their methods so that I can easily publish these results without delay.

In a week or so I am going to send a long paper to the London Mathematical Society. The results in this paper have nothing to do with those of my old results.

I have published only three short papers, two of which I have sent your uncle. For the mathematical side you may ask Mr. Seshu Iyer or your uncle whenever you meet them.

I don't write anything about the war, so that the letter may reach you safely.

I was silent so long as I had nothing to write to you. Hereafter I may tell you something about my progress as the professors here are somewhat reviving their lost interest in Mathematics.

As for my food I have no other go but to cook myself. There is no place very near this college where I can get vegetarian food and I can't go out of the college. I am getting some of the Indian things here. I will be very much obliged if you can send me some tamarind (seeds being removed) and good cocoanut oil by <u>postal parcel</u> through the cheapest route. Cocoanut oil is the best as it will be solid owing to cold and won't be spoiled. I can use lemons instead of tamarind if they are sour; but unfortunately the lemons here are not sour like our lemons and moreover they are not properly lemons at all but they are sweet Narthangaai. I can receive the things only in proper order if you send me by postal parcel, otherwise it will be very difficult for me to go to London harbour to receive the things. I beg to be excused for the trouble.

Yours sincerely
S. Ramanujan

Commentary. The lecturer that helped Ramanujan and went to war was J.E. Littlewood.

CHAPTER 4. RAMANUJAN AT CAMBRIDGE

The paper that Ramanujan submitted to the London Mathematical Society in the late fall of 1914 is his famous paper [**217**], [**226**, pp. 78-128] on highly composite numbers.

Tamarind is a fruit with a sour and tart taste and is used in South Indian cooking. As a dried fruit, it is soaked in water. The juice is then used for cooking sambar (soup) or rasam (very light soup with lentils) in place of lemon juice or essence. The tamarind tree is a beautiful large shady tree with small leaves. Tamarind trees are frequently planted along the sides of roads to give welcome shade to pedestrians.

Narthangaai, a tangy citrus fruit similar to grapefruit, is dried in the sun and salted. Afterward, the fruit is then stored in jars and used as a pickle for rice with yogurt.

The three letters below that Ramanujan wrote to Mr. E. Vinayaka Row on 11 June 1914, 24 March 1915, and 10 September 1915 have been made available through the courtesy of Professor K. Srinivasa Rao of the Institute of Mathematical Sciences, Madras. They were given to him by Mr. V. Madhava Rao, the son of E. Vinayaka Row. The latter was a lawyer and resident in Mylapore and a friend of K. Srinivasa Rao's father. The letters, parts of which have become somewhat moth-eaten, were discovered in a suitcase in Mr. Madhava Rao's attic. Doubtful or missing letters and words have been supplied in italics, or are indicated by dots ... . For some portions given in italics, undoubtedly additional words are missing. Although in some cases the original words may be uncertain or missing, the contents of the letters are clear.

E. Vinayaka Row was born on 14 July 1891 at Tanjore as the eldest son of Sri S. Ekanatha Rao and Smt. Chandrakantha (alias Meenakshi Bai). He attended Kumbakonam Town High School for boys and later graduated from Pachaiyappa's College, Madras. He was a very distinguished student in the University of Madras and was for some time a tutor in Pachaiyappa's College. Later, he studied for law in the Law College, Madras, and then joined the Bar and practiced in the Madras High Court, where he worked with Sir C.P. Ramaswamy Iyer and with Sir Alladi Krishnaswamy Iyer. He was a member of the Apollonius Circle in which Ramanujan was also a member. Ramanujan likely had contact with Vinayaka Row during the time he was a student at Pachaiyappa's College. Their correspondence confirms their mutual interest in mathematics and mathematics students and teachers of that College.

S. Ramanujan to E. Vinayaka Row   **11 June 1914**

Trinity College

My dear Vinayaka Row,

Your kind letter *to* hand. Very glad to hear that Suryanarayanan has got the Scholarship. *P*lease excuse me for the long delay in *w*riting to you. It is because that *I* felt quite uncomfortable till now.

The voyage on the whole was very pleasant though I suffered from Sea-sickness for two or three days in the Mediterranean. As soon as I arrived in England on the 14th April, Mr Neville was kind enough to come to the docks and take me to Cromwell road. After staying a few days at London I came to Cambridge on 18th evening. But as I came here *in* the middle of the year I couldn't *fin*d licensed lodging here and *stayed* in Mr Neville's house for ... ... *weeks,* and then came to the *college.*

*Liv*ing in college is more costly *than in lodg*ing houses. Cambridge *is a cost*ly place, next comes *Oxford(?)* I have no idea of the *cost. I have* to pay all the expenses *every term* unlike the lodging houses where the expenses are paid weekly. The college pays me £40 a year of which at present £20 is given to the college for my entrance and other fees.

I am staying in the college b*ecause* there will be much inconvenience *to* the professors and myself if I stay *out*side the college. I would advise *Surya*narayanan to try to live in lodg*ing*-houses and if he can fortunately *live* in an unlicensed lodging house, he will find it by far cheaper than other places.

As for the *food I would have* suffered much more had it not been for the good milk obtained here. For the first two months I felt why I had come here. Now I am alright. Suryanarayanan also will feel the same for two months from the tim*e he comes* let him not be discourag*ed* it is only a matter of two mon*ths until a* vegetarian will be home *again.* Let him select such *lodgings* where good vegetarian *food was given* previously to the *Indian occupant* he can easily get *information when* he is coming in. ... ...

There is another Suryanarayan here who was a tutor in the Pacha*i*yappa's College. He says that he knew you all. He is living in a lodging where they prepare excellent vegetarian food. During voyage there will be no *diffi*culty because most of them w*ill* be Indians and he may tell them whatever he wants. Let him be p*re*pared to take meals in an Eng*lish* Restaurant in Madras, for it will be very awkward in the beginning if he remains taking food in his house to the last moment. It is also better for him to cook one or two things for himself after coming here, if he cannot find a good house and if he <u>finds time.</u> Let him be careful not to *use fried* things as they will *just* ... fat, and not to take ... *l* ings as they may contain ... After coming here he ... cooks to fry things i*f he* wants. For veget*arian* ... eggs it will be *difficult* in the beginning ... is no difficulty ... ...

As for my studies, I am attending a few lectures and have begun to write articles and publish my results. I have two articles one *on* Definite Integrals and another *on* Elliptic functions. Another I am going to write soon on contd *fra*ctions. My hearty congratulations *to* Suryanarayanan.

Yours aff[ectionate]ly
S. Ramanujan

Ramanujan (center) with friends at Trinity College

Commentary. As a Scholar, Ramanujan would have been expected to reside in College. However, from the point of view of his health it might have been better for him to live in lodgings, particularly lodgings where, as in the case of his Indian friend, excellent vegetarian food was available.

Ramanujan mentions two persons with very similar names, which he sometimes spells in different ways. The first is **Satalur Sundara Suryanarayanan**, who was born on 5 October 1894. He won a scholarship to Merton College, Oxford, where he matriculated in October 1914. He graduated B.Sc. on 26 May 1917. At that time the Oxford B.Sc. was a postgraduate degree, more or less equivalent to the present M.Sc. or M.Litt. His Faculty was Literae Humaniores and his special subject of study was "A comparative study of the Advaita Vedanta and F.H. Bradley." The Vedanta is a system of Indian philosophy based on the Vedas, the holy books of the Hindus. It may seem surprising that this subject was combined with the name of Francis Herbert Bradley (1846–1924), who was an influential British philosopher of the time, and, moreover, a Fellow of Merton. Suryanarayanan must have known Bradley by sight, but it is perhaps unlikely that he was in any sense a pupil, since Bradley appears to have taken no part in the teaching duties of the college. There is a reference to Suryanarayanan in Hardy's letter of 20

September 1917 to Subramanian. No information is available about his later career.

The second person with a similar name is **Tenneti Surya Narayana**, who matriculated in Cambridge on 28 January 1914 as a noncollegiate student. He was an able mathematician obtaining Second Class Honours in that year in Part I of the Mathematical Tripos, and he was classed as a Wrangler in Part II of the Tripos in 1916. His subjects of study in that examination were mainly on the applied side. He graduated B.A. on 26 June 1916. He appears to have joined the Indian Civil Service in the Survey Department in 1921, but later became Professor of Applied Mathematics in the Presidency College, Madras. Two papers by him on the theory of heat were published in volumes 15 and 18 of the *Journal of the Indian Mathematical Society*. His name appears on the title page of volumes 1, 2, and 3 of the new series of that journal among the list of Collaborators. He frequently combined together his two last names into one name, as is common in India.

It is unfortunate that Ramanujan never wrote the paper on continued fractions mentioned near the end of his letter, for the methods that Ramanujan used in deriving many of his continued fractions are not known to us today. Later, he did publish a short paper [223], [226, pp. 214–215] establishing what is now known as the Rogers–Ramanujan continued fraction.

S. Ramanujan to E. Vinayaka Row   **24 March 1915**

Trinity College

My dear Vinayaka Row,

Received your letter of 18th Feb. I have received all your letters. It appears from your letter that you have not received my letter. In that letter I wrote mainly about the war.

I was not well till the beginning of this term owing to the weather and consequently I couldn't publish any thing for about 5 months. This term I have published 3 or 4 pamphlets and a long paper. Up to this time I have written on the following subjects
Definite Integrals [212], [226, pp. 53–58]. Modular equations and approximations to π (Elliptic functions) [210], [226, pp. 23–39]. Definite integrals (which come out in finite terms for the rational values of the parameter and not for the irrational values) [214], [226, pp. 59–67].
Series connected with the Eulerian Constant [220] [226, pp. 163–168], Sum of the square roots of the natural numbers (and some other allied sums) [215], [226, pp. 68–71], New expressions for the Riemann Zeta function [216], [226, pp. 72–77]. Some series having curious[?] properties [218], [226, pp. 129–132], and a very long paper on the Divisors of a number [217], [226, pp. 78–128]. All these are published in the 'Messenger of mathematics', 'Quarterly journal of mathematics' and the 'London mathematical journal'.

Except the first 3 papers all others are now being printed.

Suryanarayanan is not appearing for the I.C.S. as he has a very bad short sight. He will not be selected even if he comes in the top of the list. He will enter the Educational service. I am glad to inform you that he shines in public speech in various societies and has glorious tutorial reports in his college.

I hope you and your family are doing well and wish you success in the B.L. examination.

With best wishes
I am
Yours sincerely
S. Ramanujan

Commentary.

Hardy first reported Ramanujan's illness in the spring of 1917. In this letter, we learn that already during Ramanujan's first winter in England he was not feeling well.

Ramanujan wrote a paper on the sum of the first n square roots of a natural number that appeared in the *Journal of the Indian Mathematical Society* [211], [226, pp. 47–49]. However, Ramanujan is probably referring to [215], [226, pp. 68–71], where this topic is also examined.

The Suryanarayanan not appearing for the Indian Civil Service examination is the Oxford student about whom Ramanujan writes in his earlier letter of 11 June 1914.

S. Ramanujan to E. Vinayaka Row   **10 September 1915**

TRINITY COLLEGE,
CAMBRIDGE.

My dear Vinayaka Row,

Received your letter ... Very glad to hear of the birth of a son to you. But I am sorry that you *failed* in the examination. Tenati Suryan*arayan also* was sorry when he heard *of* your failure in D.T.M. He sat for *the I.C.S.* exam last month. I shall *let you* know about the result when it *is announ*ced. I do not know any–thing *about* S.S. Suryanarayan for these four or five *weeks. It is* being the long vacation most *likely* he may not be in Oxford at all.

Tenati Suryanarayan and *I* were exceedingly sorry to hear the sad death of Professor Singaravelu Mud*aliyar* at so early an age. We came to know of this unexpected sad news for the first time from your letter which reached here before the Indian Math. journal.

Do you want the general information about Cambridge or special information about all the colleges here? I shall wr*ite* to you in my next letter the ma*in* reasons for the rapid progress *in mathe*matics here unlike ...

Yours s*incerely,*
S. Ram*anujan*

Commentary. As mentioned earlier, Ramanujan's Cambridge friend's name was Tenneti Surya Narayana. S.P. Singaravelu Mudaliyar was Professor of Pure Mathematics at Pachaiyappa's College, Madras.

The initials D.T.M. perhaps denote Degree in Teaching Methods.

Ramanujan to his mother Komalatammal 11 September 1914

Trinity College

Salutations to the great Ramanuja! Ramanujan makes his countless prostrations to his mother and writes. Please write about your welfare. The letters you write reach me regularly. The three letters written on August 4, 10 and 11 reached me. I could not write letters for two weeks. You will henceforth be getting letters every week. There is no war in this country. War is going on only in the neighbouring country. That is to say, war is waged in a country that is as far as Rangoon is away from the city (Madras). Lakhs of persons have come here from our country to join the forces. Seven hundred Rajas have come here from our country to wage war. Ultimate victory will come only to the king of this country. You need not send any provisions. Ramachandra Rao's relative Ananda Rao, a youngster, has come to this country for study. He has not yet reached this place. He will come in October. Mr. Seshu Aiyar has told him to take with him numerous articles for being given to me. He and another youngster Sankara Rao have arrived in England.

When they reached the town of Port Said, the war commenced. Unknowingly they had sailed in the enemy ship. Their intention was to travel in an Austrian ship, get down at Austria and reach here by train. But while nearing the Island of Crete on the way to Austria after leaving Port Said, the crew of the English ship fired in the air to stop the ship and know its identity. Providentially, the ship stopped as it carried no guns. If the men of the ship had also fired, the ship would have been shot and sunk. The ship was captured and all the persons in the ship were taken prisoners and carried to Alexandria and the ship was seized. The people coming from our country and the Englishmen were put in another ship and sent here. These two youngsters reached after escaping this danger.

No war like this has raged before. The present war affects crores of people. It is not one or two crores. Germans set fire to many a city, slaughter and throw away all the people, the children, the women and the old. The small country Belgium is almost destroyed. Each town has buildings fifty to hundred times more valuable than those in Madras city.

In many towns, the people of the towns themselves blow off the supports of the bridges over rivers, leave them hanging in mid air, spread gunpowder all over the streets, lay mines and cover them up and remain ready to flee. When enemies come, the bridges fall and half of them are carried off by the current of the river and when the rest enter the city, the dwellers themselves burn the city and flee. When streets are ablaze, the enemies try to escape but the iron wires get round their legs and they perish, unable to run away.

War is waging at many places. Each place has an extent of 200 miles. This is on the land. Many ships are sunk by battles raging in mid oceans. These are of two kinds. One is to fire directly at a ship; the other is to go under water, knock at the enemy ship and sink it. Not only this. They fly in aeroplanes at great heights, bomb the cities and ruin them. As soon as enemy planes are sighted in the sky, the planes resting on the ground take off and fly at great speeds and dash against them resulting in destruction and death.

All that you sent got broken but reached me without falling off because of the supporting cloth. I get everything here. I get certain provisions from the city (Madras). You need not send anything.

Yours,
Ramanujan

Commentary. One lakh equals 100,000; one crore equals 10,000,000.

S. Ramanujan to his father Srinivasa Aiyangar   **17 November 1914**

Trinity College

Salutations to the great Ramanuja! Ramanujan makes his countless prostrations to his father and writes. Well and wish to hear the same from you. I got your letter. I have also received the letter written earlier by Tirunarayanan. I have all the pickles. I get tamarind etc. from Madras. You need not send anything. Except the kuzhuvidam (dried up precooked foodstuff made of flour) which you are sending now, do not send any other thing. My college was closed last week. It is to open in the middle of January. I am getting on well. Keep the house in such a way that it is attractive to look at. Do not allow the gutter to run as usual. Pave the place with bricks and keep it well. I am getting on well. The students who have come from our place have joined the neighbouring college.

Yours,
Ramanujan

Commentary. Ramanujan's letters to his father and mother were written in his native tongue Tamil. The translations above are due to P.K. Srinivasan; copies of the original letters in Tamil can also be found in his book [263].

Ramanujan is joking with his father about the gutter, which extended from the cooking area into the backyard and served to carry away liquid refuse from cooking.

Tirunarayanan was Ramanujan's youngest brother.

Kuzhuvidam is made from rice or tapioca and left to dry in the sun. Fried in oil, kuzhuvidam can be prepared very quickly. It can then be stored, e.g., in jars, for a long time.

RAMANUJAN'S LETTER TO HIS FATHER

S. Ramanujan to C. N. Ganapathy Iyer 17 December 1914

Trinity College

My dear Sir,

Received your letter of 2/11. My thanks to your wife for all her well wishes. It is a great mystery to me how my pamphlets reached you without my letter. You did not mention of my previous letter also. Very glad to hear that Godavery and her child are doing well. I send you herewith a copy of "Definite Integrals connected with Gauss sums". This has really no connection with Gauss's sums except that one will be reminded of Gauss's sums. I think that I am really fortunate in having made the friendship of

CHAPTER 4. RAMANUJAN AT CAMBRIDGE

such nice gentlemen as Messrs Hardy and Littlewood. They are the greatest of English pure mathematicians, and at the same time the most rigorous. Mr Neville also is a very nice gentleman and I am very sorry that our fields in Mathematics are very different and far away from each other and so have no chance of meeting him as often as with these. You must remember that the 'Orders of Infinity' which I found in your room introduced me to Messrs Hardy and Littlewood. They are doing researches in the most difficult and advanced part of pure mathematics. They have proved many properties of the zeta function (Riemann's) which defied all attempts to prove them for these so many years. I shall tell you one of the recent researches of Mr Littlewood. If you see the last paragraph in "Mathew's Theory of Numbers" you will find the statement "Hence we conclude that Gauss's approximation was too great". Mr Littlewood has proved that this is wrong beyond a certain value, in fact beyond $10^{10^{10}}$! Gauss's approximation was too great only when $x < 10^{10^{10}}$. I shall tell you one of the researches of Mr Hardy (not recent) about irrational numbers. He has shown that, if you take any irrational number, say $\sqrt{2}$, $\cos 20^o$, π or e and write these as decimals, e.g., $\sqrt{2} = 1.4142135\ldots$ to a large number of decimal places, say a billion or trillion decimal places, then the ratio of the number of 0's, 1's, 2's, 3's, ... tends to 1 as the number of decimal places becomes larger. He has shown more than this. Let $n_0, n_1, n_2, n_3, \ldots$ be the number of 0's, 1's, 2's, 3's, ... within n decimal places. Then

$\sqrt{n_0} - \sqrt{n_1}$ is finite when n becomes infinite,
$\sqrt{n_0} - \sqrt{n_2}$ ... ditto,
$\sqrt{n_0} - \sqrt{n_3}$ ... ditto,

and so on. These results are true in any scale of notation.

I am at present doing research in Arithmetical functions such as the number of divisors of N, the number of decompositions of N into two squares, etc. and functions connected with Arithmetical numbers. Here we get many sorts of functions that cannot be imagined at all. I shall show you an example. Suppose

$$F(x) = \frac{\cos\left(\sec\frac{1}{\sqrt{1-x}}\right)}{1^2} + \frac{\cos\left(\sec\frac{1}{\sqrt{1-2x}}\right) + \cos\left(\frac{1}{\sqrt{2-x}}\right)}{2^2}$$
$$+ \frac{\cos\left(\sec\frac{1}{\sqrt{1-3x}}\right) + \cos\left(\frac{1}{\sqrt{3-x}}\right)}{3^2} +$$
$$\frac{\cos\left(\sec\frac{1}{\sqrt{2-3x}}\right) + \cos\left(\frac{1}{\sqrt{3-2x}}\right)}{4^2} + \cdots.$$

I write for the denominators the squares of the natural numbers $1^2, 2^2, 3^2$, ... For the numerators first I take all rational numbers

$$1, \frac{1}{2}, \frac{1}{3}, \frac{2}{3}, \frac{1}{4}, \frac{2}{4}, \frac{3}{4}, \frac{1}{5}, \frac{2}{5}, \frac{3}{5}, \frac{4}{5},$$

and construct the numerators in order with their fractions. The series is absolutely convergent. But $F(x)$ does not exist for any rational value, say $\frac{2}{3}$, since one term will be of the form $\cos(\sec \infty)$ which is not definite. But $F(x)$ exists for all irrational values. There is another peculiarity. Suppose you approach any rational point, as $\frac{2}{3}$, from the right side, then $\sqrt{2-3x} = \sqrt{-0}$ and so

$$\cos\left(\sec\frac{1}{\sqrt{2-3x}}\right) \to \cos(\mathrm{sech}\infty) = 1.$$

But when you approach the same $\frac{2}{3}$ from the left side, then $\sqrt{2-3x} = \sqrt{+0}$, and so

$$\cos\left(\sec\frac{1}{\sqrt{2-3x}}\right) \to \cos(\sec\infty),$$

which is indeterminate. Hence $F(x)$ exists as the limit from one side and not from the other side when you approach any rational point.

Mr Hardy read before the London Mathematical Society some of my results in Arithmetical functions. It will be printed in 2 or 3 months and you will find many interesting results there. In this paper I entirely devote myself to the number of divisors of N and the functions connected with it.

I hope you and your family are doing well. I thank you very much for your encouragement to my people. My compliments to your wife.

Is Mr Patrachariar doing well? Where is he now, in Saidapet or in Madras?

Yours sincerely,
S. Ramanujan.

Commentary. We are indebted to P.K. Srinivasan for making this letter available to us. It was donated to the Ramanujan Museum in Royapuram, Madras, by C.G. Swaminathan, the son of the late Professor Ganapathy Iyer, who was Professor of Mathematics in The Presidency College, Madras. For some years the letter was kept inside an old copy of a mathematical journal, stored with other journals, books, and papers in the home of Swaminathan's grandparents in Coimbatore, and, after Swaminathan's retirement, moved to his home in Madras.

C.N. Ganapathy Iyer was born in 1889 and received B.A. and M.A. degrees from Presidency College in Madras. He was a classmate of C.V. Raman and a student of both Professors Middlemast and Littlehailes for his M.A. degree, which he received in 1909. Professor Littlehailes offered him an appointment as Lecturer in Mathematics at Presidency College in 1910; later he became Professor of Mathematics at Presidency College. While Ramanujan was at the Madras Port Trust, he and Ganapathy Iyer met regularly on the beach in the evenings to discuss mathematics. Both Ganapathy Iyer and K.S. Patrachariar later became Principal of the Government College in Kumbakonam. (The principal of a college is equivalent to the president of a college in the United States.) The college has since been renamed Kanitha Medhai Ramanujan Government College for Men in honor of its most famous student. Ganapathy Iyer died in 1973.

C.G. Swaminathan was Director of the Central Road Research Institute at New Delhi and Secretary to the Government of India. Retired, he now resides in Madras, as indicated above. His father told him that Ramanujan had scrofula, manifesting itself in lumps on his face.

It is of interest that the letter reveals that it was in Professor Ganapathy Iyer's room that Ramanujan found Hardy's tract *Orders of Infinity* [90], which is mentioned in the last paragraph of Ramanujan's letter of 16 January 1913 to Hardy.

The paper connected with Gauss sums is [214]. The work that Hardy related to the London Mathematical Society is probably that in [217].

Let
$$\text{Li}(x) = \int_2^x \frac{dt}{\log t}.$$

Gauss and Riemann conjectured that

(1) $$\text{Li}(x) - \pi(x) > 0.$$

The work of Littlewood quoted by Ramanujan is presumably in his paper [161], wherein it is proved that not only is (1) false, but that $\text{Li}(x) - \pi(x)$ changes sign infinitely often. In 1933, S. Skewes [262] showed that the first sign change of $\text{Li}(x) - \pi(x)$ occurs for some

$$x < 10^{10^{10^{34}}}.$$

The statements about the digits of numbers evidently arise from work of Hardy and Littlewood [115, pp. 185–187], [105, pp. 57–59]. The claims about the equidistribution of the digits of the cited numbers cannot be justified. However, the first assertion is true except on a set of measure 0. The second claim also cannot be justified. Hardy and Littlewood showed that

$$\sqrt{n_i} - \sqrt{n_j} = O\left(\sqrt{\log n}\right), \quad 0 \le i, j \le 9, \quad i \ne j,$$

except on a set of measure 0.

A function similar to Ramanujan's function $F(x)$ appears in his letter of 7 January 1915 to S.M. Subramanian.

S. Ramanujan to S.M. Subramanian 7 January 1915

Trinity College

My dear Subramanian,

It is very long since I heard from you. I am sorry I haven't received any reply to my letter up to this time. Are you keeping good health? How is your progress now? How long is your course? What department are you going to enter? Engineering or Archaelogical? What about your B.E. degree? Do you know what Durairajan and Govindarajan are doing?

I am doing my work very slowly. My note-book is sleeping in a corner for these four or five months. I am publishing only my present researches as I have not yet proved the results in my notebook rigorously. I am at present working in Arithmetical functions, such as the number of divisors of N, the number of ways in which N can be expressed as the sum of 2 squares and so on, and trying to get Algebraic expressions for these Arithmetical functions. I have written a very long paper on these, which will be published within a few months. I have already published 3 pamphlets on Definite integrals and Elliptic functions. I shall now tell you a very curious function. Arrange all the rational numbers thus

$$1, \frac{1}{2}, \frac{2}{1}, \frac{1}{3}\frac{3}{1}, \frac{2}{3}\frac{3}{2}, \frac{1}{4}\frac{2}{4}\frac{3}{4}\frac{4}{1}\frac{4}{4}\frac{4}{3},$$

and construct the function

$$F(x) = \frac{\cos\left(\sec\frac{1}{\sqrt{1-x}}\right)}{2} + \frac{\cos\left(\sec\frac{1}{\sqrt{1-2x}}\right)}{2^2} + \frac{\cos\left(\sec\frac{1}{\sqrt{2-x}}\right)}{2^3}$$
$$+ \frac{\cos\left(\sec\frac{1}{\sqrt{1-3x}}\right)}{2^4} + \frac{\cos\left(\sec\frac{1}{\sqrt{3-x}}\right)}{2^5} + \cdots$$

The numerators are made up of the numerators and the denominators of the rational numbers. The denominators are the regular Geometric series $2, 2^2, 2^3, 2^4, \ldots$.

Since the cosine is numerically not greater than 1, the series is absolutely convergent. If you give any rational value for x one term will be indefinite ($\cos(\sec\infty)$ is not definite). Hence $F(x)$ does not exist for all rational values of x. But $F(x)$ exists for all irrational values of x. Again there is another curiosity here. Take a rational point say $\frac{1}{2}$. If you approach from the left to this point $\frac{1}{2}$, then $\cos\left(\sec\frac{1}{\sqrt{1-2x}}\right)$ approaches the form $\cos\left(\sec\frac{1}{\sqrt{+0}}\right)$ which is indefinite. But if you approach the point $\frac{1}{2}$ from the right, then $\cos\left(\sec\frac{1}{\sqrt{1-2x}}\right)$ approaches the form

$$\cos\left(\sec\frac{1}{\sqrt{-0}}\right) = \cos\left(\operatorname{sech}\frac{1}{\sqrt{+0}}\right) = \cos 0 = 1.$$

Similarly for every rational point. Hence $F(x)$ exists if you approach every rational point from the right side but $F(x)$ is indeterminate when you approach the rational points from the left side; and $F(x)$ is obviously finite and determinate for every irrational value of x. Just imagine this function remembering that

(i) between any 2 rational points there are an infinite number of irrational points;

(ii) between any 2 irrational points there are an infinity of rational points;

(iii) every irrational point is the limiting point of some sets of rational points as well as some other sets of irrational points;

(iv) every rational point is also the limiting point of some sets of rational points as well as some other sets of irrational points.

Will you please let me know your brother's address. You have simply stated in one of your letters that he is a teacher in Bapatla. He has not written to me any letter after I left India. Either he is too sorry for his failure in the Exam. to write to me, or he has misunderstood me before I started or he is very angry with me for my having crossed the seas. I can't understand clearly. Even though I wrote to my mother to have your brother's help in packing some cookery things, even though I informed your brother of this and even though she requested your brother to help her in packing up the things properly, he did not condescend to help her. When the things reached me half lost and spoiled and when I wrote to her why she was so foolish as to pack up the things herself for sending on a long journey she replied that there was nobody to help her and the only reply she could get from your brother was "You may do everything yourself". Now as well as in future I am not in need of anything as I have gained a perfect control over my taste and can live on mere rice with a little salt and lemon juice for an indefinite time.

If you can, you may remind your brother of our old friendship.

Yours sincerely
S. Ramanujan

Commentary. The long forthcoming paper on arithmetical functions to which Ramanujan refers is probably his paper [**219**], [**226**, pp. 136–162].

Bapatla is about 300 kilometers north of Madras on the Madras–Calcutta rail line.

See our commentary after Ramanujan's letter of 22 January 1914 for the reason why crossing the seas would cause anger.

The penultimate sentence of Ramanujan's letter gives evidence that he did not keep a proper diet, which therefore may have contributed to his subsequent illness.

S. Ramanujan to S.M. Subramanian   **3 June 1915**

Trinity College

My dear Subramanian,

Received your letter of 23rd April and the Panchangam. Many thanks. Sorry to learn that you didn't receive my reply to your letter of 5th Feb. Will you please write to me definitely what books you want on architecture.

I may be conferred upon a "Research Degree" in a year but not an "Honorary Degree" as you expect. Honorary degrees are conferred upon those in a very high position in life (say Lords, dukes, maharajas etc.) here at present.

I am thinking of returning to India in a year and see you all again. I am sorry that I haven't got a photo of myself (up to this time) to send you. As for the pamphlets and dissertations of mine I shall send you some within a month. At present I do not know how many copies of my work should be sent to Vice-Chancellor, Referees, University library and a few other libraries to get my degree on any other footing. I have got only 10 or 11 copies of each in my possession. Two dozen copies are sent to me for my publication; of which 10 are distributed to the directors of studies here and 4 copies are sent to Madras.

You said in the beginning of your stay in Bombay that your work is not congenial to you owing to unpleasant surroundings. I hope you have overcome all those difficulties.

Where is your brother now at Bapatla or Kumbakonam? I am very sorry that I haven't received his letter. Is the post at Bapatla a permanent one?

Sorry to hear that your father is so ill. It is advisable that one of your brothers should remain in the house till he recovers. From your letter I understood that he was suffering from pain all over the body. I think he might have recovered in a short time.

Yours affectionately
S. Ramanujan

Commentary. A panchangam is a Hindu almanac, often published by a family. This small pamphlet normally contains dates of full moons, sunrise and sunset times, positions of stars, dates of eclipses, dates of festivals, and birthdays of important people and gods, such as Siva and Rama. A panchangam also gives the star governing the day and the auspicious periods in the day for commencing or doing specific tasks. It is consulted by every Brahmin family to find the proper day and time for specific commencements (such as purchasing, changing residences, starting journeys, etc.). For the drawing of Hindu horoscopes, the planetary positions must be known accurately. Otherwise the horoscope cannot be cast. Thus almost every Brahmin family would have a panchangam.

S. Ramanujan to S. M. Subramanian 1 July 1915

Trinity College

My dear Subramanian,

Received your letter of 29th May. I wonder why so many of my letters didn't reach you. I received all your letters except one. Three letters previous to this didn't reach you. I hope this will reach you very soon and safe owing to the overland mail.

I shall repeat briefly what I wrote to you in my previous letters.

Will you write to me definitely what kind of books on architecture you require?

RAMANUJAN WITH WESTERN CLOTHING AND HAIRCUT

A few days back a friend of mine here one Krishnamachari came to me with his camera and took my photo when I was sitting near the fireplace in my room. He gave me a copy which I am now sending you with this letter. At that time there was only dim light in the room and the photo was taken for fun. You will see in the photo many books and papers in the table as a huge white mass.

As for my dissertations you have asked me I shall send you some in a fortnight after ascertaining how many copies I may be required to send the Vice-Chancellor and the Referees to take my degree.

I am going to take a Research degree at the end of one year (hence) and I am thinking of returning to India next year this month and of coming back to England if necessary.

I am unusually slow in my work and that is why I think it may be necessary to stay here a few years more as there is no help nor references in Madras for my work.

You told me in the beginning of your stay in Bombay that you had to move with uncultured and unpleasant persons. I hope you had overcome those difficulties.

I am very sorry to learn that your brother had to leave the school without any pay for the vacation. What is he doing at present? I think it is better for him to be employed in metropolis like Madras and not in mofussils. In metropolis even if he is employed for a very short time in any department he can get a permanent job anywhere before he leaves another.

Hoping you are in good health.

Yours affectionately
S. Ramanujan

Commentary. S.M. Subramanian later became an engineer. According to his brother M. Anantharaman, Ramanujan lived with Subramanian from 1911 to 1913, while the latter was a student at the Madras Engineering College. The eldest brother was M.M. Ganapathy, who is mentioned in Ramanujan's letter of 30 March 1916. Ganapathy was a schoolteacher, and Anantharaman was a Co-operative Deputy Registrar.

The identity of Krishnamachari, which is a relatively common Indian name, is not known. However, in 1920, C. Krishnamachari, motivated by a problem posed by Ramanujan in the *Journal of the Indian Mathematical Society*, published a paper [148] on integrals and series associated with Bernoulli's and Euler's numbers.

A mofussil is a small town in an outlying area; the word is not derogatory.

Memo from S.N. Aiyar   **April 1915**

S. Ramanujan, clerk Accounts Department was given by the Madras University on 1st May 1913 a scholarship of Rs75 per mensem to carry on mathematical research. He was then given by the Chairman leave on loss of pay for two years. About the middle of March 1914, Ramanujan was prevailed upon by the Chairman to accept a scholarship of Rs10,000 from the Madras University to continue his researches at Cambridge. This he accepted and has gone to Cambridge. It is now a year since he joined the Trinity College, Cambridge. His leave expires on the 30 April 1915. On Ramanujan's behalf I request the Chairman to continue his leave on loss of pay for another year with option to regain his appointment earlier if he returns to India earlier.

The clerk who is now acting in his place may be allowed to continue to act for Ramanujan until the latter returns to Madras.

S.N. Aiyar

Commentary. Despite the increasing acclaim accorded to Ramanujan through the scholarship from Madras University and his invitation and subsequent journey to Cambridge University, S.N. Aiyar exercised a concerned caution by ensuring that Ramanujan could reassume his position with the Madras Port Trust if all means of mathematical support fail.

E.W. Barnes to Francis Dewsbury 8 November 1915

Trinity College,
Cambridge,

Dear Sir,

The work and progress of Mr. S. Ramanujan is excellent. He is entirely justifying the hopes entertained when he came here. There can be no doubt at all that his scholarship should be extended until, as I confidently expect, he is elected to a Fellowship at the College. Such an Election I should expect in Oct. 1917.

Yours faithfully,
E.W. Barnes,
Fellow and Tutor of Trinity College

Commentary. Ernest William Barnes (1874–1953) was bracketed as Second Wrangler in 1896, and placed in the first division of the First Class in Part II of the Mathematical Tripos the following year. He gained the first Smith's Prize in 1898, and was elected a Fellow of Trinity College the same year. While an undergraduate, he became President of the Cambridge Union Society (a club and debating society). He was elected a Fellow of the Royal Society in 1909. He was Senior Tutor at Trinity from 1908 to 1915 and as such was officially responsible for the well-being of Ramanujan. He was ordained as a priest in the Church of England in 1902, and was chosen by Lloyd George as Canon of Westminster in 1918. In 1924 he was selected by Ramsay MacDonald as bishop of Birmingham, from which office he retired in 1952.

His main mathematical interests were in gamma functions and other related special functions, with particular emphasis in obtaining asymptotic expansions of very general types of functions. For this purpose he made use of contour integrals of various kinds, now usually referred to as 'integrals of Barnes type.' He ceased being a professional mathematician in 1916, but he never lost interest in mathematics. His mathematical output in less than twenty years amounts to some 900 pages, an amazing total for such a short space of time. He was later the author of various religious publications and was involved in a number of serious disputes within the Anglican church.

For further information, see an obituary notice by W.N. Bailey [20].

S. Ramanujan to S.N. Aiyar **11 November 1915**

TRINITY COLLEGE
CAMBRIDGE

My dear Sir,

Your kind letter duly to hand. Very glad to hear of the kind remembrances of Sir Francis Spring. Had it not been for his special recommendation to the Government of Madras, I would not have got the scholarship so easily and quickly from Madras University. He will be glad if he knows from you about my progress.

I am sending you one of my papers read before the London Math. Society in last November. It is wrongly printed in the paper that it was read in last June. The paper that was read in last June was about some of my results in my notebook. It will take some months for me to write that paper systematically and publish it.

As for your suggestion to publish more important and general results, I have first to read French and German works and journals and to become familiar with extremely rigorous proofs and then publish my results. Now I shall be going on to write something in my own way for 6 or 7 months to come. Unless I remain here for two years more I cannot do all I have to do. If I just consider the enormous losses to this college owing to this terrible war I think I have no voice to ask the authorities of this college to do me anything if they cannot afford to retain me here of their own accord. There were about 700 students and even more than that sometimes before the war; but there are only about 150 this year. Perhaps there may be very few only next year if the war continues.

It appears that Madras University has asked this college about my work. I think the reports may reach Madras very soon after you receive this letter if they have not already reached there. If I could remain here for more than one year I should like to go over there for the coming long vacation so that I may not disappoint my people to whom I promised to return at the end of two years of my stay here.

I shall be very much obliged to you if you take so much trouble as to let Sir Francis Spring know all I have written to you. He will make all the necessary arrangements for my further stay here as well as for my voyage to India.

I am ever indebted to you and Sir Francis Spring for your zealous interest in my case from the very beginning of acquaintance.

Yours sincerely,
S. Ramanujan

P.S.—I am glad to tell you that I may be conferred upon the Research Degree next March.

Commentary. The National Archives in Delhi possesses the original letter from Ramanujan to S. N. Aiyar as well as a typed extract from this letter beginning with the sentence, "Unless I remain ... " and ending with the penultimate paragraph of the letter recorded above. The Archives also possesses an original letter of Ramanujan written on the same day. This letter is considerably shorter than the letter above but relates some of the same material. It is not known to whom the letter was addressed. However, in view of the introductory remarks made in the letter from R. Ramachandra Rao to S.N. Aiyar which follows, it seems likely that this next letter was written to Ramachandra Rao.

S. Ramanujan to R. Ramachandra Rao 11 November 1915

TRINITY COLLEGE,
CAMBRIDGE

My dear Sir,

I am glad to tell you that I may be conferred upon the Research Degree next March. I heard that Madras University was considering about my stay here but I do not know what they have decided. If I have to stay here for only one year more there is no use of my returning to Madras in the middle; but if I know for certain that I can remain here for more than one year and if circumstances favour me I am thinking of going over there for a short stay during the coming long vacation.

I am sending you a paper of mine read before the London Math Society in last November. It is wrongly printed in the paper that it was read in last June when a different paper was read.

Hoping you and your family are doing well.

Yours sincerely,
S. Ramanujan

R. Ramachandra Rao to S.N. Aiyar 6 December 1915

Nellore

Dear sir,

The enclosed letter speaks for itself. I wrote to Ramanujam after I met you at Ooty at the Government House party and the enclosed is by way of reply.

I do not know exactly what he wants but it is probably a continuance of the University Scholarship. I am not aware if the university means or can afford to do so. I can however say that Ramanujam has done a lot of work since he left—a fact which will be spoken to by Prof. Hardy and I have seen in the crude some of his researches which if valid and published will revolutionise some of the departments of mathematical thought. I have no doubt you will kindly help Ramanujan if possible.

Yours faithfully,
R. Ramachandra Rao

Commentary. Ooty is an abbreviation for Ootacamund, the most prominent "hill station" in India and the summer residence of the British governor. The town is located in south central India not far from the city of Mysore. Its high altitude and correspondingly cooler temperatures make Ooty a favorite spot for summer vacations. It thus attracted the British ruling class in Ramanujan's day and film actors and actresses in more recent times.

Francis Dewsbury to Educational Department 11 December 1915

From
> FRANCIS DEWSBURY, Esq., B.A., LL.D.,
> *Registrar, University of Madras,*

To
> The Secretary to Government
> Educational Department.

Sir,

With reference to my letter No. 523, dated 5th February, 1914, and to G.O. No. 182, Educational, dated 12th February 1914 in which the Government sanctioned the appropriation of a sum of Rs10,000 from the University Vacation Lectures Account for the grant of a scholarship to S. Ramanujam of £250 for 2 years, I am now directed by the Syndicate to request the sanction of the Government for the further appropriation of money from the same account sufficient to provide for continuing the scholarship to S. Ramanujam for another year.

2. Dr. E.W. Barnes and Mr. G.H. Hardy, both of Trinity College, Cambridge, the former being Ramanujam's Tutor and the latter the Mathematical Professor under whom he has been studying, have favoured the Syndicate with their opinion as to his work and progress. Each of them has written in very high terms of his performance and promise for the future; the latter has forwarded some of Ramanujam's published work and undertakes to furnish shortly a report for publication: meanwhile, in a private communication he states that the experiment of bringing Ramanujam to Cambridge has in his judgment been thoroughly justified and that it would be disastrous to interrupt it now, and adds that he is beyond question the best Indian Mathematician of modern times and in some ways the most remarkable Mathematician he has ever known.

Both Dr. Barnes and Mr. Hardy very strongly recommend an extension of this scholarship and anticipate the prospect of Ramanujam's election to a Fellowship in the near future.

3. The Budget Estimates for the current year for the Vacation Lectures Account show anticipated balances of Rs42,500 in securities and Rs1,663 in cash.

The scholarship to S. Ramanujam, of which an extension for one year is now proposed, would be terminated at once were he elected to a Fellowship as suggested.

> I have the honour to be,
> Sir,
> Your most obedient servant,
> Francis Dewsbury
> Registrar.

Commentary. The letters of Barnes and Hardy mentioned above have evidently been lost.

S.N. Aiyar to Sir Francis Spring　　　　　　　　**15 December 1915**

I understand that the period of the scholarship given to Mr. S. Ramanujan by the University of Madras has been extended by one year. This will carry him on to the end of March 1917. From private sources I learn that the University of Cambridge would like to keep him for another year more, i.e. to the end of March 1918.

2. Ramanujan himself has written to me that if the University of Madras should be pleased not to extend his scholarship beyond March 1917, he will be unable to keep his promise to his relatives here that he would return in April 1917. For, if he comes to India during the long vacation of 1917, he will have to return after the vacation to England for the brief period of only nine months, not leaving [an] age for [the] work that [he has] to do there. Therefore in that case he would prefer not to come to India at all during the vacation of 1917. But if his term of scholarship be extended by two years more, i.e. up to March 1918, as is desired, I understand, by mathematical Professors in Cambridge, this would enable him to come back to India during April 1916 and then to return back to Cambridge for a further stay of a year and nine months.

3. May I therefore request you to use your influence with the University of Madras to have Ramanujan's term of scholarship extended by two years more instead of by one year, as has been done by the Syndicate in their meeting on the 7th December 1915.

4. As for Ramanujan's progress at Cambridge the attached extracts from letters will speak for themselves. I also understand that Prof. G.H. Hardy of Cambridge has written to the local University Registrar in highly eulogistic terms about the progress of his work there.

> S. Narayana Aiyar

Commentary. Two copies of the letter above can be found in the National Archives in Delhi. Both are identically typed, but one has several alterations inserted in longhand. We have reproduced the amended version which evidently supersedes the other. However, some of the handwritten insertions are difficult to read. Furthermore, photocopying has deleted some words; we have put in brackets our guesses for the omitted words.

Sir Francis Spring to Francis Dewsbury 15 December 1915

The Registrar Madras University
    The Senate House. Madras

Dear Sir,

The accompanying papers refer to Mr. S. Ramanujan in whom I am interested if for no other reason than that—he was discovered while working as a Clerk in my office. The papers have reached us from my office Manager Mr. S. Narayana Aiyar M.A., L.T., who, himself a good mathematician was chiefly instrumental in bringing Mr. Ramanujan into notice.

In the interest of India's reputation as the birthplace of some good mathematicians of which Mr. Ramanujan is, I am assured, one of the most remarkable, I hope it may be possible for the University to ensure the uninterrupted continuation of his studies, but chiefly of his original research work at Cambridge, by promising him the enjoyment of his present scholarship up to the end of March 1918.

The Syndicate will, I have no doubt, give the matter their best consideration.

                                      Yours faithfully,
                                      Francis J.E. Spring

Commentary. An exact identification of the papers to which Spring refers cannot be made.

The initials L.T. denote Licentiate in Teaching.

Francis Dewsbury to Sir Francis Spring 16 December 1915

                University of Madras

                              Senate House

From
    FRANCIS DEWSBURY, Esq., B.A., LL.B.
        Registrar, University of Madras,
To
    The Hon'ble Sir Francis Spring, K.C.I.E.

Sir,

In reply to your letter, dated December 15, 1915, I have the honour to state that the University of Madras has requested sanction for the allocation of funds for the extension of the scholarship to Mr. S. Ramanujam for another year, i.e. to the end of March, 1917. Your letter and enclosures will be placed before the Syndicate for orders. The Syndicate has been given to understand that it is quite possible that Mr. Ramanujam may become independent of a scholarship before the expiry of this extension. If this proves not to be the case a further extension can be considered at a later date. In the meantime it will be difficult for the Syndicate to consider the question of providing for a return of Mr. Ramanujam to India next year, in the absence of any representation on the point from him or from those connected with his work at Cambridge. Possibly Mr. Narayana Aiyar may be able to furnish the Syndicate with more detailed information on the matter.

I have the honour to be,
Sir,
Your most obedient servant,
Francis Dewsbury
Registrar.

Commentary. The initials K.C.I.E. after Spring's name denote "Knight Commander of the Order of the Indian Empire."

S.N. Aiyar to Francis Dewsbury 20 December 1915

With reference to the concluding portion of letter no. 7923 dated the 16th December 1915 from the University Registrar, I attach an extract from Mr. S. Ramanujan's letter dated 11th November 1915.

2. He wants to arrive in Madras in April 1916 and to stay with his family during May and June, returning to England in July. If his scholarship is to end in March 1917 that leaves him only 8 or 9 months more for study at Cambridge. If he is to compete for the Fellowship mentioned by Mr. Hardy in October 1916 and if he is successful the continuance to him of the Madras University scholarship will not be necessary.

3. If the question of providing for his return to India next year can be considered only after a formal representation from him I will write to him to put in a formal application through the University authorities there. But I doubt if there is enough time for this formal representation.

S.N. Aiyar
20-12-15

Forwarded to the Registrar of Madras University in reply to his letter No. 7923 dated the 16th December 1915.

S.N. Aiyar

Commentary. Clearly, it was expected that Ramanujan would return to India in the early spring of 1916, according to original plans.

Francis Dewsbury to Sir Francis Spring   **17 January 1916**

UNIVERSITY OF MADRAS

Senate House

From
FRANCIS DEWSBURY, Esq., B.A., LL.B.,
Registrar, University of Madras,
To
The Hon'ble Sir Francis Spring, K.C.I.E.

Sir

With reference to recent correspondence on the matter of the Mathematical scholarship at Cambridge granted to Mr. Ramanujam, I have the honour to inform you that the Government has sanctioned the proposal of the Syndicate to extend this scholarship for one year. I am instructed to inform you that the Syndicate is not prepared to take any further action in the matter for the present.

I have the honour to be,
Sir,
Your most obedient servant,
Francis Dewsbury
Registrar.

S. Ramanujan to S.M. Subramanian   **30 March 1916**

Trinity College
Cambridge

My dear Subramanyan,

Your letter duly to hand. I am very sorry to hear that your scholarship has not been extended. It would have been much better if you had ascertained what was happening about your case in Madras before the time for the extension of your scholarship. Haven't you seen yet the principal of the Engineering College? You needn't go for any subordinate service. Even if you can't get any post in Madras you can get something elsewhere in India. I am thinking of returning there in June (this year) for the long vacation. We shall talk over the matter and settle it as best as we can.

Did Ganapati appear for B.A.? Has Anantaraman finished his intermediate? I hope all in your family are doing well. My kind remembrances to Ganapathi.

Yours affly,
S. Ramanujan

P.S.—I took my degree some days ago as you know full well.

RAMANUJAN AND FELLOW GRADUATES

Commentary. Ramanujan graduated with the degree, Bachelor of Science by Research, in the Cambridge Senate House on the afternoon of Saturday 18 March 1916. A photograph of him and others taken on that occasion is reproduced as the frontispiece of [227]. See also §7 of [235]. This degree was the precursor in Cambridge of the Ph.D. degree, which came into operation only in 1920.

Ramanujan was admitted as a Research Student in June 1914. Normally, Research Students were required to hold a diploma or certificate of graduation from a university. However, in exceptional cases (as for Ramanujan) they could be admitted provided they gave evidence of appropriate qualifications. When submitting his dissertation to the Registrary, the Research Student was required to pay a fee of £5, and each examiner was paid £2 from the University Chest. The title of Ramanujan's dissertation was 'Highly Composite Numbers,' and the dissertation included six other papers. When a dissertation was approved, the student received a Certificate of Research. A copy of the dissertation was deposited in the University Library, and the student could then proceed to the degree of Bachelor of Arts, provided he had kept residence at least six terms. This last condition must have been relaxed in Ramanujan's case. It is not known who were his examiners, but it is almost certain that Hardy must have acted in that capacity.

It is likely that the 'six other papers' submitted by Ramanujan were the first six of the 21 papers listed by Hardy on page xxxii of [226] as having been published in Europe by Ramanujan, with 'Highly Composite Numbers' being the seventh. Unfortunately, the dissertation, which was presumably deposited in the University Library at some stage, cannot now be traced.

Memo from S.N. Aiyar May(?), 1916

S. Ramanujan, clerk Accounts Department, who is now in England on Mathematical Research did not join his duty on the expiry of the two year's leave on <u>loss of pay</u> granted to him on the 1st May 1913. Then the C.P.T. granted him an extension of another year from the 1st May 1915 which expired on the 30 April 1916. He has not yet joined our offices. To save audit objections one of the two courses that are left to us regarding his appointment has to be adopted:

(1) His services may be dispensed with,

or

(2) His leave on loss of pay may be extended for a further period of 2 years at C.P.T.'s option. (Vide page 198 of C.S.R.).

It is submitted that course (2) may be adopted.

S.N. Aiyar

W.E.H. Berwick to S. Ramanujan 24 October 1916

University College
Bangor,
N. Wales

Dear Sir.

Very many thanks for your paper giving the singular moduli. It forms a valuable edition [addition?] to existing tables.

Some of your questions can be answered without much difficulty: others are not so easy.

First as regards references to modular equations and the values of k when $K'/K = \sqrt{D}$. Important work on these subjects will be found in the following:

<u>Sohncke</u>: Crelle, 16. (A list of errata in this paper has been given by Cayley: You will find the reference in the Index volume of Cayley's Collected Papers.)

<u>Hermite</u>. Oeuvres (3 vols. published so far); originally published in Comptes Rendus.

<u>Greenhill</u>, London Math. Soc. 19 & 21; Quarterly Journal 22.

<u>Russell</u>, London Math. Soc. 19 & 21.

<u>Mathews</u>, London Math. Soc. 21.

<u>Kronecker</u>. Berlin Sitz: 1857, 1862.

<u>Kiepert</u>. Math Annalen, 26, 32, 36, 39.

<u>Schröter</u>. Crelle 58.

There is also the very valuable work by <u>Klein</u> in Math. Ann. 14 etc. and subsequently embodied in Klein–Fricke's <u>Modulfunctionen</u>.

You will find other references in the Royal Society's Index of Scientific Papers Vol. 1 (Pure Mathematics) 1300–1900 §4050 page 326. This catalogue has been supplemented by annual issues which you will find without much difficulty in Cambridge.

Father Séguier's Multiplication Complexe contains an account of Kronecker's rather difficult theory.

When D is given the irrationality involved in k^2 for $K'/K = \sqrt{D}$ depends on the number of classes of binary quadratic forms of determinant $-D$. If h is the number of such classes (see Mathews' Theory of Numbers §63–66) the fundamental invariant

$$j = \frac{256(1 - k^2 + k^4)^3}{k^4(1 - k^2)^2}$$

is a root of an irreducible equation of degree h. [Klein's invariant is $J = \frac{j}{1728}$ and Hermite's j is connected with this by a quadratic transformation.] In certain cases k^2 can be expressed rationally in terms of j but this is not invariably true. The equation satisfied by j is always soluble by radicals and if $h = 2^m$ this equation is soluble by a chain of m quadratics. When $D = 257$ the reduced forms are 16 in number viz. $(1,0,257)$, $(2,1,129)$, $(3, \pm 1, 86)$, $(6, \pm 1, 43)$, $(9, \pm 2, 29)$, $(7, \pm 3, 38)$, $(14, \pm 3, 19)$, $(13, \pm 4, 21)$ and $(17, \pm 7, 18)$ so $j(i\sqrt{257})$ is a root of an irreducible equation of degree 128 which can be solved by a chain of 7 quadratics. In each of these cases k^2 can also be obtained by 4 and 7 quadratics respectively.

Tables of class numbers for values of D up to 3000 are given at the end of Vol. 2 of Gauss' Werke.

K can be expressed in gamma functions when $k^2 = -1$: for then

$$K = \int_0^1 \frac{dt}{\sqrt{(1 - t^4)}} = \frac{1}{4} \int_0^1 u^{\frac{1}{4}-1}(1 - u)^{+\frac{1}{2}-1} \, du = \text{ etc.}$$

There may be other cases but I do not know of any literature on the subject.

Yours faithfully
W.E.H. Berwick.

Commentary. William Edward Hodgson Berwick (1888–1944) was a member and sometime Fellow of Clare College, Cambridge. He was placed as 4th Wrangler in 1909, the last year in which candidates were listed in order of merit. At the time when the letter was written, he was a Lecturer in the University of North Wales at Bangor. After a period of five years as Reader in Mathematical Analysis at the University of Leeds, he returned to Bangor as Professor and Head of the Department. A man of highly individual character traits, he was notable among British mathematicians of his time in having an unusual mastery of all the classical theory of algebraic numbers. The Senior and Junior Berwick Prizes of the London Mathematical Society were established by his widow in his memory.

It does not seem necessary to give complete references for the journal articles and books cited in Berwick's letter. However, we shall offer some remarks to help those readers desiring to become more familiar with the literature.

Crelle was the founder and first editor of *Journal für die reine und angewandte Mathematik*. For specific information on L.A. Sohncke's work on modular equations and its relationship to Ramanujan's findings, see Berndt's book [37].

The collected papers of C. Hermite comprise four volumes.

A.G. Greenhill's work is especially valuable in addressing Ramanujan's query. "London Math. Soc." refers to the *Proceedings of the London Mathematical Society*, while the "Quarterly Journal" is *not* the journal of the same name published in Oxford today, but an English mathematical journal that ceased publication in 1927.

Ramanujan, in fact, rediscovered some of R. Russell's modular equations. See Berndt's book [37] for details.

A more complete citation for "Berlin Sitz." is S.-B. Pruess. Akad. Wiss. Phys.-Math. Kl.

The papers of L. Kiepert are also very important in addressing Ramanujan's question. The volume number 36 of *Mathematische Annalen* is incorrect and should be replaced by 37.

Although we do not know most of Ramanujan's methods in the theory of modular equations, it is certain that Ramanujan employed H. Schröter's formulas for products of theta-functions in deriving many of his modular equations. Of course, Ramanujan was unaware of Schröter's papers and so rediscovered Schröter's formulas. Schröter published two other papers on modular equations, and readers should consult Berndt's book [37] to learn how Ramanujan used Schröter's formulas in deriving many beautiful modular equations.

The book [145] is one of the most famous treatises on modular functions, while J.-A. Séguier's thesis [253] is largely unknown.

For a more extensive table of class numbers, consult D. Buell's book [56]. Buell gives the class numbers of quadratic fields with fundamental discriminant d, where $0 < \pm d < 10,000$.

More was known and more has been learned about the values of K than is provided in Berwick's letter. First, it is classical that K can be explicitly evaluated whenever iK'/K is a number belonging to one of the three imaginary quadratic fields $Q(\sqrt{-n})$, $n = 1, 2, 3$. In 1949, S. Chowla and A. Selberg [256], [257], [255, pp. 367–370; 521–545] proved the following vast extension.

THEOREM. *Let d denote the discriminant of the imaginary quadratic field* $Q(\sqrt{d})$, *and let* $h = h(d)$ *denote its class number. Let* $w = 6, 4, 2$ *according as* $d = -3, d = -4, d < -4$. *Let* (d/m) *denote the Kronecker symbol. Suppose that* iK'/K *is a number in the field* $Q(\sqrt{d})$. *Then*

$$K = \lambda\sqrt{\pi}\left\{\prod_{m=1}^{-d}\Gamma\left(\frac{m}{-d}\right)^{(d/m)}\right\}^{w/(4h)},$$

where λ *is an algebraic number.*

This formula has been considerably generalized by P. Kaplan and K.S. Williams [143] and by Williams and Z. Nan-Yue [297]. For several explicit determinations of K based on the Selberg–Chowla formula, see a paper by I.J. Zucker [300].

The evaluation of K for certain k is equivalent to the determination of $\varphi(q) := \sum_{n=-\infty}^{\infty} q^{n^2}$ for certain q. Ramanujan recorded values of $\varphi(q)$ at scattered places in his notebooks, and proofs for all of Ramanujan's original evaluations of $\varphi(q)$ can be found in a paper of Berndt and H.H. Chan [44].

S. Ramanujan to G.H. Hardy **December 1916**

It is really remarkable to get a result with an error of $O(n^{-\frac{1}{4}})$ which is as good as that with any higher powers of n in this problem. Let c_s be the s-series which is $O(s)$ and $\Omega(1)$. Suppose now that s is about $\frac{\beta\sqrt{n}}{\log n}$ where $\beta = 4\pi\sqrt{\frac{2}{3}}$. Then

$$O\sum_{O(s)}^{O(s)}\frac{|c_s|\sqrt{s}}{sn}e^{(\pi/s)\sqrt{2n/3}} = O(s\sqrt{s}n^{-\frac{3}{4}}) = O(\log n)^{-\frac{3}{2}}.$$

It would therefore follow from your arguments that the error by taking about $\beta\sqrt{n}/\log n$ terms is $O(\log n)^{-\frac{3}{2}}$. Again suppose that s is about $\frac{\alpha\sqrt{n}}{\log n}$ where $\alpha = \frac{4\pi}{5}\sqrt{\frac{2}{3}}$, then

$$\frac{|c_s|\sqrt{s}}{sn}e^{(\pi/s)\sqrt{2n/3}} = \Omega\left(\frac{n^{\frac{1}{4}}}{\sqrt{s}}\right) = \Omega\left(\sqrt{\log n}\right).$$

It therefore appears that, in order that $p(n)$ may be the nearest integer to the approximate sum, s need not be taken beyond $\beta\sqrt{n}/\log n$ and cannot be taken below $\alpha\sqrt{n}/\log n$ (for all n and very likely for <u>every</u> n too since Ω in the above equations appears to be not mere negation of o) I hope you can easily prove these. Then the problem is completely solved. Major MacMahon was kind enough to send me a typewritten copy of the 200 numbers. The approximation gives the exact number. I think you received this already from him. With kind regards to your mother and sister.

Yours sincerely S. Ramanujan

Commentary. The letter above is from a postcard and was first published with the 'lost notebook' [227, p. 132].

The content pertains to the asymptotic formula for the partition function $p(n)$, which Hardy and Ramanujan were in the process of proving [124], [105, pp. 306–339], [226, pp. 276–309]. The s-series to which Ramanujan refers is evidently $A_q(n)$ ($s = q$) in the notation of [124]. The argument given in the first part of the letter is written with more details in §6.3 of Hardy and Ramanujan's paper [124]. The second argument, which is based on the assumption that the s-series is $\Omega(1)$, is straightforward. However, this assumption does not seem to be justified, for [124, §6.22], "It should be observed in this connection that we have not even discovered anything definite concerning the order of magnitude of A_q for large values of q. We can prove nothing better than the absolutely trivial equation $A_q = O(q)$. On the other hand we cannot assert that A_q is, for an infinity of values of q, effectively of an order as great as q, or indeed even that it does not tend to zero (though of course this is most unlikely)."

The "200 numbers" refers to a table of the partition function $p(n)$, $1 \leq n \leq 200$, prepared by MacMahon.

One can infer from the last sentence of the card that Ramanujan had met Hardy's mother, née Sophia Hall (1845–1917), and his sister Gertrude Edith Hardy (1879–1963).

Francis Dewsbury to Educational Department 7 March 1917

From
> FRANCIS DEWSBURY, Esq., B.A., LL.D.,
> *Registrar, University of Madras,*

To
> The Secretary to Government
> Home (Education) Department.

Sir,

With reference to this office letter No. 7780, dated 10th December 1915, and to G.O. No. 1441, Educational, Mis., dated 20th December 1915, sanctioning the further appropriation from the University Vacation Lectures Fund of an amount sufficient to provide for the continuance for another year of the grant of scholarship to Mr. S. Ramanujam, I am now directed by the Syndicate to request similar sanction for the further appropriation of money from the same account to provide for another extension of the scholarship for the ensuing year.

In July last the Syndicate received from G.H. Hardy Esq., M.A., F.R.S., a detailed report of Mr. S. Ramanujam's work at Cambridge. This report has since been published and a copy of it is enclosed herewith. The attention of the Government is specially invited to paragraphs on pages 1, 4, 7, 9 and 13 containing expressions showing Mr. Hardy's appreciation of Mr. Ramanujam's extraordinary ability.

Subsequently in February last the Syndicate received through Mr. Hardy a formal certificate from Mr. R.V. Laurence, Mr. Ramanujam's tutor, 'that Mr. Ramanujam has resided at Trinity College, Cambridge, during the past year and that his conduct and progress in researches have been in every way satisfactory to the authorities of the college.' Mr. Hardy enclosed with this certificate a private and confidential letter, in which he urged that Mr. Ramanujam should be allowed to remain at Cambridge for at least another year. This letter is enclosed in a sealed cover for the information of the Government and return to this office. In view of Mr. Hardy's representations, the Syndicate at its meeting on the 6th March 1917 unanimously decided, subject to the approval of the Government, to continue the scholarship to Mr. Ramanujam for another year and directed me to take immediate action in the matter. I have to request very early orders so that immediate intimation together with necessary funds may be cabled to Mr. Ramanujam whose scholarship expires on the 31st March 1917.

> I have the honour to be,
> Sir,
> Your most obedient servant,
> Francis Dewsbury
> REGISTRAR.

Commentary. Hardy's report and confidential letter, as well as Mr. Laurence's certificate, have evidently been lost.

Francis Dewsbury to S.N. Aiyar   14 May 1917

Kindly refer urgently to Mr. S. Ramanujan's register in your office, and let me know the date of his birth and the name of his father.

Please see your office No. G. 30 dated 14-4-1913. Mr. Ramanujan was a clerk in the accounts department of the Port Trust office and is now in England studying Mathematics in the University of Cambridge.

S.N. Aiyar to Francis Dewsbury   14 May 1917

We have no records in this office to show Ramanujan's date of birth nor the name of his father.

But I believe he was born in 1890 and his father's name is Srinivasa Aiyangar. His people are living at 17, Sarangapani Covil Sannadhi Street Kumbakonam.

Commentary. In fact, the two passages above are transcriptions of telephone conversations. Since Ramanujan was born on 22 December 1887, S.N. Aiyar's reply is further evidence that one's date of birth is not as important nor frequently recorded as in the West. S.N. Aiyar had evidently helped Ramanujan compose his first letter to Hardy, and his recollection of Ramanujan's (incorrect) age probably arose from its mention in that letter.

The next two pairs of correspondence are also transcriptions of telephone conversations.

Francis Dewsbury to S.N. Aiyar   14 May 1917

Can you please tell me if Mr. Srinivasa Iyengar, father of Mr. Ramanujan, is alive and whether he was or is a District munsiff?

S.N. Aiyar to Francis Dewsbury   14 May 1917

Mr. Srinivasa Aiyangar is alive. He was a petty clerk on about Rs20 under a merchant. He does not know English and was never a District Munsiff.

Commentary. Appointed by the Collector, a munsiff is an officer or head of a small town or district. The munsiff kept land records for the town.

Francis Dewsbury to S.N. Aiyar   14 May 1917

Could you please let me know in what year Mr. S. Ramanujan son of Srinivasa Aiyangar was matriculated, and whether he applied for the first or intermediate examination?

S.N. Aiyar to Francis Dewsbury   14 May 1917

I know he is a matriculate of this University and failed in F.A. in all the subjects. I do not know the year in which he matriculated.

Commentary. It has often been inaccurately reported that Ramanujan failed all of his subjects; he did not fail his mathematics examination.

Chapter 5
Ramanujan is Ill

S. Ramanujan to G.H. Hardy

Please mention in a foot note above the result (1.61) that there was a mistake in Comptes Rendus concerning the error of $P(61)$ over $p(61)$. I think it was given there as -44.641 instead of the correct value -34.641. Please ascertain if there was really a mistake: the figures might not have been mentioned in Comptes Rendus.

In yesterday's letter instead of writing

$$\frac{\pi^2}{\sin \pi\alpha \sin \pi\beta}, \qquad \frac{\pi^4}{\sin \pi\alpha \sin \pi\beta \sin \pi\gamma}$$

I inverted them in haste. Is it not better to construct a small table showing the maximum value $M_s(n)$ and minimum $m_s(n)$ of $|C_s(n)|$ in the partition paper like this

s	1	2	3	4	...	18
$M_s(n)$	1	1	$2\cos\frac{\pi}{18}$	$2\cos\frac{\pi}{8}$	...	
$m_s(n)$	1	1	$2\sin\frac{\pi}{18}$	$2\sin?$	...	

Commentary. This letter was probably written in mid-1917 shortly after Ramanujan entered a nursing hostel on Thompson's Lane in Cambridge. Undoubtedly, interrupted mail service due to World War I was responsible for Ramanujan's not yet seeing the final version of his joint paper with Hardy [123], [105, pp. 274-276], [226, pp. 239-241], published on 2 January 1917. This paper does not contain any numerical differences between the values of $p(n)$ and $P(n)$, the approximation to $p(n)$ achieved by taking only the main term in the asymptotic series for $p(n)$ found by Hardy and Ramanujan. Consequently, the mistake that Ramanujan feared did not exist. In fact, Ramanujan's opening sentence is slightly erroneous; replace 61 by 62. Evidently, at that time, Hardy and Ramanujan were writing the final draft of their epic paper [124] providing a proof of their asymptotic series for $p(n)$, two terms of which are given by (1.61). Above (1.61), Hardy and Ramanujan state, for $n = 61, 62$, and 63, the exact values of $p(n)$, the approximate values $P(n)$, and their differences.

It appears that $C_s(n)$ equals $A_s(n)$ in [124], for which Hardy and Ramanujan computed an extensive table (Table II). However, the values for $m_s(n)$, when $s = 3$ and (possibly) 4, stated in the letter are not correct.

A facsimile of the original letter appears in [227, p. 371]. We have not been able to discern with certainty the argument of the last sine appearing in the table above.

Commentary. In two letters from Hardy to Mittag-Leffler there are references to Ramanujan, which are given below. Gösta Mittag-Leffler (1846–1927) was a distinguished Swedish mathematician, who worked on the theory of functions of a complex variable. He was founder and, for forty-five years, the chief editor of *Acta Mathematica*. The letters primarily refer to the publication of papers by Hardy in that journal.

G.H. Hardy to G. Mittag-Leffler   **20 August 1917**

Trinity College
Cambridge

... You will be sorry to hear that Mr Ramanujan is seriously ill, and that we are very much alarmed about him ...

G.H. Hardy to G. Mittag-Leffler   **12 October 1917**

Trinity College
Cambridge

... I am glad to be able to say that Mr Ramanujan is decidedly better. But it is, I fear, almost certain that he is infected with tuberculosis in some form. He is a Brahman, and a strict vegetarian. This, in a cold climate, is a terrible difficulty in the way of securing an appropriate diet. Two months ago the doctors almost despaired of saving him: now they are distinctly more hopeful ...

G.H. Hardy to S.M. Subramanian   **20 September 1917**

Trinity College,
Cambridge

Dear Sir,

I was very glad to hear from you about Ramanujan. He has been seriously ill but is now a good deal better. It is very difficult to get him to take proper care of himself; if he would only do so we should have every hope that he would be quite well again before very long.

It was only a few months ago—when he was for a time in a Nursing Hospital here—that we discovered that he was not writing to his people nor apparently, hearing from them. He was very reserved about it, and it appeared to us that there must have been some quarrel. As at that time he was really very ill indeed I wrote (on the advice of Mr. Sureyaranan of Merton College, Oxford) to Mr. R. Ramachandra Rau, Collector, Madras. It would be a good thing if you could communicate with him. But it is important to let him know that the prospects as regards Ramanujan's health are now decidedly better than when I wrote first to him. I wrote two letters, the first of which was (by the kindness of Mr. Montagu) sent by special despatch, while the second went by the ordinary post. The first, I fear, will give him an unduly pessimistic idea of the state of affairs.

I showed your letter to Ramanujan (who is now in Cambridge) and got him to promise to write to his people. We are most anxious that any trouble which may have arisen should be cleared away.

I am
Yours sincerely
G.H. Hardy

Commentary. Sureyaranan is the same person, Suryanarayanan, mentioned in Ramanujan's letter of 11 June 1914 to Vinayaka Rao.

The Nursing Hostel in Thompson's Lane, Cambridge, was partly supported by Trinity College. There evidently were financial problems, and it was closed sometime before 1921. The building was then converted to students' lodgings. It was demolished in 1981 and replaced by a block of flats overlooking the River Cam.

G.H. Hardy to S. Ramanujan   **6 February 1918**

T. C. C.

Dear Ramanujan

I have managed to find a way of summing the 'partition series' for

$$(1 + 2x + 2x^4 + \cdots)^r \quad r \text{ odd}.$$

The conclusion seems to be that it <u>does</u> give $r_r(n)$ exactly for 1, 3, 5, 7 (for 1 of course it is divergent).

The series is (I take $r = 5$)

$$\frac{\pi^{5/2}}{\Gamma(\frac{5}{2})} \sum_{p,q} \frac{U_{p,q}^5}{q^5} f\left(xe^{-p\pi i/q}\right), \qquad \left(f(x) = \sum n^{\frac{3}{2}} x^n\right)$$

where $q > 1$; $-q < p < q$; p, q have no common factor; one is odd and one is even: also

$$U_{p,q} = \sum_{0}^{q-1} e^{s^2 p\pi i/q}.$$

Now it follows from our previous results (e.g. those in the paper of yours which is now being printed; or by the formulae in Tannery and Molk) that

$$(U_{p,q})^4 = q^2(-1)^p.$$

Hence

$$(U_{p,q})^5 = (-1)^p q^2 \sum_0^{q-1} e^{s^2 p\pi i/q}.$$

Also it can be shown by contour integration that

$$f(x) = \Gamma\left(\frac{5}{2}\right) \sum_{-\infty}^{\infty} \frac{1}{\left(\log \frac{1}{x} + 2n\pi i\right)^{5/2}}$$

(if the values of the roots are chosen in the proper and most obvious manner). Hence

$$\text{part}^n \text{series} = S = \sum_{p,q} \sum_n \frac{(-1)^p U_{p,q}}{\sqrt{q}} \frac{1}{\{[(p+2nq) - q\tau]i\}^{5/2}}.$$

We can write this

$$S = \sum_{p,q} \frac{(-1)^p U_{p,q}}{\sqrt{q}} \frac{1}{[(p - q\tau)i]^{5/2}},$$

the condition that $-q < p < q$ being now standard. Multiplying by

$$\frac{1}{1^2} + \frac{1}{3^2} + \cdots = \frac{\pi^2}{8}$$

we get

$$\frac{\pi^2 S}{8} = \sum \frac{(-1)^p U_{p,q}}{\sqrt{q}} \frac{1}{[(p-q\tau)i]^{5/2}}$$

where now p and q need no longer be prime to one another: in fact p, q are any odd & even, or even and odd integers, of which $q > 0$.

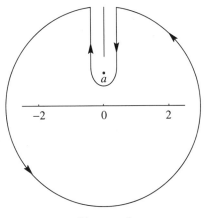

FIGURE 1

Thus

$$\frac{\pi^2 S}{8} = \sum_q \frac{1}{\sqrt{q}} \sum_{s=0}^{q-1} \sum_p \frac{e^{s^2 p\pi i/q}}{[(p-q\tau)i]^{5/2}}$$

if q is odd $p = 0, \pm 2, \pm 4, \ldots,$

if q is even $p = \pm 1, \pm 3, \ldots.$

The p series can now be summed by contour integration. We must write

$$s^2 = \mu q + \Lambda \qquad (0 \leq \Lambda < q)$$

and use the integrals

$$\int \frac{e^{\pi i \theta z}}{e^{\pi i z} \pm 1} \frac{dz}{[(z-a)i]^{5/2}}.$$

$a = q^2$, $\theta = \Lambda/q$, $\pm$ according as q is even or odd. [e.g. with q odd we consider

$$\int \frac{e^{\pi i \theta z}}{e^{\pi i z} - 1} \cdots$$

round [the contour above] and make the circular part $\to \infty$: this requires $0 \leq \theta < 1$.] I get finally

$$\frac{\pi^2 S}{8} = \frac{32}{3} \left\{ \sum_{\substack{1,3,5,\ldots \\ (q)}} \frac{1}{q^2} \sum_{(\Lambda)} \sum_{\substack{0 \\ (n)}}^{\infty} (nq+\Lambda)^{3/2} x^{nq+\Lambda} \right.$$

$$\left. - \sum_{\substack{2,4,6,\ldots \\ (q)}} \frac{1}{q^2} \sum_{\Lambda} (-1)^{\mu} \sum_0^{\infty} (-1)^n (nq+\Lambda)^{3/2} x^{nq+\Lambda} \right\}$$

(to explain the meaning of $\sum_\Lambda$ and $(-1)^\mu$:
$q = 6$

$$\begin{array}{cccccc} 0^2 & 1^2 & 2^2 & 3^2 & 4^2 & 5^2 \\ 0 & 1 & 4 & 9 & 16 & 25 \\ 0 & 1 & 4 & 3 & 4 & 1 \end{array} \quad \text{mod } 6$$

i.e. $\Lambda = \quad 0 \quad 1 \quad 4 \quad 3 \quad 4 \quad 1$
the corresponding μ's $= \quad 0 \quad 0 \quad 0 \quad 1 \quad 2 \quad 4$

Each Λ is to be taken as often as it occurs.). Now I have verified that, when $n = 1$, this is $\underline{10}$. The coefficient of x is

$$\frac{32}{3}\left\{1 + \sum_{3,5,\ldots} \frac{N_q}{q^2} - \sum_{2,4,\ldots} \frac{N_q'}{q^2}\right\}$$

where N_q = no. of incongruent solutions
$\qquad\qquad$ (mod q) if $x^2 \equiv 1 \pmod{q}$,
$N_q' = \sum(-1)^\mu$ extended over all these solutions.

And

$$\begin{aligned} N_q &= 2^j, & q &= p_1^{\alpha_1} p_2^{\alpha_2} \ldots p_j^{\alpha_j} \\ &\text{or} & q &= 2 p_1^{\alpha_1} p_2^{\alpha_2} \ldots p_j^{\alpha_j} \\ &= 2^{j+1}, & q &= 2^2 p_1^{\alpha_1} p_2^{\alpha_2} \ldots p_j^{\alpha_j} \\ &= 2^{j+2}, & q &= 2^\alpha p_1^{\alpha_1} p_2^{\alpha_2} \ldots p_j^{\alpha_j} \quad (\alpha \geq 3) \end{aligned}$$

This is

$$\frac{32}{3}\left[\left(\frac{1}{1^2} - \frac{1}{2^2} - \frac{2}{4^2} - \frac{0}{2^6} - \frac{0}{2^8} - \cdots\right)\right.$$
$$\times \left(\frac{1}{1^2} + \frac{2}{3^2} + \frac{2}{3^4} + \cdots\right)$$
$$\left.\times \left(\frac{1}{1^2} + \frac{2}{5^2} + \frac{2}{5^4} + \cdots\right)\cdots\right]$$

and this reduces to

$$\frac{32}{3}\left[\left(1-\frac{1}{4}-\frac{1}{8}\right)\frac{1-\frac{1}{4}\zeta^2(2)}{1+\frac{1}{4}\zeta(4)}\right]=10$$

This seems conclusive. Further in general it is clear that the solution will depend on series like

$$\sum \frac{1}{q^2}\left(\frac{n}{q}\right)$$

—and it is known that $r_5(n)$ does so.

I mean to make a great effort to find some direct proof that, when $r < 8$, the partition series must give $r_r(n)$ exactly. Mordell found one for 2, 4, 6, 8: not for the partition series, indeed, but for something which I can identify with it. But this involves the theory of modular functions

$$\sum\sum \frac{1}{(m+n\tau)^r} \qquad (r = \underline{\text{integer}})$$

and I don't believe there is any such theory when $r = \frac{1}{2}+$ integer.

The fact that

$$S = (\text{exactly}) \quad (1 + 2q + \cdots)^r$$

<u>must</u> depend on
(1) coeff. of x^n in $S = r_r(n) + O(n^{\frac{r}{2}-\frac{1}{2}})$—of course we can only <u>prove</u> $O(n^{\frac{r}{4}})$.
(2) $\frac{r}{4} - \frac{1}{2} < 0$ if $r < 8$.
(3) if an <u>integer</u> is $o(1)$ it is 0 exactly.

If I get any more I will write to you again. I wish you were better and back here—there would be some splendid problems to work at. I don't know if you feel well enough to think about such difficult things yet.

I hope it is not so horribly cold now as it was.

All kind regards,

Yours most sincerely
G.H. Hardy.

But at present you must do what the doctors say. However you might be able to think about these things a little: they are very exciting.

G.H.H.

Commentary. This letter describes work communicated the following month to the London Mathematical Society; see [93], which is not in Hardy's *Collected Papers*. A fuller version appeared later in [94], [105, pp. 340-344].

In the letter it is tacitly assumed that $x = e^{\pi i \tau}$, and Hardy writes

f, p, q, U, x, Λ, and r(number of squares)
where the published paper has
F, h, k, S, q, j, and s,
respectively.

Beside the diagram showing the contour of integration, Hardy has added the following note in pencil in the left-hand margin:

When r is even $f\left(xe^{-p\pi i/q}\right)$ is an <u>elementary</u> function. The same sort of method applies but is much easier. Hence we see <u>why</u> the odd case is so much harder. When r is even we get a different proof of the results you proved in your paper.

The paper to which Hardy refers is evidently [219], [226, pp. 136–162].

In a later and more important paper by Hardy [96], [105, pp. 345–374] the problem is examined by different methods. In that paper, and in the joint work of Hardy and Littlewood on Waring's problem, the term 'singular series' was first introduced, replacing what Hardy calls the 'partition series' in his letter.

The reference to Ramanujan's paper "now being printed" seems to refer to [221], [226, pp. 179–199], although Gauss sums are not explicitly mentioned there.

Mordell's paper [166] is mentioned in the letter, but not his subsequent paper [167], which shows that, contrary to Hardy's belief, a theory for $r = \frac{1}{2}+$ integer does exist.

G.H. Hardy to J.J. Thomson

Trinity College
Cambridge

Dear Thomson,

I should like to say more clearly what the position about Ramanujan is. I wouldn't have done so if you hadn't asked me: I am much afraid of prejudicing him by persistent pushing, in view of his double candidature for a Fellowship and the Royal Society.

If he had not been ill I would have deferred putting him up a year or so: not that there is any question of the strength of his claim, but merely to let things take their ordinary course. As it is, I felt no time must be lost. I hardly hoped that he would have much chance this year—I don't know whether he has or not. If he has no doubt his illness will have much to do with it, and I should tell you precisely how (so far as I can know) matters stand.

Batty Shaw found out, what other doctors didn't know, that he had undergone an operation about 4 years ago. His worst theory was that this had really been for the removal of a malignant growth, wrongly diagnosed. In view of the fact that Ramanujan is no worse than 6 months ago, he has now abandoned this theory—the other doctors never gave it any support.

Tubercle has been the provisionally accepted theory, apart from this, since the original idea of gastric ulcer was given up. There is a perfectly regular rise of temperature every night, and persistent weakness. But there is certainly no crude lung trouble: no coughing or spitting. And Dr Kincaid, who runs Matlock, told me he was quite doubtful still, thought it might be some obscure Oriental germ trouble imperfectly studied at present.

For my own part, I think <u>probably</u> he will be alive in a year's time, and that he <u>may</u> recover completely. If he were like Jeans—a man who made a scientific fight to get right—he would have a much better chance. Like all Indians, he is fatalistic, and it is terribly hard to get him to take care of himself. Everyone, too, is frightened of the continual illness and solitude affecting his mind.

As I said, I am nervous about trying to rush him: and I am aware that for the time being I am not an ideal supporter. And I realise that the Royal Society has many other things to consider. But there is no doubt that (especially after his disappointment in the Fellowships) any striking recognition now might be a tremendous thing for him. It would make him feel that he was a success, and that it was worth while going on trying. It is this much more than the fear of the Royal Society losing him entirely which seems to me important.

I write on the hypothesis that his claims are such as, in the long run in any case, could not be denied. This is to me quite obvious. There is an absolute gulf between him and all the other mathematical candidates.

Yours sincerely,
G.H. Hardy

P.S. I forget whether I ever sent you a copy of my report on him. In case I didn't I send another.

I also send a copy of some suggestions I made to the Cambridge Philosophical Society Council about its publication. They were referred to a Committee of the Officers. As so much of the Proceedings is Physics, your opinion would be most helpful.

Commentary. The letter is undated but would have been written before 21 February 1918, probably early in that month, for the candidates for election to the Royal Society were considered by the Council firstly on 14 February and again on 21 February, when the list, including Ramanujan's name, was drawn up. This provisional list was read at the meeting on 28 February, the names were published in *Nature*, and the election finally took place on 2 May 1918.

Sir John Joseph Thomson (1856–1940), the distinguished physicist, Nobel prize winner, and discoverer of the electron, was President of the Royal Society and was admitted as Master of Trinity College on 5 March 1918.

Harold Batty Shaw (1867–1936) was a London specialist on consumption and chest diseases. It was 9, not 4, years earlier that Ramanujan underwent a surgical operation to remove a hydrocele.

Sir James Hopwood Jeans (1877–1946) was a distinguished scientist who made many contributions to astronomy and theoretical physics. He was awarded the Order of Merit in 1939. He was 3rd Wrangler in 1898, the same year that Hardy was placed 4th, and they both were awarded Smith's Prizes in 1901. It is of interest that Jeans was a member of the Council of the Royal Society at the time of Hardy's letter to Thomson. Between the years 1902 and 1903, Jeans contracted tuberculosis of the joints and spent time in various sanatoria from which he emerged fully recovered.

Ramanujan to G.H. Hardy

Matlock House
Matlock

Dear Mr Hardy,

My words are not adequate to express my thanks to you. I did not even dream of the possibility of my election. When I opened your telegram I read thrice Fellow Philosophical Society instead of Royal Society—I came to know only very recently of my election to Cambridge Philosophical Society—and I was very much puzzled why you sent a telegram from Piccadilly for that. It is only after some time that I read your telegram correctly.

Please convey my heartfelt thanks to Major MacMahon and Mr Littlewood. I am sorry I didn't write to them as I did not know their addresses.

Ever yours
S. Ramanujan

Commentary. This letter was probably written on Friday, 1 March 1918, or on the previous day.

Ramanujan was elected a Fellow of the Cambridge Philosophical Society on 18 February 1918. His recommendation form was signed by Hardy (from personal knowledge) and supported by the signatures of (i) Arthur Everett Stanley (1861–1927), a zoologist and Master of Christ's College, Cambridge, who was later Vice-Chancellor of the University, and was created a G.B.E. in 1920; (ii) Arthur Stanley Eddington of Trinity College, the distinguished Plumian Professor of Astronomy, who was knighted in 1930; (iii) Edward Alexander Newell Arber (1870–1918) of Trinity College, who was a botanist; and (iv) John Edward Purvis (1862–1930) of St. John's College, a University Lecturer in Chemistry and Physics in relation to hygiene and preventive medicine, who later became Mayor of Cambridge. This information is reproduced on page 103 of R. Bharathi's volume [45] with some errors in the transcription of the persons mentioned.

Percy Alexander MacMahon (1854–1929) had a distinguished career as a soldier in the Royal Artillery, retiring in 1898 with the rank of major. He made numerous contributions to combinatorics and combinatorial number theory, in particular, the theory of partitions. He was elected a Fellow of the Royal Society and served as President of the London Mathematical Society from 1894 to 1896. He became a member of St. John's College, Cambridge, in 1904, and received many academic honours, although he was never a university student in the ordinary sense. For further information about MacMahon, see [22].

Excerpt from Records of the Royal Society 18 December 1917

SRINIVASA RAMANUJAN
   Trinity College,
      Cambridge.

Research student in Mathematics. Distinguished as a pure mathematician, particularly for his investigations in elliptic functions and the theory of numbers. Author of the following papers, amongst others:—"Modular Equations and Approximations to π" (Quart. Journ., vol. xlv.); "New Expressions for Riemann's Functions $\zeta(s)$ and $\Xi(t)$" (ibid., vol. lxvi.); "Highly Composite Numbers" (Proc. Lond. Math. Soc., vol. xiv) "On Certain Arithmetical Functions" (Trans. Camb. Phil. Soc., vol. xxii.); "On the Expression of a Number in the Form $ax^2 + by^2 + cz^2 + dt^2$" (Proc. Camb. Phil. Soc., vol. xix.). Joint author with G.H. Hardy, F.R.S., of the following papers:—"Une formule asymptotique pour le nombre des partitions de n" (Comptes Rendus, January 2, 1917); "Asymptotic Formulae for the Distribution of Integers of various Types" (Proc. Lond. Math. Soc., vol. xvi); "The Normal Number of Prime Factors of a Number n" (Quart. Journ. vol. xlvii.); "Asymptotic Formulae in Combinatory Analysis" (Proc. Lond. Math. Soc., awaiting publication).

G.H. Hardy.	P.A. MacMahon.	J.H. Grace.	Joseph Larmor.
T.J.I'A. Bromwich.	E.W. Hobson.	H.F. Baker.	J.E. Littlewood.
J.W. Nicholson.	W.H. Young.	*E.T. Whittaker.*	*A.R. Forsyth.*
A.N. Whitehead.			

Commentary. The last three names on the citation for the Fellowship of the Royal Society are in italics, indicating that these sponsors had 'General Knowledge' of the candidate's work, rather than 'Personal Knowledge.' With the exception of Major MacMahon, all thirteen of those who subscribed had been Wranglers in the Cambridge Mathematical Tripos, and six of them had been Senior Wranglers. It is interesting to see among them the names of E.W. Hobson and H.F. Baker, who had not replied to letters written by Ramanujan from India.

Francis Dewsbury to G.H. Hardy 5 March 1918

University of Madras
SENATE HOUSE

From
FRANCIS DEWSBURY, Esq., B.A., LL.B.,
Registrar, University of Madras

To
G.H. Hardy, Esq., M.A.,
Trinity College, CAMBRIDGE.

Sir,

Your letter, dated 25th January 1918, and your cablegram of the 28th February have both reached me today, 5th of March.

I have to thank you for the very clear and explicit statement on the position of Mr. Ramanujam. Unfortunately the Syndicate has just met and will not meet again until the second week of April. I shall then place the matter before it for orders and I think you may anticipate a sympathetic consideration.

Mr. Ramanujam's friends and admirers here will be gratified to learn that he has been elected a Fellow of the Royal Society.

I shall write to you again immediately after the Syndicate has met. Your letter I shall treat as personal and confidential.

I have the honour to be,
Sir,
Your most obedient servant,
Francis Dewsbury
REGISTRAR.

Commentary. Hardy's letter of 25 January 1918 to Dewsbury has evidently not survived.

Article from Madras Mail 7 March 1918

Intimation has been received by cablegram from Mr. G.H. Hardy, the Cayley Lecturer in Mathematics at the University of Cambridge, that Mr. S. Ramanujam, the holder of a special scholarship granted by the University of Madras for the study of Mathematics at Cambridge, has recently been elected a Fellow of the Royal Society. This is the first time, we believe, that a student of the Madras University has received such distinguished recognition, and for an almost wholly self-taught scholar to achieve an F.R.S., so soon after coming under expert direction, must be very exceptional indeed.

Memoranda circulated between Sir Francis Spring and S. Narayana Aiyar

F. Spring to S.N. Aiyar 9 March 1918

1. Mr Narayana Aiyar has doubtless seen in I think in [*sic*] the Madras Mail of March 7th that Mr Ramanujan has been made a Fellow of the Royal

Society, one of the highest honours that can be conferred in the whole of India. Sir Thomas Holland, Dr Bourne and Dr Gilbert Walker are F.R.S. and some few others no doubt.

2. What exactly is Mr Ramanujan's nominal connection with this office? Is there any practical use in keeping up the connection if there is one? It seems inconceivable that we should ever be of any use to him.

3. What has he now got to live on? A Madras University Scholarship, a subvent[ion] or a grant allowance? or what?

F.J.E.S.

S.N. Aiyar to F. Spring 12 March 1918

1. I read this in the newspapers.
2. Ramanujan was granted leave on loss of pay on 1st May 1913 for two years. This period was extended by another three years, the maximum limit admissable under C.S.R. [Civil Service Regulations]. His service with the Port Trust will therefore terminate on the 30th April 1918.
3. I understood from Ramanujan's mother, who was here in Madras last November at the request of the Home Secretary to the Local Government, that the University Scholarship was continued to him and that he would get a University Professorship on his return to Madras on not less than Rs 400 a month.

SN

I am very pleased to hear this. We are all very proud I think of an F.R.S.

F.J.E.S.

Commentary. Spring's handwriting is difficult to read, so that some of the words cannot be read with certainty.

P.V. Seshu Aiyar to S.N. Aiyar 17 March 1918

18, PYCROFT'S ROAD
TRIPLICANE
Madras

Dear Mr. Narayana Aiyar,

Some of our young friends are anxious to have a congratulatory party on S. Ramanujan's election to a Fellowship of the Royal Society. I suggested to them that Sir Francis Spring will be the fit person to preside at the party since he did so much for Ramanujan. Please talk to Mr. Spring and get him to consent to preside. The meeting may be held under the auspices of the Maths. Association of the Presidency College of which he is an honorary member or if you prefer under the auspices of "The Madras Members of the Indian Mathematical Society."

Yours sincerely,
P.V. Seshu Aiyar

Commentary. Peruvemba Venkateswara Aiyar Seshu Aiyar was born into a poverty-stricken family in January 1872. His father was a petty farmer, and his mother was blind from smallpox. During his entire life, Seshu Aiyar's health was not robust. In 1891, he received a Bachelor's degree from Madras Christian College but had to terminate his studies for the M.A. degree due to a breakdown in health. He returned in 1893 to his native Palghat to become Headmaster of the high school there, but soon thereafter he began service at the Government College at Kumbakonam as Lecturer. At the Government College, he was Ramanujan's mathematics instructor during the one year that Ramanujan studied there. Sometime thereafter, Seshu Aiyar joined Presidency College in Madras, where for 15 years he lectured and eventually became Professor of Applied Mathematics. He returned to Kumbakonam to become Principal of the Government College until his retirement. He was one of the original members of the Indian Mathematical Society and eventually served as Secretary and President. He died in October 1935. For further information, see [237].

Indian Mathematical Society to G.H. Hardy   22 March 1918

To
     G.H. Hardy Esq. M.A. F.R.S.   Cayley lecturer,
     Cambridge University

Sir,

We have the honour to convey to you by direction, the grateful thanks of the Madras Members of the Indian Mathematical Society and the Mathematical Associations of the Presidency and Christian Colleges Madras, for the aid and guidance you have been providing to Mr. S. Ramanujan in his work. A copy of the minutes of the meeting conveyed for the purpose is herein enclosed.

> We have the honour to be,
> Sir,
> Your most obedient servants,
> P.V. Seshuiyar
> Representative of the Madras section
> of the Indian Math. Society.
> R. Ramaniah
> Secretary, Mathematical Association,
> Presidency College.
> L.V. Subramanian
> Secretary, Mathematical Association,
> Christian College.

P.S.

May I also add my personal thanks to you for the parental care you have been bestowing on him during these months when his health has not been good?

P.V. Seshuiyar

Francis Dewsbury to Educational Department   **9 April 1918**

From
> FRANCIS DEWSBURY, Esq., B.A., LL.D.,
> *Registrar, University of Madras,*

To
> The Secretary to Government
> Home (Education) Department

Sir,

With reference to G.O. No. 345, dated 13th March 1917, and previous correspondence in connection with the Scholarship granted to Mr. S. Ramanujam at Cambridge, I have the honour to inform you that the Syndicate has resolved that this Scholarship be renewed for another year, and has directed me to request the sanction of the Government for the appropriation of funds sufficient for the purpose from the University Vacation Lectures Account.

I enclose in a confidential cover a letter, dated 25th January 1918, received here on 5th March 1918, from Mr. G.H. Hardy which gives full particulars of Mr. Ramanujam's present circumstances and future prospects. The Syndicate is of opinion that Mr. Ramanujam should not in any circumstances, and more especially in his illness, be left dependent upon the benevolence of others.

The Government is no doubt aware that a cablegram was received on 5th March 1918 from Mr. Hardy announcing Mr. Ramanujam's election as a Fellow of the Royal Society.

I have to request that Mr. Hardy's letter may be returned after perusal, and that early orders may be passed upon this application. Mr. Ramanujam's Scholarship expired on 31st March 1918.

I have the honour to be,
Sir,
Your most obedient servant,
Francis Dewsbury
Registrar.

Commentary. Both Hardy's report and letter have evidently been lost.

Francis Dewsbury to G.H. Hardy   **10 April 1918**

University of Madras
Senate House

Dear sir,

In continuation of my letter No. 1556, dated 5th March 1918, I now write to say that the Syndicate has resolved that the scholarship to Mr. S. Ramanujam be extended for one additional year, and I have received instructions to obtain the sanction of the Government for the appropriation of the necessary funds. As soon as such sanction is received, I shall cable to you and at the same time make a remittance to Mr. Ramanujam's account also by cable. As I anticipated, the Syndicate gave very sympathetic consideration to your representations about Mr. Ramanujam. It is probable that before the termination of this extension of his scholarship the Syndicate will have before it further proposals as to Mr. Ramanujam's future unless intimation is received before then that this has been decided by the action of the authorities at home. I shall be much obliged if you will keep me posted as to his prospects in England so that the Syndicate may be in a position to judge the necessity for further action or otherwise as the case may be.

I trust that Mr. Ramanujam's health continues to improve. News of him from time to time will be welcomed here.

Yours faithfully,
Francis Dewsbury

G.H. Hardy Esq., M.A.,   Trinity College, Cambridge.

Francis Dewsbury to G.H. Hardy   **16 April 1918**

University of Madras
SENATE HOUSE

From
    FRANCIS DEWSBURY, Esq., B.A., LL.B.,
        Registrar, University of Madras

To
    G.H. Hardy Esq., M.A.,
        Trinity College, CAMBRIDGE.

Sir,

In continuation of my letter No. 2722, dated 10th April 1918, I have now the honour to inform you that the Government have sanctioned the necessary appropriation of funds to provide for the continuance of the scholarship to Mr. S. Ramanujam for another year. I am arranging to cable to Mr. Ramanujam's bankers the value of his scholarship for the first quarter of the year.

I have the honour to be,
Sir,
Your most obedient servant,
Francis Dewsbury
REGISTRAR.

S. Ramanujan to the Royal Society 17 May 1918

Matlock Sanatorium
Matlock
Derbyshire

Dear Sir,

I regret very much that, owing to my ill health, I am not able to travel to London at present.

Will you be kind enough to apply to the Council for an extension of the time before admission?

I remain
Sir,
Yours faithfully
S. Ramanujan

Commentary. On 2 May 1918, Ramanujan became the second Indian to be elected to the Royal Society. The first Indian Fellow of the Royal Society was Ardaseer Cursetjee (1808–1877). He was elected F.R.S. in 1841, and was a Parsee and a distinguished marine engineer in Bombay. Cursetjee was responsible for introducing gas to Bombay. Being the first Indian native to be placed over Europeans, he held the post of Principal Inspector of Steam Machinery at the Bombay Steam Factory from 1840 to 1858. The third and fourth Indians elected to the Royal Society were physicists Sir Jagadish Chandra Bose (1858–1937), elected in 1920, and Sir C. V. Raman (1888–1970), elected in 1924.

This letter was evidently written to an official of the Royal Society to request that Ramanujan's induction be delayed because of ill health. The letter appears in a monograph published by the Association of Mathematics Teachers in India commemorating the centenary of Ramanujan's birth [17].

S. Ramanujan to A.S. Ramalingam 19 June 1918

Matlock House
Matlock

My dear Ramalingam,

The whole of last night I had fever and my temperature this morning was about 102 deg. The old cook has left this place the day before yesterday. The present cook spoiled yesterday all the *appalappu* by scorching some of them and leaving some raw (without stirring them perhaps). The curried rice was just like *akshata* and as hard as uncooked rice. Yesterday I had no dinner. At least I had some breakfast and plain boiled rice before. Even these she can't prepare properly now. And so please don't send me anything except *avarakkai, vendakkai* and *nilakkadalai urundai*. Don't send anything to the matron with instructions or any such thing.

Did you meet Dr. Ram yesterday? I redirected a letter to you yesterday. Perhaps it may be a reply from Mr. Field.

Ever yours
S. Ramanujan

(Don't send me oil, powder, *appalappu*. You may give me all of these things after I leave the sanatorium.)

Commentary. Ramanujan met Alambadi Subbaraya Ramalingam (1891-1953) at the Indian Students Hostel at 21 West Cromwell Road in London shortly after Ramanujan arrived in England in 1914. Evidently, they did not meet again until Ramanujan became confined to nursing homes. From the letters which follow, it is clear that Ramalingam, or Ram, as he was known, was a very close and caring friend who was greatly concerned about Ramanujan's comfort and health. Ram gives some information about himself in his letter to Hardy, and it has been possible to supplement this from other sources.

A.S. Ramalingam was born on 26 August 1891. He registered as a student at King's College, University of London, in 1910 and obtained a B.Sc. in civil engineering in 1913. Although his curriculum consisted primarily of scientific subjects, it is of interest to note that he had a keen interest in religion. Upon graduation, he worked as an engineer for the North Staffordshire Railway before joining the 6th Royal Fusiliers Division of the British Army at Dover. While in the army, he became known as A.S. Ram. The Secretary of State for India appointed Ramalingam as Assistant Engineer, Imperial Service, in August 1918, and, when he retired from the service in March 1944, he had the status of an Executive Engineer. He returned to India in August or September of 1918, almost a year before Ramanujan returned. The fact that there is no record of any meeting in India between Ramanujan and A.S. Ram had led to the erroneous conjecture that Ram perished on the sea voyage back to India; the real reason for this lack of contact would appear to be the fact that, at the time, Ram was serving far away from Madras in the Public Works Department in Bengal. He died in Madras on 6 May 1953.

Notes left in Ramalingam's small diary indicate that he apparently attended a lecture by Sir Francis Spring on Madras Harbour on 2 May 1918 in Jarrow by Newcastle upon Tyne. Dr. Ram, who attended Ramanujan at Matlock House, is also mentioned, and the words "tubercular meningitis" appear in pencil.

The Tamil words for the various foodstuffs mentioned have been transliterated and appear in italics. Ram's abbreviations have, in most cases, been converted into complete words. Appalappu are small pappadums, called aplams by Ram, made from rice flour and fried in oil or ghee. Avarakkai are broad beans, vendakkai are okra, also known as lady's fingers, and nilakkadalai urundai are groundnut balls made from roasted groundnut and jaggery.

With the following letter to Hardy, Ram included a letter to Ramanujan for perusal before forwarding. Because the letter has survived, we can surmise

A. S. RAMALINGAM

that Hardy did not forward it to Ramanujan, possibly because he felt that Ram's admonition was perhaps too harsh.

A.S. Ram to G. H. Hardy 23 June 1918

From: A.S. Ram, To: Prof. Hardy, F.R.S.
25 Croft Terrace, Jarrow on Tyne. Cambridge

Dear Sir,

I am very much obliged to you for your kind reply of 26th ult, informing me about the whereabouts and bad or illhealth of my friend Mr. S. Ramanujan.

By appointment I visited Ramanujan last Sunday [16 June] and stayed with him at the sanatorium itself till Tuesday afternoon. I must say I was shocked and horrified to find him in the thin, weak and emaciated state I found him in.

I desire to note a few observations on his state, health, food, mind etc., and also venture to make a few suggestions to you, so as to improve his health. Probably before I do these, I had better state a word or two as to the "Who is Who" of myself.

I am a native of Cuddalore, S. of Madras. In 1910 I came to England, as a student of Engineering Science. I graduated at London (King's Coll) and was serving my time as an assistant Engineer on the North Staffordshire Rly at Stoke on Trent. About April 1914, when I was in London, at 21, Cromwell Road, on a short visit, I met Ramanujan on his arrival there fresh from Madras. I bid him welcome and made him as at home as possible in a country then strange to him.

I regret I had not seen him, since then, till last week. While on the N.S. Railway at Stoke in September 1914, I joined H.M.'s Forces. After 16 months service in the Army, a small part of which was spent amongst the Indian Contingent, I was "temporarily released" to work "on munitions" at Messrs Palmers Shipbuilding and Iron Co. Ltd. Jarrow, where I still continue to be employed as an engineer and an undischarged soldier. Lately I had a bit of luck and was one of those selected for service (civilian) in the Indian Public Works Department. Conditioned by my passing a test in Horseriding, which will be held at Woolwich on 31st July, I have been promised, by the India Office, my release from the Army. I believe I am expected to sail for India before or about the end of September next. I am (have to) still sticking to Palmers and also manage to have some riding practice at Newcastle on Tyne. No doubt I am looking forward to my total discharge papers from my unit; before the receipt of these, I dare not leave Palmers, except with Palmers consent perhaps.

I may add, that I have been all these years, in touch with the Indian students' Departments of the India Office and also 21 Cromwell Road.

Before I resume re Ramanujan, I desire to add, that in view of rationing scheme in England, I had by last December arranged that my people should be sending me, fairly regularly, parcels of eatables for my use. I also must state that I am a vegetarian, but took to eggs in England. I may however say that apart from my not being a vegetarian by caste—Ramanujan being a Brahmin is a very strict vegetarian—I have not been, both by circumstances and otherwise, not been a vegetarian staunch and strict. My people, I mean, the members of my family are not all meateaters.

Anyway, my good people seem to have got somewhat alarmed at my going short of food and have been for the last nearly 2 months piling parcels upon parcels on me at Jarrow, so much so that it is still a problem with me as to how I am going to dispose of these; I say *still*, because, in view of my leaving England shortly, I sent a cable home on 23rd May "Stop, sending, food, Ram".

Both to enquire of Ramanujan's health and also to see if he would care to share some of these eatables with me, I wrote to him about the middle of May, to the care of your address. Having not had a reply, I had to enquire of you.

As soon as I received your kind reply, I wrote to Ramanujan and pressed for an answer. After all I had a reply—a meagre one though—asking me

CHAPTER 5. RAMANUJAN IS ILL

to send some ghee—nearest Indian equivalent to butter—and a few Madras made pickles. Straightaway I forwarded to Matlock, two full bottles of ghee and one of gingelly oil for frying purposes. I might say, that 2 parcels containing the above 3 bottles, were simply forwarded to him *intact* as sealed by my people at Madras. You are no doubt aware of the difficulty and bother involved in packing a bottle of oil for carriage by post. Therefore I simply readdressed these 2 parcels to Ramanujan to Matlock. In the same week, I sent to him in a pigeonhole box, sample quantities of 3 different pickles or chutneys and of various other Indian preparations, and also cashew nuts or really kernels.

In indenting for and acknowledging these, I was able to elicit only 2 letters and a p.c. from him. Owing to changes in my work at Palmers, I found myself unable to postpone my holidays (annual privilege) to 31st July to suit my riding examination, and without choice I had to take the last fortnight. This holiday came rather unexpectedly. I did not feel like taking the holiday so early. I tried to do something particular during this fortnight but I was disappointed and couldn't get a start. Hence on Friday morning (14th inst) I sent a reply paid wire to Ramanujan as follows "*Could I call Sunday morning. Write immediately stating needs, Ram*". I received a reply "*certainly with pickles*". On Saturday I looked in vain for a letter from Ramanujan.

Soon after I finished my riding practice on Saturday 15th inst., I set off by the night train to Matlock, where after a cold weary journey, I arrived at about 8 a.m.

There were fortunately a few vacancies there and the doctor had kindly permitted me to stay in one of the rooms of the sanatorium. The Matron was kind enough to let me have tea with Ramanujan in his room and I had the other meals in the dining hall.

I was conversing with Ramanujan all sorts of topics, personal, political, war, Indian social, Christian missions, etc. etc. etc. I had not the slightest reason to suspect or think that he was "going off" at all, as one or two were presumptuous enough to think.

I also was careful to observe his temperature whenever the nurse measured it and also to see what food he was getting and how he was relishing the food.

Well, it is with regard to food that I have to write somewhat harshly and tersely and at a good length. For breakfast he was having scrambled eggs or toast and tea. Well this could hardly be improved perhaps, though I can make something better to suit Ramanujan's palate, but still for his health's sake, no improvement is desirable in this meal.

For lunch he was having plain boiled rice, chillies and mustard fried in butter, lately began having cucumber and lemon fruit. The doctor was kind enough to provide just for a day or two, some green peas specially for him. I maintain this meal could very considerably be improved.

Tea: More or less a repetition of breakfast with green onions substituted for eggs at times.

MATLOCK HOUSE

Dinner. More or less a repetition of lunch + a glass of milk.
I believe a glass of milk is given about lunch time as well.
On Monday I consulted the doctor if he recommended or prohibited any particular items of diet e.g. whether he objected to pungent stuff like curry. The doctor advised that Ramanujan could take anything he liked best.

Ramanujan told me on my enquiry, that he did not like porridge; nor did he care for bread and milk, on which diet he had been formerly, for some months, he told me.

Naturally therefore I asked him why he did not think of going to a vegetarian sanatorium. I was told, the India Office was unable to recommend to him any vegetarian sanatorium. The India students Department of the India Office, with its mighty influence and with knowledge renowned all over India, not to be able to suit Ramanujan, was an unwelcome surprise to me.

Ramanujan has wrong notions of what is meant by Sanatorium and Nursing Home.

Besides Ramanujan seems to have been somewhat cowed down by Dr Ram, who seems to have told him "As long as you are a patient and not well you are not free and the doctor has control over your movements". Besides I do not know how Ramanujan got into his head the impression that Dr Ram had told him or given him to understand that Ramanujan had by some agreement to stay in Matlock sanatorium at least 12 months. Anyway before

CHAPTER 5. RAMANUJAN IS ILL

I could think of shifting Ramanujan I was advised by Ramanujan to consult both Dr Ram and Prof. Hardy (*i.e.* yourself).

At once the desire to consult Dr Ram, myself, sprang in me and it was strengthened by one or two incidents.

I called on the matron and consulted her if I myself could cook something for Ramanujan. "No you can't go to the kitchen" was the answer. Then I offered to write out a recipe and asked them if they would cook to the recipe. Then there was some trouble about the old cook leaving. I was rather keen on having my stuff and recipe tried while I was there, so that I could see it was properly cooked and also if Ramanujan liked it. Well the opportunity was really denied to me till Tuesday 1 p.m.

On Monday there was prepared some mixture of macaroni custard and something else for pudding and on enquiry I found—this was before I went to lunch—that it was macaroni and something else. I at once requested that Ramanujan should be served with whatever pudding that is going in the dining room and it must be left to Ramanujan to eat it or leave it. Ramanujan afterwards told me that he liked macaroni prepared with cheese. It seems, unfortunately, that the sanatorium people couldn't get cheese. Well goodness I thought that the sanatorium had a priority claim for such foodstuffs. Anyway I had some last week at Jarrow and so also this week. As a matter of fact, I managed to spare some for Ramanujan and sent on to him about one lb of cheese on Friday afternoon (21st inst.)

Also, I understood from Ramanujan that the Matlock Sanatorium people couldn't get bananas for him, whereas on Tuesday afternoon, I saw a fellow traveller from Sheffield who was in the panel or ward of Dr Ram, carrying a bunch of about a dozen bananas. On enquiry I gathered they cost 4d each. Still they are obtainable.

Nextly Ramanujan complained that he used to get once e.g. when Dr Ram visited him, delicious rice pudding, sago pudding, etc and that he was not getting them lately. Also I understood from Ramanujan that the sanatorium people couldn't afford butter to fry his potatoes in.

Therefore you see, Prof. Hardy, that there is plenty of scope for improving his diet and I saw little chance of improving it so long as he was in that sanatorium. Also Ramanujan is pretty keen on moving to somewhere near London but feels helpless to carry out his desire. As a matter of fact Ramanujan desires to stay in London but I strongly advised him not to entertain that idea at all. His object is the easiness with which he could obtain Indian dishes and Indian condiments. I certainly do not agree with him.

However I am inclined to think that it will be better if he could be shifted to a vegetarian sanatorium or nursing home, down South or S.W. or at least to a private place under the supervision of a doctor where he could get more individual attention with regard to cooking.

Therefore I made up my mind, in consultation with Ramanujan, to break my return journey to Newcastle at Swinton, so as to enable me to visit Dr

Ram and discuss the question fully with him. I rang him up on the phone and asked for an appointment for Tuesday. Dr. Ram was very kind and told me almost anytime would suit him. I left Ramanujan directly after lunch on Tuesday and with a bit of luck I happened to catch the train which I got in at Belper and did not change at all till I got down to Swinton where I got out.

I was glad to see Dr Ram and his family at Mexborough. I first made myself sure that Dr Ram had made no agreement whatever as to the length of time he should stay at Matlock with Dr Kincaird. Dr Ram also gave me some hope of Ramanujan's life and added that food alone could save him. I quite believe it.

Dr Ram asserts that there are very few doctors who would care to have Ramanujan in their nursing homes and fewer still who would bother with humouring Ramanujan's palate. It is only after I saw Dr Ram that I was somewhat impressed with the nonpliability and the inadaptability of Ramanujan's palate. Also Dr Ram placed before me the enormous difficulties that are involved in running a separate say 4 roomed cottage with perhaps a special nurse and a special cook and probably a servant in addition.

In any case when I left Dr Ram on Tuesday at about 9 p.m. I was better acquainted with the difficulties of Ramanujan's case; also that Ramanujan had too particular a palate to be easily pleased.

Now then I will turn to the solution of the problem. Though Dr Ram thinks that Dr Kincaird is the only kind and charitable doctor that will tolerate Ramanujan, I am hopeful—I have consulted a few of my friends and doctors out here—that I could find one or two nursing homes which will suit Ramanujan better. Though my desire is to move him to near the sunny S. or S.W. the few homes that I got the addresses of, so far are all situated in S.E. or near London. I desire to avoid these because of the air raids. However I will try and place my results before you shortly.

This question appeals to me in 3 main items.

(a) The desirability of an operation by a skilled surgeon and specialist like Dr Boyd of Newcastle, thorax specialist. The operation should not be undertaken just for practice or for small chances. The purpose and location should be clearly defined beforehand. If there is an ulcer, well if it is not too far gone, there is a chance of its being removed by operation; if there is no ulcer, the idea of operation should be thought out beforehand not only more than once but very soundly and several times.

(b) Merely to have the same treatment viz. fresh air treatment—as he is having now. In this case, the possibility of moving him to a place, preferably vegetarian, where he will be better looked after.

(c) Both of the above are conditioned by Ramanujan's being in a state fit to be transported and no doubt of the availability of transport.

Well I do not know if (a) is decided upon, it can be carried out at Matlock. It means the specialist surgeon will have to be at or go to Matlock.

So that I could consult my friends—both doctors and others over here—I jotted down a few particulars about Ramanujan's disease and the theories as diagnosed by the four or five doctors he came under. Only what I put down on paper was Ramanujan's statement. But my friends over here would rather have a medical statement by say Dr Kincaird, describing all the particulars and also summing them up for judgment by my friends.

I however think it won't be fair if *I* asked Dr Kincaird for such a report. You pay the doctor and it is reasonable that you have a far better right to ask for such a report, than myself. Therefore if you would rather that I continued my efforts, I should be obliged for placing such a report in my hands at your earliest convenience.

I need hardly add, that I will treat the report as confidential and will not use it in any way objectionable.

Well as a matter of fact I desired to consult Dr Kincaird privately i.e. outside the reach of Ramanujan's hearing—but Dr Kincaird was so busy, I believe, that he forgot to fix me up with 5 minutes.

Still a soldier as I am, I am confident of getting away to visit Ramanujan or help Ramanujan with transport. I do not know if it is anything of value but I desire to place it before you for what it is worth. I have passed the second year's examination in ambulance or first aid work and for Ramanujan's sake, I do not mind qualifying in sick nursing.

When you send for Dr Kincaird's medical statement I do not know if it is advisable that you should ask to be enlightened about this point. I personally think it is desirable that some light should be thrown on this procedure. It is this. Dr Kincaird *distinctly* told me that Ramanujan could eat anything that he liked and that nothing was injurious or objectionable.

On the other hand I hear from my friends in Jarrow that eatables like pickles, chillies etc should be avoided by consumptive patients. There is one reason however in which case these can be excused. When the patient has gone too far to be remedied, and when it is a question of only a few weeks or months, it matters little, as long as the patient feels happy and comfortable in the last days of his life. In permitting me to give anything to Ramanujan to eat, is this the object of Dr Kincaird?

It is anguishing and breaking one's heart to feel that Ramanujan, with his wonderful capabilities and valuable contributions, should be given up for such as hopeless. War with all its horrors might have made us callous to the whole-sale slaughter and loss of lives but surely should Ramanujan be given up?

Now I desire to place before you certain proposals. Before I do that I desire to call your attention to this fact. Ramanujan is not only a vegetarian but is one with a strong Madras palate; what is worse he is suspicious of anything that is offered to him being meat. I will write to him and advise him that he should choose between killing—no, controlling will do—his palate and killing himself. He is thinking of his vegetarianism even at the expense of his health and life. But one cannot but think of him as cranky or headstrong when he refused cream and say plums on it; he also dislikes porridge or oatmeal. The solution is to provide him with Madras dishes; but is it practicable; are they commendable for his health?

Nextly might I call your attention to his letter (original) of 19th inst., since I left him. It is enclosed herewith. The stuff—call them *Aplams* if you like—I have marked blue on his letter is good for him. It is neither hot nor pungent nor acidy nor sour nor sweet. It simply wants frying in oil which I did provide. It is fried exactly like chips. As a matter of fact I was surprised to find, when I went to Matlock, that the bottles of oil and ghee had not been opened into or used at all. Ramanujan's explanation was that he did not then feel like eating these preparations or did not feel well enough to eat these. This is not true because he had eaten the dried vegetables (marked red on his letter)—call it Avarakai—raw without being cooked. The latter won't do because uncooked avarakai only throws additional work on the digestive organs.

This is why on Thursday I fried here (at Jarrow) some aplams and sent them on to him, together with some avarakai. Aplams once fried are eatable and alright cold but certainly far better warm—not warmed up.

Now then I come to 3 somewhat bold proposals impracticable perhaps but not impossible. All these three are dependent upon influence exerted by you thro' the India Office and also Ramanujan's ability as an F.R.S.

(i) The very first is the possibility of shifting Ramanujan to Southern France or Italy. Of course India will be better but that is impractible. In case he is permitted to be moved to S. France or S. Italy, he can be moved on a Red Cross carriage or wagon and a hospital ship. The sea voyage is so comparatively short. The climate of S. Italy is a very effective remedy for Ramanujan.

(ii) In addition to (i) or even without (i), the "temporary release" of a soldier cook (Indian) to work for Ramanujan for making his Indian diet. I mention a soldier cook because a soldier cook appreciates his freedom from military discipline and is more or less faithful to his civilian job. It may be argued that this is misusing soldiers. So many soldiers on munitions there are at present like myself. It is more or less taking advantage of the fact that we are soldiers and putting us to where they want us and not necessarily to where we are fit for.

I dare say I can get hold of one or two lascars—coloured workers on mercantile marine—knocking about on the Tyneside. Amongst them are some good cooks; but they are extremely unreliable. They will desert Ramanujan as quickly as they have deserted their ships.

(iii) Even in case of moving Ramanujan within England, it will be much better to get permission for transporting him on a Red Cross van or Rly carriage.

There is little else that I can usefully add now; but I am busy trying to get some valuable information and I will place these before you shortly.

Do you think you can still fall on your first offer from Bournemouth? I mean the one that Ramanujan had to go to, if Dr Ram had not brought him to Mexborough.

I desire to meet you at Matlock when you will call on Ramanujan at the end of this session. Therefore I left a stamped and addressed telegraph form in Ramanujan's writing box, so that he loses no time in informing me of your arrival.

In case I wire to you and you confuse my name with that of Dr Ram, please look for place of origin. Or if the telegram is from London, say, I will sign myself *Sram* to denote S Ram. Dr Ram is L Ram.

I regret to have made my letter so very long but I could hardly help it. One would think that such a question cannot be thrashed out on paper. But still I have little other choice.

I look forward to your kind advice and guidance.

Believe me, Dear Sir,
I am,
Yours very sincerely,
A S Ram.

P.S.

I am enclosing my letter to Ramanujan herewith. Please peruse it and post it straight away. I trust you follow my position and case clearly.

ASR

A.S. Ram to Ramanujan 23 June 1918

A.S. Ram, 25 Croft Terrace,
Jarrow on Tyne. Sunday

My dear Ramanujan,

I was exceedingly grieved to have your painful letter of 19st. Sorry to hear the new cook is a failure so far as you are concerned. How did you like the aplams cold? I mean those that I sent on to you on Thursday last (20th inst.).

Please ask matron to boil those avarakais for you. They shouldn't find it difficult to do that. It only means another sauce–pan on the fire.

Also you told me you liked macaroni prepared with cheese. I forwarded about 1 lb—possibly less—of cheese on Friday 21st inst. Please instruct matron to have these prepared to your taste.

I am starting work tomorrow in the shop as a mechanic. I have to turn up at 6 a.m. for the first time for the last over 18 months. Well I am glad it won't be long; at the same time it is an opportunity to train my hands. I had a reply from Mr Field—that was not the letter you so kindly and promptly forwarded. The India Office does not recommend private arrangements. Therefore my plan to go as a marine engineer falls thro'.

Possibly tomorrow I will be forwarding to you a parcel containing (a) ground nut balls (b) vegetable fingers and (c) Avarakai. These were the only 3 you have asked for in yours 19th/18 .

By the bye, I am forwarding your letter 19th/18 to Prof. Hardy.

I am writing to Prof. Hardy today and making a sound case for you. I hope to have his advice soon.

Now then I will have to be a bit harsh with you.

Both from my talk with Dr Ram and after my second thoughts, I am impressed with your being so particular about your palate. Well you will have to choose between killing or controlling your palate and killing yourself. You must try and get yourself to like porridge or oatmeal, cream etc. My friends have strongly advised me not to let you indulge in pickles and chillies. You know you can train your palate. Dr Ram told a patient who expressed unwillingness to take milk

"You must take milk or you will go to Hell".

I am not going to the extremes and asking you to take beef tea or bovril, though considering your life, my asking you to take such a thing is quite excusable nay desirable and even unavoidable.

Be reasonable and don't be bigoted. Practise self restraint and not stick to curries and pickles. I am conserving all my ghee and oil for you.

Please provide me a weekly report stating among other things your normal morning and evening temperatures. State what food you are getting and what food you would like me to send.

Tinned Maize is very hard to get but I have not given up my search. There is little hope of my getting any dessicated cocoanut. Unfortunately cocoanut cakes and cocoanut biscuits cannot be well made with ordinary moist cocoanut kernel.

From the last few lines of your letter 19th/18 I gather you are confident of being moved from Matlock Sanatorium. Even if it were otherwise possible, are you in a state fit to be moved? Certainly not.

Well it is up to you therefore, to eat or drink plenty of porridge, eat tomatoes, bananas, cheese, cream, macaroni etc. Bananas and cream go together lovely. Eat these and fatten yourself and then alone, the possibility of your moving will be considered.

Dr Ram has certainly made no agreement whatever as to the length of time you should stay or will stay at Matlock House.

He gives you plenty of hope to recover but he makes it entirely dependent upon food. Have you any objection to taking foods like cod liver oil, sanatogen which aim at strength building and also fattening.

Now then hurry up and begin eating or devouring plenty and get fat. I shall take you with me to India. That is a good boy.

I am extremely busy. You are constantly in my thoughts and prayers.

Hoping to take you on my visit to London next month.

I am,
Yours
A S Ram

P.S.

Johnny Guardplate—I mean guard against evil spirits etc,—that I left with you, doing you any good? Please keep it carefully or send it on to me when you feel inclined.

A.S.R.

I say Ramanujan, my cadet friend J B Madgshon is taking his final examination for his commission at Cambridge about 28th inst and will be going over to Newcastle soon after. I could ask him to fetch your stuff—I mean Indian eatables and grain etc—and then I could take them and present to Dr Ram if you like.

However if you want anything done in this respect you must hurry up and write to me by return. You must give full instruction as to where he could get the key etc and where exactly you have kept these.

A.S.R.

Commentary. The Mr. Field mentioned in Ramanujan's letter and in Ram's reply was Robert Edwin Field, a minor staff clerk in the department of the Secretary for Indian students at the India Office. The Department of the Educational Adviser to Indian Students was housed in 21 Cromwell Road, London. Immediately after his arrival in London aboard the S S Nevasa, Ramanujan was taken there and it was there that he first met A.S. Ram.

In his letter to Hardy, Ram refers several times to Dr. Kincaird. The correct spelling is Kincaid. The thoracic specialist mentioned in the letter may be John William Hugh Boyd, who graduated with M.B. and B.S. degrees from the University of Durham in 1899 and practiced in Newcastle upon Tyne. There is no evidence that he was ever consulted about Ramanujan.

The Matlock House Sanatorium in Rutland Street, Matlock, Derbyshire, was advertised in 1904, not as a sanatorium, but as a hydropathic establishment specializing in thermotherapeutic treatment and recommended as an all-the-year-round resort with a continual round of entertainments, outdoor and indoor. By 1918, it had become a more clinical establishment under the medical superintendence of Dr. Frederick Kincaid, who was still in charge in 1920, according to the Medical Register of that year, but who emigrated to Vancouver seven years later. The building has not been used for medical purposes for some time, and its name has been changed to Rutland Court. Part of the building is now used as a Local Taxation Office by the Inland Revenue, and the remainder of the building has been converted into flats.

It is known that, for a brief period (probably in November 1917) before he went to Matlock House, Ramanujan was a patient at Hill Grove, a large house, now ruined, three miles northeast of Wells in the Mendip Hills in Somerset. This establishment was run for patients with tuberculosis by Dr David Jacob Aaron Chowry-Muthu, who had been a fellow-passenger with Ramanujan on the voyage from India to England in 1914. Chowry-Muthu

Hill Grove

was known affectionately in the district as "the black doctor" and had novel ideas for the treatment of tuberculosis, in which his patients were required to wear masks at the beginning of their treatment; this may have proved uncongenial to Ramanujan and may therefore have been the reason for his short stay.

There is a suggestion at the end of the letter that Ramanujan might have gone at an earlier stage to Bournemouth for treatment. We have found no other reference to this possibility, nor do we know why it should have been Dr. Ram who "brought him to Mexborough." L. Ram was a medical graduate of the University of Edinburgh in 1904 and practiced continuously in Mexborough from 1907 onwards. He appears to have had no connection with Cambridge, but may have been recommended by one of Ramanujan's friends. He was the author of a paper on abdominal dropsy and presumably, being in charge of the ward at Matlock House, was skilled in the treatment of chest diseases.

Unfortunately, Hardy's replies to A.S. Ram's letter have not survived, so that it is not possible to determine what action may have been taken to meet Ram's various suggestions, or whether Ram was responsible for Ramanujan's admission to Fitzroy House or Colinette Road.

The following excerpts are taken from an article by Bollobás [**50**, pp. 84,85]. The complete letters and their dates are unavailable to us.

S. Ramanujan to G.H. Hardy

Matlock

I have been here a month and I have not been allowed fire even for a single day. I have been shivering from cold many a time and have not been able to take my meals sometimes. In the beginning I was told that I could not possibly have any except the welcome fire I had for an hour or two when I entered this place. After a fortnight of stay they told me that they received a letter from you about one and promised me fire on those days in which I do some serious mathematical work. That day hasn't come yet and I am left in this dreadfully cold open room. Even if I do any serious mathematics in future I am not going to ask them for fire on that day.

The bath rooms are nice and warm. I shall go to the bath room with pen and paper every day for about an hour or so and send you two or three papers very soon. This thought did not strike me before. Else I would have written something already. In a week or so you may perhaps have a complaint against me from the doctor that I am having bath every day. But I assure you beforehand that I am not going to bathe but to write something.

S. Ramanujan to G.H. Hardy

Matlock

It is true that I promised my mother that I was going home at the end of 2 years; I wrote them several letters $1\frac{1}{2}$ years ago that I was coming over there for the long vacation; but I had many letters of protest from my mother to the effect that I ought not to come to India till I took my M.A. degree. So I gave up the idea of going there.

It is not true that I am getting letters from my wife or brothers–in–law or anybody. I had only very few formal letters from my wife just explaining to me why she had to leave my home and asking me why my mother should trouble her by [not] speaking to her when they met anywhere by chance. When I do not know the whereabouts of my wife my mother's grievance is that I have left my wife in some secret place somewhere in India and that she is waiting for me frequently to come to that place without my mother's knowledge and that I am listening to her words all the time ...

The initial S. in my name stands for Srinivasa which is my father's name. I haven't got a surname, really speaking.

S. Ramanujan to G.H. Hardy

Matlock
Sunday

Dear Mr. Hardy,

In one of my letters I wrote about the least number of terms which will give the nearest integer to the actual coefficient in $1/g_3$ problem. It will be extremely difficult to prove such a result. But we can prove this much as follows.

$$\sum \alpha_n x^n = \frac{1}{1 - 504\left(\frac{1^5 x}{1-x} + \frac{2^5 x^2}{1-x^2} + \cdots\right)}$$

(1)
$$\alpha_n = c\left\{e^{2n\pi} + (-1)^n \frac{e^{n\pi}}{2^4} + 2\cos\left(\frac{2n\pi}{5} + 8\arctan 2\right)\frac{e^{2n\pi/5}}{5^4} + \cdots \right.$$
$$\left. + \left\{2\cos\left(\frac{2n\pi}{65} + 8\arctan\ldots\right) + 2\cos\left(\frac{2n\pi}{65} + 8\arctan\ldots\right)\right\}\frac{e^{2n\pi/65}}{(65)^4} + \cdots\right\}$$

where

$$c = \frac{32}{9}\frac{\{\Gamma(\tfrac{3}{4})\}^{16}}{\pi^4} = .94373\ldots$$

We shall first prove that, if we take only

(2)
$$\left[\frac{\tfrac{3}{4}n(1-\epsilon)}{(\log n)^{3/2}\sqrt{(1-\tfrac{1}{5^2})(1-\tfrac{1}{(13)^2})(1-\tfrac{1}{(17)^2})\cdots}}\right]$$

terms in the right hand side of (1), ϵ being any positive number less than 1 and 5, 13, 17, ... are primes of the form $4n+1$, then it is possible to find an infinity of values of n for which α_n is <u>not</u> the nearest integer to the sum of the asymptotic series up to (2) terms. Suppose that $\omega(\lambda)$ is the number of sums of any two squares such as 1, 2, 4, 5, 8, 9, 10, 13, 16, ... not exceeding λ and that $p(\lambda)$ is the number of sums of two squares that are prime to each other such as 1, 2, 5, 10, 13, 17, 25, ... not exceeding λ. Then it is easy to see that

(3) $$p(\lambda) = \omega(\lambda)\left(1 - \frac{1}{2^2}\right)\left(1 - \frac{1}{3^2}\right)\left(1 - \frac{1}{7^2}\right)\left(1 - \frac{1}{(11)^2}\right)\cdots$$

where 2, 3, 7, ... are the primes excluding those of the form $4k+1$ and the right side is to be understood as $\omega(\lambda) - \omega(\lambda/4) - \omega(\lambda/9) + \omega(\lambda/36) - \cdots$ But Landau has proved that

(4) $$\omega(\lambda) \sim \frac{\lambda}{\sqrt{\log \lambda}}\frac{1}{\sqrt{2(1-\tfrac{1}{3^2})(1-\tfrac{1}{7^2})(1-\tfrac{1}{(11)^2})\cdots}}\overset{*}{,}$$

3, 7, 11, ... being the primes of the form $4k+3$.

CHAPTER 5. RAMANUJAN IS ILL

---

*Please see if this constant is correct. It may be somewhat interesting to note that

$$\left(1+\frac{1}{7}\right)\left(1+\frac{1}{11}\right)\left(1+\frac{1}{19}\right) = \sqrt{2\left(1-\frac{1}{3^2}\right)\left(1-\frac{1}{7^2}\right)\left(1-\frac{1}{(11)^2}\right)\left(1-\frac{1}{(19)^2}\right)}$$

(The first four terms in the constant)

We can easily prove from (3) and (4) that

$$p(\lambda) \sim \frac{\lambda}{\sqrt{\log \lambda}} \frac{(1-\frac{1}{2^2})(1-\frac{1}{3^2})(1-\frac{1}{7^2})\cdots}{\sqrt{2(1-\frac{1}{3^2})(1-\frac{1}{7^2})(1-\frac{1}{(11)^2})\cdots}}$$

that is to say

(5) $$p(\lambda) \sim \frac{3\lambda}{2\pi\sqrt{\log \lambda}} \frac{1}{\sqrt{(1-\frac{1}{5^2})(1-\frac{1}{(13)^2})(1-\frac{1}{(17)^2})\cdots}}$$

5, 13, 17, ... being the primes of the form $4k+1$.

Suppose we take only (2) terms and that the last term is of the form

$$c(2\cos\cdots + 2\cos\cdots + \cdots)\frac{e^{2n\pi/\lambda}}{\lambda^4}$$

where λ is a number belonging to the class

$$1, 2, 5, 10, 13, \ldots$$

and c is the same as in (1). Then obviously

$$p(\lambda) = (2).$$

It follows from (2) and (5) that

(6) $$\lambda \sim \frac{\pi n}{2\sqrt{\log n}}(1-\epsilon).$$

Now choose a prime p of the form $4k+1$ such that

$$p > \lambda \quad \text{and} \quad p \sim \lambda.$$

This is quite possible since consecutive primes of the form $4k+1$ are asymptotically equivalent to one another just like ordinary primes.

Consider the asymptotic series for $\alpha_n, \alpha_{n-1}, \alpha_{n-2}, \ldots, \alpha_{n-p+1}$ remembering that we take only

$$\left[\frac{\frac{3}{4}n(1-\epsilon)}{(\log n)^{3/2}\sqrt{(1-\frac{1}{5^2})(1-\frac{1}{(13)^2})\cdots}}\right], \left[\frac{\frac{3}{4}(n-1)(1-\epsilon)}{\{\log(n-1)\}^{3/2}\sqrt{(1-\frac{1}{5^2})(1-\frac{1}{(13)^2})\cdots}}\right], \ldots$$

terms in $\alpha_n, \alpha_{n-1}, \alpha_{n-2}, \ldots, \alpha_{n-p+1}$, respectively.

A term
$$c \cdot 2\cos\left(\frac{2n\pi}{p} + 8\arctan\theta\right)\frac{e^{2n\pi/p}}{p^4}$$
will occur in α_n beyond the terms we have taken. Similarly
$$c \cdot 2\cos\left(\frac{2(n-1)\pi}{p} + 8\arctan\theta\right)\frac{e^{2(n-1)\pi/p}}{p^4}$$
will occur in α_{n-1} beyond the terms we took. And so on. In all these θ and p remain the same throughout. There is only one term $2\cos$ instead of many $\cos$'s since p can be expressed as the sum of 2 squares in one and only way. Now take a number K such that
$$0 < K < 2.$$
Then out of the p numbers
$$2\cos\left(\frac{2n\pi}{p} + 8\arctan\theta\right), 2\cos\left(\frac{2(n-1)\pi}{p} + 8\arctan\theta\right),$$
$$\ldots 2\cos\left(\frac{2(n-p+1)\pi}{p} + 8\arctan\theta\right)$$
we can certainly find a number
$$2\cos\left(\frac{2(n-s)\pi}{p} + 8\arctan\theta\right) > K$$
and
$$2\cos\left(\frac{2(n-t)\pi}{p} + 8\arctan\theta\right) < -K.$$
The corresponding term in α_{n-s} is greater than $Kce^{2(n-s)\pi/p}$ and that in α_{n-t} is less than $-Kce^{2(n-t)\pi/p}$.

Now we know that s and t are not negative and do not exceed p and
$$p \sim \frac{\pi n}{2\log n}(1-\epsilon).$$
Hence,
$$Kc\frac{e^{2(n-s)\pi/p}}{p^4} = Kc\frac{16(\log n)^4}{\pi^4 n^4}\exp\left(\frac{4\log n}{1-\epsilon}\{1+o(1)\}\right)$$
$$= Kc\frac{16}{\pi^4}(\log n)^4 n^{4\epsilon/(1-\epsilon)+o(1)}.$$
Precisely in the same way
$$-Kc\frac{e^{2(n-t)\pi/p}}{p^4} = -Kc\frac{16}{\pi^4}(\log n)^4 n^{4\epsilon/(1-\epsilon)+o(1)}.$$
The right hand sides surpass all limits as $n \to \infty$.

Thus we see that the difference between α_n and the sum of the asymptotic series for α_n to (2) terms is not even bounded as $n \to \infty$.

CHAPTER 5. RAMANUJAN IS ILL

We shall now prove that, if ϵ is any positive number and if we take

(7) $$\left[\frac{n(1+\epsilon)}{(\log n)^{3/2}\sqrt{(1-\frac{1}{5^2})(1-\frac{1}{(13)^2})(1-\frac{1}{(17)^2})\cdots}}\right]$$

terms, then the nearest integer to the sum of (7) terms will be α_n for all sufficiently large values of n. As before let the last term we have taken viz. (7)th term be

(8) $$c(2\cos\cdots + 2\cos\cdots + \cdots)\frac{e^{2\pi n/\lambda}}{\lambda^4}$$

where λ is a number belonging to the class

$$1, 2, 5, 10, 13, 17, \ldots$$

and let the subsequent numbers be $\lambda', \lambda'', \lambda''', \ldots$. Now the number of cos's in (8) is the number of ways in which λ can be expressed as the sum of two squares, it follows that in (8)

$$|2\cos\cdots + 2\cos\cdots + \cdots| \leq d(\lambda)$$

where $d(\lambda)$ is the number of divisors of λ. Hence the absolute value of the sum of all the remaining terms to ∞ is less than

$$c\left\{\frac{d(\lambda')}{\lambda'^4}e^{2\pi n/\lambda'} + \frac{d(\lambda'')}{\lambda''^4}e^{2\pi n/\lambda''}\cdots\right\}$$

$$= O\left\{\lambda'^{-4+o(1)}e^{2\pi n/\lambda'} + \lambda''^{-4+o(1)}e^{2\pi n/\lambda''} + \cdots\right\}$$

$$= O\left\{e^{2\pi n/\lambda}\left(\lambda'^{-4+o(1)} + \lambda''^{-4+o(1)} + \cdots\right)\right\}$$

$$= O\left\{e^{2\pi n/\lambda}\left((\lambda+1)^{-4+o(1)} + (\lambda+2)^{-4+o(1)} + \cdots\right)\right\}$$

$$= O\left(\lambda^{-3+o(1)}e^{2\pi n/\lambda}\right)$$

But $p(\lambda) = (7)$ so that, in virtue of (5),

(9) $$\lambda \sim \frac{2\pi n}{3\log n}(1+\epsilon)$$

Hence

$$O\left(\lambda^{-3+o(1)}e^{2\pi n/\lambda}\right) = O\left(n^{-3\epsilon/(1+\epsilon)+o(1)}\right) = o(1).$$

Thus we have completed the proof.

Yours
S. Ramanujan

$$\sum p_n x^n = \frac{1}{1 - 504\left(\frac{1^5 x}{1-x} + \frac{2^5 x^2}{1-x^2} + \cdots\right)}.$$

$$C = 1 + 480 \left(\frac{1^7}{e^{2\pi} - 1} + \frac{2^7}{e^{4\pi} - 1} + \cdots \right).$$

If n is zero or a positive integer, then

(i) $\quad \frac{1}{2} C p_n = \frac{c_1(n)}{1^4} e^{2n\pi} + \frac{c_2(n)}{2^4} e^{n\pi} + \frac{c_5(n)}{5^4} e^{2n\pi/5} + \cdots,$

(ii) $\quad \frac{1}{2} C u_n = \frac{c_1(n)}{1^3} e^{2n\pi} + \frac{c_2(n)}{2^3} e^{n\pi} + \frac{c_5(n)}{5^3} e^{2n\pi/5} + \cdots,$

(iii) $\quad \frac{1}{2} C v_n = \frac{c_1(n)}{1^2} e^{2n\pi} + \frac{c_2(n)}{2^2} e^{n\pi} + \frac{c_5(n)}{5^2} e^{2n\pi/5} + \cdots,$

where $c_1(n) = 1$, $c_2(n) = (-1)^n$, $c_5(n) = 2\cos(\frac{4n\pi}{5} + 8\arctan 2), \ldots$ and

$$\sum u_n x^n = \frac{\pi}{3} \frac{1 - 24 \left(\frac{x}{1-x} + \frac{2x^2}{1-x^2} + \frac{3x^3}{1-x^3} + \cdots \right)}{1 - 504 \left(\frac{1^5 x}{1-x} + \frac{2^5 x^2}{1-x^2} + \frac{3^5 x^3}{1-x^3} + \cdots \right)},$$

$$\sum v_n x^n = \frac{\pi^2}{9} \frac{\left\{ 1 - 24 \left(\frac{x}{1-x} + \frac{2x^2}{1-x^2} + \frac{3x^3}{1-x^3} + \cdots \right) \right\}^2}{1 - 504 \left(\frac{1^5 x}{1-x} + \frac{2^5 x^2}{1-x^2} + \frac{3^5 x^3}{1-x^3} + \cdots \right)}.$$

If n is a negative integer, then

(i) $\quad 0 = \frac{c_1(n)}{1^4} e^{2n\pi} + \frac{c_2(n)}{2^4} e^{n\pi} + \frac{c_5(n)}{5^4} e^{2n\pi/5} + \cdots,$

(ii) $\quad 0 = \frac{c_1(n)}{1^3} e^{2n\pi} + \frac{c_2(n)}{2^3} e^{n\pi} + \frac{c_5(n)}{5^3} e^{2n\pi/5} + \cdots,$

(iii) $\quad 0 = \frac{c_1(n)}{1^2} e^{2n\pi} + \frac{c_2(n)}{2^2} e^{n\pi} + \frac{c_5(n)}{5^2} e^{2n\pi/5} + \cdots.$

All these can be proved using the formula:
If $ad - bc = 1$, $q = \exp\left(i\pi \frac{c+id}{a+ib} \right)$, then

(I) $\quad \frac{\pi}{3} \left\{ 1 - 24 \left(\frac{q^2}{1-q^2} + \frac{2q^4}{1-q^4} + \frac{3q^6}{1-q^6} + \cdots \right) \right\} = a^2 + b^2.$

There is <u>only one</u> more formula like (I) which is given at the end of the next page. Besides (I) and (II) there is no more possibility.

$$\sum p_n x^n = \frac{1}{1 + 240 \left(\frac{1^3 x}{1-x} + \frac{2^3 x^2}{1-x^2} + \cdots \right)},$$

$$C = 1 + 504 \left(\frac{1^5}{e^{\pi\sqrt{3}} + 1} - \frac{2^5}{e^{2\pi\sqrt{3}} - 1} + \frac{3^5}{e^{3\pi\sqrt{3}} + 1} - \cdots \right).$$

CHAPTER 5. RAMANUJAN IS ILL

If n is 0 or a positive integer, then

(i) $\frac{1}{3}Cp_n = (-1)^n \left\{ \frac{c_1(n)}{1^3}e^{n\pi\sqrt{3}} + \frac{c_3(n)}{3^3}e^{n\pi/\sqrt{3}} + \frac{c_7(n)}{7^3}e^{n\pi\sqrt{3}/7} + \cdots \right\}$,

(ii) $\frac{1}{3}Cu_n = (-1)^n \left\{ \frac{c_1(n)}{1^2}e^{n\pi\sqrt{3}} + \frac{c_3(n)}{3^2}e^{n\pi/\sqrt{3}} + \frac{c_7(n)}{7^2}e^{n\pi\sqrt{3}/7} + \cdots \right\}$,

where $c_1(n) = 1$, $c_3(n) = -1$, ... and

$$\sum u_n x^n = \frac{\pi}{2\sqrt{3}} \frac{1 - 24\left(\frac{x}{1-x} + \frac{2x^2}{1-x^2} + \frac{3x^3}{1-x^3} + \cdots\right)}{1 + 240\left(\frac{1^3 x}{1-x} + \frac{2^3 x^2}{1-x^2} + \cdots\right)}.$$

If n is a negative integer then

(i) $0 = \left\{ \frac{c_1(n)}{1^3}e^{n\pi\sqrt{3}} + \frac{c_3(n)}{3^3}e^{n\pi/\sqrt{3}} + \frac{c_7(n)}{7^3}e^{n\pi\sqrt{3}/7} + \cdots \right\}$,

(ii) $0 = \left\{ \frac{c_1(n)}{1^2}e^{n\pi\sqrt{3}} + \frac{c_3(n)}{3^2}e^{n\pi/\sqrt{3}} + \frac{c_7(n)}{7^2}e^{n\pi\sqrt{3}/7} + \cdots \right\}$.

These are proved using the formula:

If $ad - bc = 1$, $q = \exp\left(i\pi \frac{c+\omega d}{a+\omega b}\right)$, $\omega = \frac{-1+i\sqrt{3}}{2}$, then

(II) $\frac{\pi}{2\sqrt{3}} \left\{ 1 - 24\left(\frac{q^2}{1-q^2} + \frac{2q^4}{1-q^4} + \frac{3q^6}{1-q^6} + \cdots\right) \right\}$

$= a^2 - ab + b^2$.

$$\sum p_n x^n = \frac{1}{1 - 504\left(\frac{1^5 x}{1-x} + \frac{2^5 x^2}{1-x^2} + \cdots\right)}$$

$p_0 =$ 1
$p_1 =$ 504
$p_2 =$ 270648
$p_3 =$ 144,912096
$p_4 =$ 77599,626552
$p_5 =$ 41,553943,041744
$p_6 =$ 22251,789971,649504
$p_7 =$ 11,915647,845248,387520
$p_8 =$ 6380,729991,419236,488504
$p_9 =$ 3,416827,666558,895485,479576
$p_{10} =$ 1829,682703,808504,464920,468048
$p_{11} =$ 979779,820147,442370,107345,764512
$p_{12} =$ 524,663917,940510,191509,934144,603104

$$524{,}663917{,}940510{,}190119{,}197271{,}938395.329$$
$$+\ 1390{,}736872{,}662028.140$$
$$+\ 2680.418$$
$$+\ .130$$
$$-\ .014$$
$$-\ .003$$
$$\overline{524{,}663917{,}940510{,}191509{,}934144{,}603104.000}$$

Commentary. This letter was published with the "lost notebook" [227, pp. 97–101].

From the opening sentence of the letter, it is evident that Ramanujan had written Hardy earlier letters from nursing homes that have not been preserved.

As usual, in the theory of elliptic functions,

$$(1') \qquad 216\left(\frac{\omega_1}{\pi}\right)^6 g_3 = 1 - 504 \sum_{n=1}^{\infty} \frac{n^5 q^n}{1-q^n}, \qquad |q| < 1,$$

where ω_1 is a fundamental period. In the notation of Ramanujan's paper [219], [226, pp. 136–162], and in the notation of his notebooks [225], the right side of $(1')$ is denoted by R and N, respectively.

In (1), Ramanujan states, in different notation, the primary theorem of his paper with Hardy [125], [226, pp. 310–321], [105, pp. 294–305]; see Theorem 3 and the beginning of §7. In particular, $\alpha_n = p_n$, in the notation of [125].

The theorem of Landau quoted by Ramanujan in (4) is the same theorem communicated on page (3) of Ramanujan's first letter to Hardy.

The remainder of Ramanujan's argument is self-contained.

The curious arithmetical observation set off by horizontal bars was recorded in both Ramanujan's second and third notebooks, pages 309, 363 in the pagination of [225].

Following the letter given above, we have reproduced a three-page fragment which is found in the collection at Trinity College, Cambridge, and in the publication of the "lost notebook" [227, pp. 102–104] immediately after the letter. These three pages were obviously written on a later date, for, in contrast to the letter, the notation coincides with that in the published paper [125]. The first two pages give analogues of the main theorem in [125], and it is surprising that these results are not even mentioned in [125]. The calculation comprising the third page, however, concludes the paper [125].

S. Ramanujan to G.H. Hardy   28 June 1918

Dear Mr. Hardy,

I sent you the proofs. In (1.16) you say that the partition method gives exact values only in the case of 4, 6 and 8 squares. What about 3, 5, 7 squares? Are they not exact? I am sending you the analogous results in case of g_2. Please mention them in the paper without proof. After all we have got only <u>two neat</u> examples to offer, viz. g_2 and g_3. So please don't omit the results.

You can write many things at the foot notes instead of the body of the paper so that they will occupy very much less space. I made a slight mistake in the proof of the least number of terms; instead of writing

$$2\cos\left(\frac{2kn\pi}{p} + 8\arctan\theta\right), 2\cos\left(\frac{2k(n-1)\pi}{p} + 8\arctan\theta\right),$$
$$2\cos\left(\frac{2k(n-2)\pi}{p} + 8\arctan\theta\right), \ldots$$

where k is a positive integer less than p, I omitted k in my letter. I don't think the arguments are affected. Are they?

Recently I received my scholarship from the Madras University. I am sending you with this letter a cheque for £100 for the present.

Ever Yours
S. Ramanujan

The result in the case of g_3 is

$$C = 1 + 480\left(\frac{1^7}{e^{2\pi}-1} + \frac{2^7}{e^{4\pi}-1} + \cdots\right),$$

$$ad - bc = 1, \qquad \mathbf{q} = \exp\left(i\pi\frac{c+id}{a+ib}\right),$$

$$\frac{\frac{1}{2}C}{1 - 504\left(\frac{1^5 q^2}{1-q^2} + \frac{2^5 q^4}{1-q^4} + \cdots\right)} = \frac{1}{1-q^2 e^{2\pi}} + \frac{1}{2^4}\frac{1}{1+q^2 e^\pi} + \cdots$$
$$= \sum \frac{1}{(a+ib)^8}\frac{1}{1-q^2/\mathbf{q}^2}.$$

The analogous result in the case g_2 is

$$C = 1 + 504\left(\frac{1^5}{e^{\pi\sqrt{3}}+1} - \frac{2^5}{e^{2\pi\sqrt{3}}-1} + \frac{3^5}{e^{3\pi\sqrt{3}}-1} - \cdots\right),$$

$$ad - bc = 1, \qquad \omega = \frac{-1+i\sqrt{3}}{2}, \qquad \mathbf{q} = \exp\left(i\pi\frac{c+\omega d}{a+\omega b}\right).$$

(1) $$\frac{\frac{1}{3}C}{1 + 240\left(\frac{1^3 q^2}{1-q^2} + \frac{2^3 q^4}{1-q^4} + \cdots\right)} = \frac{1}{1+q^2 e^{\pi\sqrt{3}}} - \frac{1}{3^3}\frac{1}{1+q^2 e^{\pi/\sqrt{3}}} + \cdots$$
$$= \sum \frac{1}{(a+\omega b)^6}\frac{1}{1-q^2/\mathbf{q}^2}.$$

$$C = \frac{32}{11}\left\{1 - 504\left(\frac{1^5}{e^{2\pi\sqrt{3}}-1} + \frac{2^5}{e^{4\pi\sqrt{3}}-1} + \frac{3^5}{e^{6\pi\sqrt{3}}-1} + \cdots\right)\right\}$$

$$= (2\sqrt{3})^3 e^{-\pi\sqrt{3}/2}\left\{(1+e^{-\pi\sqrt{3}})(1-e^{-2\pi\sqrt{3}})(1+e^{-3\pi\sqrt{3}})\cdots\right\}^{12}$$

$$= 2(4\sqrt{3})^3 e^{-\pi\sqrt{3}}\left\{(1-e^{-2\pi\sqrt{3}})(1-e^{-4\pi\sqrt{3}})\cdots\right\}^{12}$$

$$= \left(\frac{\pi}{3}\right)^{3/2}\frac{2^3}{\{\Gamma(\frac{5}{6})\}^9}$$

$$= \left(\frac{\pi}{3}\right)^6 \frac{2^9}{\{\Gamma(\frac{2}{3})\}^{18}}.$$

$$\sum p_n x^n = \frac{1}{1 + 240\left(\frac{1^3 x}{1-x} + \frac{2^3 x^2}{1-x^2} + \cdots\right)}.$$

(2)
$$\frac{1}{3}Cp_n = (-1)^n\left\{c_1 e^{n\pi\sqrt{3}} + \frac{c_3}{3^3}e^{n\pi/\sqrt{3}} + \frac{c_7}{7^3}e^{n\pi\sqrt{3}/7} + \frac{c_{13}}{(13)^3}e^{n\pi\sqrt{3}/13} + \cdots\right\}$$

$$= (-1)^n \sum \frac{c_\lambda}{\lambda^3}e^{n\pi\sqrt{3}/\lambda},$$

where λ is a number of the form $a^2 + ab + b^2$, a and b being prime to each other. That is to say, λ is of the form

$$3^{a_3} \cdot 7^{a_7} \cdot 13^{a_{13}} \cdot 19^{a_{19}} \cdots,$$

where $a_3 = 0$ or 1 and $a_7, a_{13}, \ldots$ are zeros or positive integers, 7, 13, 19, ... being primes of the form $6k + 1$. The first 20 values are

$$1, 3, 7, 13, 19, 21, 31, 37, 39, 43,$$
$$49, 57, 61, 67, 73, 79, 91, 93, 97, 103.$$

$$c_1 = 1, \; c_3 = -1, \; c_7 = 2\cos\left(\frac{2n\pi}{7} + 6\arctan 3\sqrt{3}\right),$$

$$c_{13} = 2\cos\left(\frac{6n\pi}{13} + 6\arctan 2\sqrt{3}\right),$$

$$c_{19} = 2\cos\left(\frac{4n\pi}{19} - 6\arctan 5\sqrt{3}\right),$$

$$c_{21} = -2\cos\left(\frac{4n\pi}{7} + 6\arctan 3\sqrt{3}\right).$$

If n is a negative integer the right hand side of (2) is <u>zero</u>. If $|q| \geq 1$, the right hand side of (1) is <u>zero</u>. The proofs for these statements are exactly similar to those for g_3.

CHAPTER 5. RAMANUJAN IS ILL

Commentary. A facsimile of the original letter has been published with the "lost notebook" [**227**, pp. 113-115].

In the opening sentence, Ramanujan refers to his joint paper with Hardy [**125**]. In particular, Ramanujan questions the following assertion made in the paragraph underneath (1.6): "Thus, the sum of the series (1.6) is equal to $r_s(n)$ if s is 4, 6, or 8, but not in any other case." Indeed, finding a formula for odd s is considerably more difficult, and this was accomplished by Hardy in his paper [**96**], [**105**, pp. 345-374].

Ramanujan's plea that his result for g_2 be at least mentioned in their paper [**125**] was not heeded, evidently because Hardy felt that the required alterations were too substantial to be made in galley proofs.

See problems 3 and 4 on page 176 of [**177**] for a characterization of those positive integers representable in the form $a^2 + ab + b^2$.

S. Ramanujan to G.H. Hardy

Matlock House
Monday

Dear Mr. Hardy,

Will my letters be forwarded to you during the long vacation if I write to you to the Trinity College?

I considered another problem connected with g_3. Suppose that

$$\sum p_n x^n = \frac{1}{1 - 504\left(\frac{1^5 x}{1-x} + \frac{2^5 x^2}{1-x^2} + \cdots\right)}$$

and that

$$p_n = T_1(n) + T_2(n) + T_5(n) + T_{10}(n) + \cdots,$$

where the T's are the terms in the asymptotic expansion. Now $T_1(n) + T_2(n) + T_5(n) + \cdots$ is an integer for integral values of n. The question arises: If we separate the odd and the even suffixes and consider the two separate series

$$T_1(n) + T_5(n) + T_{13}(n) + \cdots$$

and

$$T_2(n) + T_{10}(n) + T_{26}(n) + \cdots,$$

what is the nature of the values of each of these series? I did not succeed in solving this problem but at any rate I have arrived at the following result which follows from the identities I wrote to you in my last letter.

I have already written to you that if

$$\sum a_n x^n = \frac{1 - 24\left(\frac{x}{1-x} + \frac{2x^2}{1-x^2} + \cdots\right)}{1 - 504\left(\frac{1^5 x}{1-x} + \frac{2^5 x^2}{1-x^2} + \cdots\right)}$$

then

$$a_n = \frac{3}{\pi}\left(T_1(n) + 2T_2(n) + 5T_5(n) + 10T_{10}(n) + \cdots\right).$$

Suppose now that
$$\sum b_n x^n = \frac{1}{\left\{1 + 24\left(\frac{x}{1-x} + \frac{3x^3}{1-x^3} + \cdots\right)\right\}^2};$$
then
$$b_n = 18\left\{\left(n + \frac{3}{\pi}\right)T_2(n) + \left(n + \frac{15}{\pi}\right)T_{10}(n) + \left(n + \frac{39}{\pi}\right)T_{26}(n) + \cdots\right\}.$$

It follows that
$$18\left\{\left(n + \frac{3}{2\pi}\right)T_1(n) + \left(n + \frac{15}{2\pi}\right)T_5(n) + \left(n + \frac{39}{2\pi}\right)T_{13}(n) + \cdots\right\}$$
is also an integer as we already know that $T_1(n) + T_2(n) + \cdots$ is an integer and $\frac{3}{\pi}\{T_1(n) + 2T_2(n) + 5T_5(n) + \ldots\}$ is an integer and consequently
$$2\left\{\left(n + \frac{3}{2\pi}\right)T_1(n) + \left(n + \frac{3}{\pi}\right)T_2(n)\right.$$
$$\left. + \left(n + \frac{15}{2\pi}\right)T_5(n) + \left(n + \frac{15}{\pi}\right)T_{10}(n) + \cdots\right\}$$
is an integer. Hence the result.

I am now considering another problem I told you viz., the expression of the various powers of $\Gamma(\frac{1}{4})$ and $\Gamma(\frac{1}{6})$ as products in primes just like the powers of $\Gamma(\frac{1}{2})$ i.e. the powers of π. For instance we know that

$$\pi^2\left(1 - \frac{1}{2^2}\right)\left(1 - \frac{1}{3^2}\right)\left(1 - \frac{1}{5^2}\right)\left(1 - \frac{1}{7^2}\right)\cdots = 6,$$

$$\pi^4\left(1 - \frac{1}{2^4}\right)\left(1 - \frac{1}{3^4}\right)\left(1 - \frac{1}{5^4}\right)\left(1 - \frac{1}{7^4}\right)\cdots = 90,$$

$$\pi^6\left(1 - \frac{1}{2^6}\right)\left(1 - \frac{1}{3^6}\right)\left(1 - \frac{1}{5^6}\right)\cdots = 945,$$

$$\pi^8\left(1 - \frac{1}{2^8}\right)\left(1 - \frac{1}{3^8}\right)\left(1 - \frac{1}{5^8}\right)\cdots = 9450,$$

$$\pi^{10}\left(1 - \frac{1}{2^{10}}\right)\left(1 - \frac{1}{3^{10}}\right)\left(1 - \frac{1}{5^{10}}\right)\cdots = 93555,$$

and so on, the right-hand side being a rational number. I find that

$$\left\{\frac{\Gamma(\frac{1}{4})}{2\pi^{1/4}}\right\}^4 \left(1 - \frac{1}{3^2}\right)\left\{\left(1 - \frac{1}{(1+2i)^2}\right)\left(1 - \frac{1}{(1-2i)^2}\right)\right\}$$
$$\times \left(1 - \frac{1}{7^2}\right)\left(1 - \frac{1}{11^2}\right)\left\{\left(1 - \frac{1}{(3+2i)^2}\right)\left(1 - \frac{1}{(3-2i)^2}\right)\right\}$$
$$\times \left\{\left(1 - \frac{1}{(1+4i)^2}\right)\left(1 - \frac{1}{(1-4i)^2}\right)\right\}\left(1 - \frac{1}{19^2}\right)\cdots = 4.$$

CHAPTER 5. RAMANUJAN IS ILL

Here we take the odd primes 3, 5, 7, 11, 13, 17, 19, ... and retain the primes of the form $4k - 1$ viz. 3, 7, 11, 19, ... and instead of the primes of the form $4k + 1$ viz. 5, 13, 17, ... we write the conjugate factors $1 + 2i$ and $1 - 2i$; $3 + 2i$ and $3 - 2i$; $1 + 4i$ and $1 - 4i$ and so on. It should be remembered that the odd number must be the real part and the even number the imaginary one in the complex factors. For instance, $29 = 2^2 + 5^2$ and $37 = 1^2 + 6^2$. The conjugate factors corresponding to 29 and 37 must be written as $5 + 2i$ and $5 - 2i$; $1 + 6i$ and $1 - 6i$ and not as $2 + 5i$ and $2 - 5i$.

With the same convention I find that

$$\left\{\frac{\Gamma(\frac{1}{4})}{2\pi^{1/4}}\right\}^8 \left(1 - \frac{1}{3^4}\right) \left\{\left(1 - \frac{1}{(1+2i)^4}\right)\left(1 - \frac{1}{(1-2i)^4}\right)\right\}$$
$$\times \left(1 - \frac{1}{7^4}\right)\left(1 - \frac{1}{(11)^4}\right) \cdots = 12;$$

$$\left\{\frac{\Gamma(\frac{1}{4})}{2\pi^{1/4}}\right\}^{12} \left(1 - \frac{1}{3^6}\right) \left\{\left(1 - \frac{1}{(1+2i)^6}\right)\left(1 - \frac{1}{(1-2i)^6}\right)\right\}$$
$$\times \left(1 - \frac{1}{7^6}\right)\left(1 - \frac{1}{(11)^6}\right) \cdots = 40;$$

$$\left\{\frac{\Gamma(\frac{1}{4})}{2\pi^{1/4}}\right\}^{16} \left(1 - \frac{1}{3^8}\right) \left\{\left(1 - \frac{1}{(1+2i)^8}\right)\left(1 - \frac{1}{(1-2i)^8}\right)\right\}$$
$$\times \left(1 - \frac{1}{7^8}\right)\left(1 - \frac{1}{(11)^8}\right) \cdots = 140;$$

$$\left\{\frac{\Gamma(\frac{1}{4})}{2\pi^{1/4}}\right\}^{20} \left(1 - \frac{1}{3^{10}}\right) \left\{\left(1 - \frac{1}{(1+2i)^{10}}\right)\left(1 - \frac{1}{(1-2i)^{10}}\right)\right\}$$
$$\times \left(1 - \frac{1}{7^{10}}\right)\left(1 - \frac{1}{(11)^{10}}\right) \cdots = 480;$$

$$\left\{\frac{\Gamma(\frac{1}{4})}{2\pi^{1/4}}\right\}^{24} \left(1 - \frac{1}{3^{12}}\right) \left\{\left(1 - \frac{1}{(1+2i)^{12}}\right)\left(1 - \frac{1}{(1-2i)^{12}}\right)\right\}$$
$$\times \left(1 - \frac{1}{7^{12}}\right)\left(1 - \frac{1}{(11)^{12}}\right) \cdots = 1650;$$

and so on. I am confident that I will certainly get similar results for the powers of $\Gamma(\frac{1}{6})$ and $\Gamma(\frac{1}{6})$ is substantially $\Gamma(\frac{1}{3})$ since

$$\Gamma\left(\frac{1}{6}\right)\sqrt{\frac{\pi}{3}}\sqrt[3]{2} = \left\{\Gamma\left(\frac{1}{3}\right)\right\}^2.$$

But I am not sure whether I'll succeed in dealing with the powers of $\Gamma(\frac{1}{8})$ or not. Could you tell me if any such results were known before?

Yours ever
S. Ramanujan

Commentary. A facsimile of the original letter is printed with the publication of the "lost notebook" [227, pp. 105–109]. Following the letter in [227], three pages from the "lost notebook" pertaining to the content of the second part of the letter are reproduced.

Ramanujan does not define $T_j(n)$, $1 \leq j < \infty$. However, $T_j(n)$ is clearly the expression involving $e^{2n\pi/j}$ in (1) of his first letter to Hardy from Matlock House. The claim "$T_1(n) + T_2(n) + T_5(n) + \cdots$ is an integer for integral values of n," as well as similar statements, must be reinterpreted, for the series is a divergent asymptotic series. More precisely, we should say that $T_1(n) + T_2(n) + T_5(n) + \cdots$ is an asymptotic representation of the integer p_n.

The generating function and asymptotic series for a_n, $n \geq 0$, are found in the fragment after the first letter from Matlock House. In the notation of that fragment, $\frac{3}{\pi}u_n = a_n$. On the other hand, the generating function and asymptotic expansion for b_n, $n \geq 0$, are not found in this fragment. It might be noted that

$$1 + 24\sum_{n=0}^{\infty} \frac{(2n+1)x^{2n+1}}{1-x^{2n+1}} = 2L(x^2) - L(x),$$

where

$$L(x) := 1 - 24\sum_{n=1}^{\infty} \frac{nx^n}{1-x^n}.$$

Ramanujan concludes page 2 and the first part of his letter with "Hence the result." This is somewhat mysterious, for he did not indicate earlier in the letter that this result was his goal.

Ramanujan commences the second portion of his letter by quoting the values of $\zeta(2n)$, $1 \leq n \leq 5$. The remainder of the letter has been thoroughly discussed by K.G. Ramanathan [197]. In particular, Ramanujan offers six values of a Hecke zeta-function attached to a certain Grössencharakter. The "similar results for the powers of $\Gamma(\frac{1}{6})$" are established in Ramanathan's paper [197]. Ramanathan also confirms the skepticism expressed in the penultimate sentence of Ramanujan's letter. However, a certain Hecke zeta-function evaluated at positive integral arguments $n \equiv 0 \pmod 4$ can be evaluated in terms of powers of $\Gamma(\frac{1}{8})\Gamma(\frac{3}{8})$.

S. Ramanujan to G.H. Hardy

Matlock House
Tuesday

Dear Mr. Hardy,

I have been unwell since you saw me except for 2 or 3 days. It is only today I feel somewhat better.

I considered about g_2, g_3. It presents some interesting features. In the first place it is easy to prove the following identities.

Suppose that

$$\phi(x) = 1 + 24\left(\frac{x}{1-x} + \frac{3x^3}{1-x^3} + \frac{5x^5}{1-x^5} + \cdots\right).$$

Then

(1) $$\phi(\sqrt{x}) + \phi(-\sqrt{x}) = 2\phi(x),$$

(2) $$\phi(\sqrt{x})\phi(-\sqrt{x})\phi(x) = 1 - 504\left(\frac{1^5 x}{1-x} + \frac{2^5 x^2}{1-x^2} + \cdots\right),$$

(3) $$5\{\phi(x)\}^2 = 1 + 240\left(\frac{1^3 x}{1-x} + \frac{2^3 x^2}{1-x^2} + \cdots\right) \\ + 4\left\{1 + 240\left(\frac{1^3 x^2}{1-x^2} + \frac{2^3 x^4}{1-x^4} + \cdots\right)\right\},$$

(4) $$7\{\phi(x)\}^3 = -\left\{1 - 504\left(\frac{1^5 x}{1-x} + \frac{2^5 x^2}{1-x^2} + \cdots\right)\right\} \\ + 8\left\{1 - 504\left(\frac{1^5 x^2}{1-x^2} + \frac{2^5 x^4}{1-x^4} + \cdots\right)\right\},$$

(5) $$\frac{2}{3}\left\{\frac{1}{\phi(\sqrt{x})} + \frac{1}{\phi(-\sqrt{x})}\right\} - \frac{1}{3\phi(x)} = \frac{1 + 240\left(\frac{1^3 x}{1-x} + \frac{2^3 x^2}{1-x^2} + \cdots\right)}{1 - 504\left(\frac{1^5 x}{1-x} + \frac{2^5 x^2}{1-x^2} + \cdots\right)},$$

(6) $$\frac{11}{24}\left\{\frac{1}{\phi(\sqrt{x})} + \frac{1}{\phi(-\sqrt{x})}\right\} + \frac{1}{12\phi(x)} = \frac{1 + 240\left(\frac{1^3 x^2}{1-x^2} + \frac{2^3 x^4}{1-x^4} + \cdots\right)}{1 - 504\left(\frac{1^5 x}{1-x} + \frac{2^5 x^2}{1-x^2} + \cdots\right)}.$$

Also we have

(7)
$$15(1 - 2x + 2x^4 - 2x^9 + \cdots)^8 = -\left\{1 + 240\left(\frac{1^3 x}{1-x} + \frac{2^3 x^2}{1-x^2} + \cdots\right)\right\}$$
$$+ 16\left\{1 + 240\left(\frac{1^3 x^2}{1-x^2} + \frac{2^3 x^4}{1-x^4} + \cdots\right)\right\},$$

(8)
$$240x(1 + x + x^3 + x^6 + \cdots)^8 = \left\{1 + 240\left(\frac{1^3 x}{1-x} + \frac{2^3 x^2}{1-x^2} + \cdots\right)\right\}$$
$$- \left\{1 + 240\left(\frac{1^3 x^2}{1-x^2} + \frac{2^3 x^4}{1-x^4} + \cdots\right)\right\}.$$

Suppose now that
$$\sum a_n x^n = \frac{1 + 240\left(\frac{1^3 x}{1-x} + \frac{2^3 x^2}{1-x^2} + \cdots\right)}{1 - 504\left(\frac{1^5 x}{1-x} + \frac{2^5 x^2}{1-x^2} + \cdots\right)}.$$

Then
$$a_n = \frac{2}{C}\left\{e^{2n\pi} - \frac{(-1)^n}{2^2}e^{n\pi} + \frac{2\cos\left(\frac{4n\pi}{5} + 4\arctan 2\right)}{5^2}e^{2n\pi/5} + \cdots\right\}$$
$$= \frac{2}{C}\sum \frac{1}{(a+ib)^4}\exp\left(-2in\pi\frac{c+id}{a+ib}\right),$$
where $ad - bc = 1$ and
$$C = 1 + 240\left(\frac{1^3}{e^{2\pi} - 1} + \frac{2^3}{e^{4\pi} - 1} + \cdots\right).$$

Say then
$$a_n = T_1 + T_2 + T_5 + T_{10} + T_{13} + \cdots,$$
where $T_1, T_2, T_5, \ldots$ are the terms in the asymptotic series. The following interesting identities follow from the identities (1)–(8).
If
$$\sum b_n x^n = \frac{1}{1 + 24\left(\frac{x}{1-x} + \frac{3x^3}{1-x^3} + \frac{5x^5}{1-x^5} + \cdots\right)},$$
then
$$b_n = -3(T_2 + T_{10} + T_{26} + \cdots),$$
where 2, 10, 26, etc. are the even suffixes occurring in $a_n = T_1 + T_2 + T_5 + T_{10} + \cdots$.
If
$$\sum g_n x^n = \frac{1 + 240\left(\frac{1^3 x^2}{1-x^2} + \frac{2^3 x^4}{1-x^4} + \cdots\right)}{1 - 504\left(\frac{1^5 x}{1-x} + \frac{2^5 x^2}{1-x^2} + \cdots\right)},$$
then

CHAPTER 5. RAMANUJAN IS ILL

$$g_n = \frac{11}{16}(T_1 + T_5 + T_{13} + \cdots) - \frac{1}{4}(T_2 + T_{10} + T_{26} + \cdots).$$

If

$$\sum k_n x^n = \frac{(1 - 2x + 2x^4 - \cdots)^8}{1 - 504\left(\frac{1^5 x}{1-x} + \frac{2^5 x^2}{1-x^2} + \cdots\right)},$$

then

$$k_n = \frac{2}{3}(T_1 + T_5 + T_{13} + \cdots) - \frac{1}{3}(T_2 + T_{10} + \cdots).$$

If

$$\sum \ell_n x^n = \frac{x(1 + x + x^3 + \cdots)^8}{1 - 504\left(\frac{1^5 x}{1-x} + \frac{2^5 x^2}{1-x^2} + \cdots\right)},$$

then

$$\ell_n = \frac{1}{768}(T_1 + T_5 + T_{13} + \cdots) + \frac{1}{192}(T_2 + T_{10} + \cdots).$$

If n is a negative integer then

$$T_1 + T_2 + T_5 + T_{10} + T_{13} + \cdots = 0;$$

$$T_1 + T_5 + T_{13} + \cdots = 0; \qquad T_2 + T_{10} + T_{26} + \cdots = 0.$$

Of all these the coefficients in

$$\frac{1}{1 + 24\left(\frac{x}{1-x} + \frac{3x^3}{1-x^3} + \cdots\right)} = \sum b_n x^n$$

are the most interesting and the numerical approximations must be more strikingly remarkable than in $1/g_2 \sim 1/g_3$ problems since $b_n = -3(T_2 + T_{10} + T_{26} + \cdots)$ and T_{10}, T_{26} are very small.

Ever yours
S. Ramanujan

Commentary. The original letter has been published in facsimile along with the "lost notebook" [**227**, pp. 116–118]. Pages 119–120 of [**227**] offer two pages of related results from the "lost notebook."

Note that the coefficients a_n and b_n do not have the same meanings as in the previous letter.

At the end of [**125**], the authors promise "But we must reserve the discussion of these peculiarities for some other occasion." Ramanujan's illness and imminent death precluded the writing of this promised paper, and so the results described in Ramanujan's four letters from Matlock House were never published. Several years later, in 1940, H.S. Zuckerman [**302**] extended the methods developed by Hardy and Ramanujan [**125**], but he did not address any of the contents of these four letters.

Most of the claims made by Ramanujan in the four letters above have been proved by P. Bialek in his doctoral dissertation [**46**].

S. Ramanujan to G.H. Hardy

> Fitzroy House
> 16 Fitzroy Square
> London
> Friday

Dear Mr Hardy,

My heartfelt thanks for your kind telegram. After you succeeded in getting me elected by the Royal Society my election at Trinity probably became very much less difficult this year.

My tooth extraction was to be done this morning. But the dentist was not able to come as he was indisposed. I have not been very well since you saw me and my temperature has been as irregular and high as [it] was at Matlock. Dr Bolton thinks that my feverish attacks and my rheumatic pain are both due to my teeth. His idea explains very well the rheumatic pain and no other doctor gave any cause for the pain. But I can't see any connection between the teeth and the feverish attacks which reduce my body and which existed long before anything was wrong in my teeth. I shall just remove one or two teeth for the present and I do not want to trouble with the other ones now.

> Yours ever,
> S. Ramanujan

Commentary. Ramanujan's election to his Fellowship at Trinity was not as straightforward as he imagined; see page 136 of [48]. The election took place on Thursday, 10 October 1918, so that Ramanujan's letter to Hardy would have been written on the following day.

S. Ramanujan to G.H. Hardy

> Fitzroy House
> 16. Fitzroy Square
> Monday

Dear Mr. Hardy,

Please tell Mr Littlewood and Major MacMahon that I thanked them very much. Had it not been for your pains and their encouragement I would be neither the fellow of the one nor that of the other. Have the tutors and the Master of Trinity more voice in the election than the other fellows? I heard that in some colleges that there are two kinds of fellowships: one lasting for two or three years and the other for five or six years; if that is so in Trinity, is mine the first or the second kind?

I have considered more or less exhaustively about the congruency of $p(n)$ and in general that of $p_r(n)$ where

$$\sum p_r(n)x^n = \frac{1}{\left\{(1-x)(1-x^2)(1-x^3)\cdots\right\}^r},$$

by four different methods. Each method has its own advantages. In the case of $p(n)$ the results are roughly these!

I method—Very special and very simple. Gives only
$$p(5n+4) \equiv 0 (\mathrm{mod}\, 5), \qquad p(7n+5) \equiv 0 (\mathrm{mod}\, 7).$$
This you are publishing now.

II ϑ-function method. This is general theoretically but becomes practically unworkable due to tediousness of calculation with the exception of few cases. I have simplified the proof very much and made it quite elementary as the first method. From this the divisibility by $5, 5^2, 5^3$ and $7, 7^2, 7^3$ is established quite easily. I have not considered the case of 11 due to tediousness. This method gives many more results like (17) and (18) in the proof sheet.

III Dirichlet's series method. This is very general and practically workable. The conjectured products of Dirichlet series corresponding to $\left\{(1-x)(1-x^2)(1-x^3)\cdots\right\}^n$ when $n = 2, 4, 6, 8, 12, 26$ were proved by Mr Mordell. I have now conjectured analogous results for some more values of n viz. 10, 14, 16, 18, 20, 22, 28, 30, 32, 36, 48. All these may be proved by Mr Mordell. If we assume these results this method gives information about the divisibility of $p(n)$ by
$$5, 5^2, 7, 7^2, 11, 13, 17, 19, 23, 29, 31 \text{ and } 37.$$

IV g_2, g_3 method. This is very general. This gives all the previous results except the case of $5^3, 7^3$, and the divisibility by 11^2, as well. Thus the divisibility by $5^a 7^b 11^c$ when $a = 0, 1, 2, 3; b = 0, 1, 2, 3; c = 0, 1, 2$ amounting to $4 \times 4 \times 3 - 1$ or 47 cases of the conjectured theorem are proved.

Ever yours
S. Ramanujan

Commentary. A photocopy of the original letter may be found in [227, pp. 121–122].

Ramanujan was made a Fellow of Trinity College in the fall of 1918; he was the first Indian to be so elected. The second Indian to be elected was S. Chandrasekhar in 1933.

Ramanujan evidently left Matlock House sometime between July 1918 (since one of the four letters from Matlock House is dated 28 June 1918) and early October 1918 (since Ramanujan was elected as a Fellow of Trinity College on 13 October 1918). On 2 May 1918, he had been chosen as a Fellow of the Royal Society. His Trinity Fellowship was for a period of six years, as is made clear by Hardy's letter to Francis Dewsbury of 26 November 1918. See Kanigel's book [142, pp. 291–300] for more details on Ramanujan's elections to these fellowships.

For proofs of general congruences satisfied by $p_r(n)$, see M.I. Knopp's book [146].

The two congruences mentioned under Method I were proved in a very simple, elegant manner by Ramanujan in [222], [226, pp. 210–213]. The proofs were reproduced by Hardy in his book on Ramanujan's work [110, pp. 87–90]. For further discussion of Ramanujan's conjectures on congruences satisfied by $p(n)$, see our commentary following D.H. Lehmer's letter of 2 November 1937 to Hardy.

It is not certain what the "theta–function method" entails. If it involves modular equations and identities for products of eta–functions and theta–functions, then the method has been quite successful, for Watson [291] and A.O.L. Atkin [18] employed such ideas in proving Ramanujan's general conjectures modulo powers of 5, 7, and 11. See also Knopp's book [146].

In the third method which utilizes Dirichlet series, one considers positive integral powers of the Dedekind eta–function, which is defined by

$$\eta(x) = x^{\frac{1}{24}} f(x) = x^{\frac{1}{24}} \prod_{m=1}^{\infty} (1 - x^m) \qquad (x = e^{2\pi i z}, \quad \text{Im } z > 0).$$

When n is a positive integral divisor of 24, L.J. Mordell [168] proved that the coefficients of η^n are multiplicative, i.e., he proved that in each of these cases the associated Dirichlet series has an Euler product. See also Rankin's paper [236] for a discussion of Mordell's method. In the letter, Ramanujan appears to assert that these properties continue to hold for the ten values of n between 10 and 48 that he lists. This is not strictly the case. For example, when $n = 48$, $\eta^{48} = \Delta^2$, the square of the discriminant function Δ, and no Euler product exists.

However, Ramanujan's brief statement in the letter may only be a rough approximation to what he knew, then or later, to be the case, namely that Euler products exist, but involve other functions, such as Eisenstein series E_k, in addition to powers of η. This is corroborated by what B.J. Birch has called Fragment V, an unpublished manuscript of Ramanujan described by Birch [47]. This manuscript was transcribed by G.N. Watson and is found in Watson's copy of the second notebook; the original no longer exists. Watson's copy was reproduced with the publication of the "lost notebook" [227, pp. 233–237]. Fragment V consists of a list of over two dozen functions having Euler products. Their weights k satisfy $k = n/2 \leq 17$, and each function is of the form

$$a\eta^n + b\eta^m E_{(n-m)/2},$$

for certain positive integral values $m < n$. The coefficient b is zero only for $n = 2, 4, 6$, and 8, i.e., for values of n not in Ramanujan's second list. This demonstrates Ramanujan's remarkable insight into a part of the theory of modular forms only rigorously established nearly twenty years later through the works of E. Hecke. For proofs of the formulas in Fragment V, see [228] and [192].

Further corroboration comes from Ramanujan's unpublished manuscript reproduced on pages 135–177 and 238–243 of [227]; in the sequel, we refer to this as M. A small part of M was published by Hardy [224], [226, pp. 232–238]. The methods described in this manuscript are almost certainly those used by Ramanujan to derive the congruences for the partition function $p(m)$ mentioned in paragraph III of his letter.

Very roughly, his method is to take a fixed prime $q > 3$ and write $q^2 - 1 = 24r$, so that r is an integer. Then

$$\{f(x)\}^{q^2} = f(y) + qJ,$$

where $y = x^{q^2}$ and J is a power series in q with integral coefficients. The coefficients of $f(y)$ were found by Euler (the Euler pentagonal number theorem), and those of $1/f(x)$ are the partition function $p(m)$. Thus,

$$x^r \{f(x)\}^{24r} \equiv x^r f(y)/f(x) \pmod{q}.$$

For certain small values of r, such as $r = 1, 2, 5$ (corresponding to $p = 5, 7, 11$), the coefficient $a(m)$ of x^m on the left-hand side is congruent modulo q to a function of the form $m\,d(m)$, where $d(m)$ is a combination of simple divisor functions and nonnegative powers of m with integral coefficients. E.g., for $r = 1$, the coefficient $a(m)$ is Ramanujan's function $\tau(m)$, and the congruence is

$$\tau(m) \equiv m\sigma_1(m) \pmod{5},$$

given by Ramanujan [227, p. 136]. In all cases, it follows that the coefficient of x^{mq} is divisible by q, and from this we deduce, by examining the right-hand side, that

$$p(mq - r) \equiv 0 \pmod{q}.$$

Observe that, for $r = 2$ and 5, $a(m)$ is not multiplicative.

In the examples above, the modular forms are Δ, Δ^2, and Δ^5, respectively, but Ramanujan also considers cases when the form is a product of E_k (for $k = 2$ and 4) and a power of Δ, and he was aware that the Fourier coefficients of these forms need not be multiplicative. Thus, in §17 of M, he mentions, to use more modern terminology, that the cusp forms $E_4\Delta^2$ and $E_6\Delta^2$ are not newforms, but each is expressible as a difference of two newforms. This is certainly the case, since in each case the dimension of the relevant space of cusp forms is 2.

The congruence for $\tau(m)$ modulo 5 given above is only one example of many other congruences for cusp form coefficients in terms of divisor functions found by Ramanujan; see, for example, §2 of [231] and page 92 of [233].

It is not clear to what extent Ramanujan's method IV overlaps method III and whether it involves the methods of M. Certainly g_2 and g_3 appear nowhere in M in that notation, although they do occur *passim* (together with P) as Q and R.

The two parts of M were first published in 1988 [227]. However, J.M. Rushforth, who was a research student of Watson, wrote his doctoral thesis [248] and one paper [249] on results appearing in M, which was then in his supervisor's possession. For a description of the manuscript, see [231] and §4.7 of [235].

Ramanujan's general conjecture on the congruences of $p(n)$, and further discussion, can be found in the commentary after D. H. Lehmer's letter of 2 November 1937 to Hardy.

S. Ramanujan to G.H. Hardy

Fitzroy House
Saturday

Dear Mr. Hardy,

You told me that you can prove at present that all numbers can be expressed as the sum of 33 fourth powers. I wonder whether you can get a better result by starting with the following function instead of $1 + 2x + 2x^{16} + 2x^{81} + \cdots$.

We have

$$\sum_{-\infty}^{\infty} x^{\mu^2 + \mu\nu + \nu^2} = 1 + 6\left(\frac{x}{1-x} - \frac{x^2}{1-x^2} + \frac{x^4}{1-x^4} - \frac{x^5}{1-x^5} + \cdots\right).$$

It follows that if $\phi(x) = x + x^4 + x^9 + x^{16} + \cdots$ then

$$\sum_{-\infty}^{\infty} x^{(\mu^2 + \mu\nu + \nu^2)^2}$$
$$= 1 + 6\left\{\phi(x) - \phi(x^4) + \phi(x^{16}) - \phi(x^{25}) + \phi(x^{49}) - \phi(x^{64}) + \cdots\right\}$$

where 1, 4, 16, 25 are the squares of natural numbers without the multiples of 3. Suppose now that you can prove that for a certain value of s all the coefficients after a certain point in the expansion of

(1) $$\left\{1 + 6\left(\phi(x) - \phi(x^4) + \phi(x^{16}) - \phi(x^{25}) + \cdots\right)\right\}^s$$

are all greater than 0 by using the partition method. Then since

$$\mu^4 + \nu^4 + (\mu + \nu)^4 = 2(\mu^2 + \mu\nu + \nu^2)^2$$

it follows that all even numbers can be expressed as the sum of $3s$ fourth powers after a certain point. Since any odd number is the sum of a fourth power and an even number it follows that all odd numbers can be expressed as the sum of $3s + 1$ fourth powers after a certain point onwards.

Now the asymptotic value of the coefficient of x^n in (1) according to the partition method is

(2) $$C^s \frac{n^{\frac{1}{2}s-1}}{\Gamma(\frac{1}{2}s)} \{T_1 + T_2 + T_3 + T_4 + \cdots\}$$

where T's are the terms arising out of the partition method; e.g.,

$$T_1 = 1; \qquad T_2 = (-1)^{n+s}\left(\frac{1}{2}\right)^s;$$

$$T_3 = 2\cos\left(\frac{\pi s}{2} - \frac{2\pi n}{3}\right)\left(\frac{1}{3}\right)^{s/2};$$

$$T_4 = 2\cos\left(s\arctan 3 - \frac{\pi n}{2}\right)\left(\frac{5}{8}\right)^{s/2};$$

$$T_6 = 2\cos\left(\frac{\pi s}{2} - \frac{\pi n}{3}\right)\left(\frac{1}{12}\right)^{s/2};$$

$$T_{12} = \left\{2\cos\left(s\arctan\frac{2}{3} - \frac{\pi n}{6}\right) + (-1)^s 2\cos\left(s\arctan\frac{2}{3} - \frac{5\pi n}{6}\right)\right\}\left(\frac{13}{48}\right)^{s/2}.$$

The constant

$$C = 3\sqrt{\pi}\left(1 - \frac{1}{2} + \frac{1}{4} - \frac{1}{5} + \cdots\right)$$

$$= \pi\sqrt{\frac{1}{3}\pi}.$$

If the asymptotic value of the coefficient of x^n in $(1+2x+2x^4+2x^9+\cdots)^s$ be

$$\frac{\pi^{\frac{1}{2}s} n^{\frac{1}{2}s-1}}{\Gamma(\frac{1}{2}s)} \{t_1 + t_2 + t_3 + \cdots\}$$

then but for the cosines

$$t_1 = 1, \quad t_2 = 0, \quad t_3 = \left(\frac{1}{3}\right)^{s/2}, \quad t_4 = \left(\frac{1}{2}\right)^{s/2}, \quad t_6 = 0, \quad t_{12} = \left(\frac{1}{6}\right)^{s/2}$$

and so there is a slight increase in some of the T's over the t's which can naturally be expected. Now in the square problem the value of $s = \underline{5}$ in order that the coefficients are positive using the partition method (though $\underline{4}$ is the right number). This being so the value of s in the present problem must certainly be greater than 5 due to the slight increase in the value of T's (using the partition method). If it is possible to use your arguments to this function I don't think the value of s will be so high as $\underline{10}$.

$$s = 10 \text{ gives } \begin{cases} \text{even numbers } - 30 \text{ fourth powers,} \\ \text{odd numbers } - 31 \text{ fourth powers.} \end{cases}$$

$s = 5$ is certainly impossible; for if we had $s = 5$ then all even numbers after a certain point would be expressible as the sum of 15th fourth powers which is certainly false. As a matter of fact it appears that all even numbers without exception can be expressed as the sum of 18 fourth powers while the odd numbers without exception can be expressed as the sum of 19 fourth powers. There cannot be a better result than this (for all numbers) since $78 = 4 \cdot 2^4 + 14 \cdot 1^4$ and $79 = 4 \cdot 2^4 + 15 \cdot 1^4$ though it may be likely that all numbers after a certain point can be expressed as the sum of 16 fourth powers. From the formula

$$\mu^4 + \nu^4 + (\mu + \nu)^4 = 2(\mu^2 + \mu\nu + \nu^2)^2$$

it is quite easy to prove that the number of ways in which a number N can be expressed as the sum of 3 fourth powers is greater than

$$2^{(\frac{1}{2}-\epsilon)\log N/\log\log N}$$

for an infinity of values of N.

If
$$a_\lambda = 1, \qquad \lambda = \mu^2 + \nu^2$$
and
$$a_\lambda = 0, \qquad \lambda \neq \mu^2 + \nu^2$$
then it is known that
$$a_1 + a_2 + \cdots + a_n \sim \frac{cn}{\sqrt{\log n}}$$

where c is a constant. Could you tell me if there are results like this to other powers that is to say if
$$a_\lambda = 1, \qquad \lambda = \mu^s + \nu^s$$
and
$$a_\lambda = 0, \qquad \lambda \neq \mu^s + \nu^s$$
then is the order of $a_1 + a_2 + \cdots + a_n$ known?

Ever yours
S. Ramanujan

Commentary. A facsimile of this letter can be found on pages 93–96 of [**227**].

Evidently, Ramanujan's ideas in attacking Waring's problem for fourth powers were never utilized by him, Hardy, or any other mathematician. Now the problem has been completely solved. As usual, let $g(k)$ be the minimum number n such that every positive integer can be expressed as the sum of n kth powers (including 0^k), and let $G(k)$ be the minimum number N such that every sufficiently large positive integer is expressible as the sum of N kth powers (including 0^k). Hardy had informed Ramanujan that he (and Littlewood) had shown that $G(4) \leq 33$. This, in fact, is a special case of the theorem

$$G(k) \leq 2^{k-1}k + 1,$$

published in 1920 by Hardy and Littlewood [**117**], [**105**, pp. 382–403]. Ramanujan's conjecture that $G(4) = 16$ is, indeed, correct, for in 1939 H. Davenport [**68**], [**70**, pp. 946–962] proved that $G(4) = 16$. Ramanujan's conjecture about the value of $g(4)$ is also correct, for in 1989 R. Balasubramanian, F. Dress, and J.-M. Deshouillers [24], [25], [71], [72], [73], [74] proved that $g(4) = 19$.

For proofs of theorems connected with Waring's problem and related problems, see Vaughan's book [269].

The formula

(*) $$\sum_{j,k=-\infty}^{\infty} q^{j^2+jk+k^2} = 1 + 6\sum_{n=0}^{\infty} \left(\frac{q^{3n+1}}{1-q^{3n+1}} - \frac{q^{3n+2}}{1-q^{3n+2}} \right)$$

is very interesting. If $r(n)$ denotes the number of representations of the positive integer n by the quadratic form $j^2 + jk + k^2$ and if $d_{m,3}(n)$ ($m = 1, 2$) denotes the number of positive divisors of n of the form $3\ell + m$, then (*) implies that

(**) $$r(n) = 6(d_{1,3}(n) - d_{2,3}(n)).$$

We are not certain who first explicitly stated (**). However, it is a consequence of a general theorem of P.G.L. Dirichlet [75] on the number of representations of a positive integer by a representative set of inequivalent positive definite binary quadratic forms. See a paper of Berndt [39] wherein (**) is proved via results on theta–functions well known to Ramanujan. The identity (*) is stated without proof by J.M. and P.B. Borwein [51].

Near the end of his letter, Ramanujan again quotes a theorem of Landau that Ramanujan communicated on page (3) of his first letter to Hardy.

G.H. Hardy to Francis Dewsbury 26 November 1918
Trinity College,
Cambridge

Dear Sir,

I have been meaning for some days to write to you again about Ramanujan but have been prevented by stress of work. I think it is now time that the question of his temporary return to India and of his future, generally, should be reconsidered.

There is at last, I am profoundly glad to say, a quite definite change for the better. I think we may now hope that he has turned the corner, and is on the road to a real recovery. His temperature has ceased to be irregular, and he has gained nearly a stone in weight. The consensus of medical opinion is that he has been suffering from some obscure and only partially diagnosed source of blood poisoning, which has now dried up: and that it is reasonable to expect him to recover his health completely and if all goes well fairly rapidly. He would even now be almost fit to make the journey if accompanied by a careful friend. Moreover, the other reasons which made his continuous stay in England desirable (his candidature for the Royal Society for a Fellowship) have now ceased to have importance.

At various times we have felt considerable anxiety about his mental state. I do not think there is really anything seriously amiss with it. But the long illness, and spells of comparative solitude have undeniably had an effect and he has been subject to fits of depression and been difficult to manage. This (with a man of his rather nervous temperament and abnormal quickness of mind) is only natural and almost inevitable. But I think that (assuming his physical condition to have been improving as it has lately) a return for a while to his own country would be a very good thing. His tenure of his fellowship is in no way affected. It involves no duties nor any obligations to residence.

He has apparently been approached (with a view to return) directly by several friends. It is possible, I think that the suggestion has not been made in the most tactful way possible at any rate it seems to have turned him rather against the idea of going. My own view is that the suggestion would best be made more or less officially and by letter simultaneously to him and to me. His Fellowship, of course, makes him financially independent (when once he gets well enough to live by himself). But, if I were assured officially (as I have been unofficially) that it is the intention of Madras to make permanent provision for him, in a way which will leave him free to do research and to visit England from time to time, I would support the proposal for his return, and no doubt he would be willing to go.

There has never been any sign of any diminution in his extraordinary Mathematical talents. He has produced less, naturally during his illness but the quality has been the same.

Possibly you would be kind enough to communicate this letter or the substance of it to Mr. Ramachandra Rao, with whom I have corresponded previously about Ramanujan, and who has taken great interest in his welfare. He is, I believe, a Collector in the Government of Madras, unfortunately I cannot lay my hands on one of his letters or any note of his address, at the moment.

His fellowship is worth £250 a year for 6 years. The first payment does not come till Xmas 1919, but it is possible to anticipate some of it. Until August 1918, i.e. through the first 15 months or so of his illness, his Madras Scholarship, Trinity Exhibition and some £80–90 that was raised for him, enabled him to pay his way in spite of hospital medical expenses. During the last few months, he has been in London, he has seen several specialists and his expenses must have been heavy. On the other hand he has, I fancy, a substantial reserve of his own in the bank—his tastes are frugal and he saved a good deal of money during his first few years here.

He will return to India with a scientific standing and reputation such as no Indian has enjoyed before, and I am confident that India will regard him as the treasure he is. His natural simplicity and modesty has never been affected in the least by success—indeed all that is wanted is to get him to realize that he really is a success.

Yours very truly,
G.H. Hardy

S. Ramanujan to Francis Dewsbury   11 January 1919

> Colinette House,
> 2 Colinette Road,
> Putney, SW 15.

To the Registrar of the University of Madras

Sir,

I beg to acknowledge the receipt of your letter of 9th December 1918, and gratefully accept the very generous help which the University offers me.

I feel, however, that after my return to India, which I expect to happen as soon as arrangements can be made, the total amount of money to which I shall be entitled will be much more than I shall require. I should hope that, after my expenses in England have been paid, £50 a year will be paid to my parents and that the surplus, after my necessary expenses are met, should be used for some educational purpose, such in particular as the reduction of school-fees for poor boys and orphans and provision of books in schools. No doubt it will be possible to make an arrangement about this after my return.

I feel very sorry that, as I have not been well, I have not been able to do so much mathematics during the last two years as before. I hope that I shall soon be able to do more and will certainly do my best to deserve the help that has been given me.

> I beg to remain, Sir,
> Your most obedient servant
> S. Ramanujan

Commentary. Ramanujan was not formally admitted to his Trinity Fellowship until 22 February 1919, after he had signed the Statutable Declaration. The declaration is normally made in the College Chapel in the presence of the Master and the College Council, and obliges the new Fellow to observe the Statutes, Ordinances, and good customs of the College, and in all things to endeavour to promote its welfare. However, as Ramanujan was too ill to come to Cambridge from the Nursing Home in Colinette Road, the College Council agreed, very exceptionally, that he could make the declaration by signing a statement in the presence of a witness. Hardy wrote out the declaration, which was signed by Ramanujan and witnessed by Hardy.

Ramanujan's offer to give his excess earnings to poor students evidently generated hostility in his parents, since they thought that this money should be given to them. It is unlikely that his instructions were carried out.

The letter above is taken from S.R. Ranganathan's book [**229**]. Ramanujan must have been very ill when this letter was written, because the letter is in Hardy's handwriting.

Francis Dewsbury to Educational Department 13 March 1919

From
>   FRANCIS DEWSBURY, Esq., B.A., LL.D.,
>>   *Registrar, University of Madras,*

To
>   The Secretary to Government
>>   Home (Education) Department

Sir,

With reference to correspondence ending with G.O. No. 475, dated the 11th April 1918, on the subject of the scholarship granted to Mr. S. Ramanujam, F.R.S., at Cambridge, I have the honour, by direction, to inform you that in November last a letter was received from Mr. G.H. Hardy to the effect that Mr. Ramanujam had been elected a Fellow of the Trinity College, Cambridge. His Fellowship is of the value of about £250 a year for six years and to it are attached no duties or conditions either of work or of residence. On receipt of this letter the Syndicate resolved to recommend to the Senate that a grant be made from the University Vacation Lectures Account to Mr. Ramanujam of a further sum of £250 a year for five years from the termination of his present scholarship on the 1st April 1919 together with his actual expenses incurred in returning to India and on such passages from India to Europe and back as the Syndicate may approve of during the five years. This, the Syndicate hoped, would sufficiently supplement Mr. Ramanujam's income to enable him to continue his mathematical work free from financial anxiety and to keep in touch with European thought. This resolution of the Syndicate was in accord with suggestions previously received from Mr. Hardy, and anticipated a request from him for financial assistance to Mr. Ramanujam which was received on the 2nd January 1919. Mr. Hardy was informed of the proposals of the Syndicate and he replied "the whole plan is everything I could have wished."

2. I am directed to request that the Government will be pleased to sanction the Syndicate's proposals for the appropriation of funds from the Vacation Lectures Account to meet the expenditure involved. The balance in that account of Rs28,500 in Government Securities at $3\frac{1}{2}$ per cent should be sufficient.

3. At the meeting of the Senate held on the 7th March 1919, sanction was accorded to the Budget Estimates for 1919–20 which specifically included an allocation of a sum of Rs5,000 from this account to meet the proposed expenditure.

>   I have the honour to be,
>>   Sir,
>>>   Your most obedient servant,
>>>>   Francis Dewsbury
>>>>   REGISTRAR.

Chapter 6
Ramanujan Returns to India

Commentary. Through the cooperation of T.V. Rangaswami and K. Srinivasa Rao, we have obtained a photocopy of Ramanujan's complete passport and record the relevant information below. Note that the birthdate is incorrect.

DESCRIPTION OF BEARER

Age   30   *Profession*   research scholar
Place and date of birth   Erode, India Dec. 1888
Height   5 *feet* 6 *inches*
Forehead medium   *Eyes*   normal, dark
Nose   broad   *Mouth*   normal
Chin   normal   *Colour of Hair*   black
Complexion   olive   *Face*   oval
Any special peculiarities   smallpox marks
National Status British–India-born subject–Hindu Brahmin
   son of Mr Srinivasa Aiyangar of Kumbakonam, Tanjore

Upon Ramanujan's return to India, *The Madras Mail* and *The Madras Times* published articles on Ramanujan, which, except for the opening paragraph, are identical. We transmit below the version appearing in the *Times*; the *Madras Mail* article appeared one day earlier.

The Madras Times **30 March 1919**
Mr. S. Ramanujam

The Registrar, University of Madras, writes:—In view of the return of Mr. S. Ramanujam, F.R.S., to India and the description of him which has been lately appearing in the papers "as a Government State Scholar," I am directed by the Syndicate to communicate to your paper the accompanying official statement as to Mr. Ramanujam's career, for favour of publication:—

Mr. S. Ramanujam, F.R.S., who was recently elected to a Fellowship at the Trinity College Cambridge, is not, and has never been, a Government State Scholar, but for the last six years has been in receipt of special scholarships granted to him by the Syndicate of the University of Madras from the

University funds. The facts of his case are briefly as follows:—

In February, 1913, the Director General of Observatories, Sir Gilbert Walker, drew the attention of the Syndicate to the mathematical work of Mr. S. Ramanujam, at that time a clerk in the Accounts Department in the Madras Port Trust and previously an undergraduate of the Madras University. In March of the same year Mr. Arthur Davies, the Secretary to the Madras Students' Advisory Committee, forwarded a letter from the Secretary for Indian Students in London, also inviting enquiry into the work of Mr. Ramanujam on a suggestion made by a tutor of the Trinity College, Cambridge, with whom Mr. Ramanujam had been in correspondence. Mr. Davies added that Mr. Ramanujam was unwilling to leave India.

The Board of Studies in Mathematics recommended a special scholarship of Rs. 75 per mensem to enable Mr. Ramanujam to devote his whole time to the study of Mathematics. As there were no University Regulations enabling the Syndicate to grant this scholarship, a reference was made to the Government for the necessary permission to do so. This permission was received and Mr. Ramanujam was offered and accepted a scholarship tenable from the 1st May of that year, 1913. This scholarship was paid from the University Fee Fund.

In the early part of the following year Mr. E.H. Neville, a Fellow of Trinity College, Cambridge, visited Madras as a special lecturer in Mathematics, and interested himself in Mr. Ramanujam's work upon which he reported to the Syndicate and suggested that Mr. Ramanujam should be afforded an opportunity of being trained in modern methods and of coming into contact with European scholars. Mr. R. Littlehailes, the Professor of Mathematics in the Presidency College, also at the same time recommended to the Syndicate that Mr. Ramanujam, who had overcome his unwillingness to leave India, should be sent to Trinity College, Cambridge, by means of a special scholarship of £250 a year. The Syndicate decided to adopt Mr. Littlehailes' suggestion and requested the Government's permission to utilise for this purpose part of the balances of the grant previously made to the University to provide Vacation Lectures to teachers. The Government sanctioned the appropriation, from this University fund, of a sum not exceeding Rs. 10,000 for a period of two years, free passage and outfit. Mr. Ramanujam accepted the scholarship and proceeded to Cambridge in March of the same year. Two years later, in 1916, this scholarship was extended for another year, and again in 1917 and for a third time in 1918, the necessary funds being provided, with the permission of the Government, from the same source.

In March last the Syndicate communicated to the Senate its intention to make a further grant to Mr. Ramanujam of £250 a year for five years from April next, together with his expenses on approved journeys to and from Europe in the hope that this would sufficiently supplement the amount of his Fellowship to enable him to pay visits to India and also to keep in touch with Western Mathematicians. The Senate accepted an allocation, in the

University Budget, for this purpose, of Rs. 5,000 for the ensuing year. The sanction of the Government to the appropriation of this sum from the grant above mentioned is awaited.

Mr. Ramanujam, who has been very ill in England, arrived in India by the s.s. Nagoya on the 27th instant. Mr. G.H. Hardy, F.R.S., who periodically reported to the Syndicate of the University of Mr. Ramanujam's condition and work and has befriended him in every possible way, describes Mr. Ramanujam in one of his letters as being "in some respects the most remarkable mathematician in the world."

Commentary. There is no evidence that in 1913 Ramanujan had been in correspondence with anybody in Trinity College other than Hardy. Accordingly, Hardy must have been the "tutor of Trinity College, Cambridge" referred to in the third paragraph of the article, although he never held that office officially.

The following article is one of the most extensive newspaper accounts of Ramanujan's life. It appeared in both the Saturday evening and Sunday morning editions. We are grateful to Mr. A. Sivasailam, Chairman, Amalgamations Limited, Madras, and the son of the Proprietor of *The Madras Times*, for providing us with a copy of this article.

The Madras Times 5 April 1919

A FAMOUS MADRAS MATHEMATICIAN

MR. S. RAMANUJAN, F. R. S.

It is the belief that Madras Presidency may well be proud to have produced perhaps the greatest—within certain definite lines—pure mathematician of this or any age, we venture to day to offer to our readers such notes as we have been able to collect, showing how Mr. S. Ramanujan obtained recognition, in spite of many unfavourable surroundings, and so has been enabled, as explained in the note by the Registrar, Madras University, which we published on March 30th. to offer to the world the fruits of his marvelous genius. The notes for this article have chiefly been collected from papers in the possession of the Madras Port Trust.

EARLY INTEREST IN MATHEMATICS.

Mr. S. Ramanujan is a native of Kumbakonam in the Tanjore District of Madras Presidency. He was born in 1888 of poor, and so far as English goes, illiterate parents, of the Vaishnava sect of Brahmans. His father, whose name is Sreenivasa Iyengar, lives at 17, Sarangapani Coil Sannadhi Street, Kumbakonam. He has two of his younger brothers of 20 and 14 years of age living. Three sisters died while young. In the year 1900, when in his 12th year, he found himself progressing in his mathematical studies faster than his teacher could carry him. A mathematical friend of his gives the following terse notes of his studies at this period:—

"Arithmetic and geometric series. Was curious to know what Trigonometry was. So he borrowed from a student a Trigonometry. This happened to be Part II of Loney's. He went through the whole book and was thus fortunate to learn at the outset that sine, cosine and other functions represented algebraic series and not the ratios as is generally taught to young schoolboys. 1902 learnt from another how to solve cubic equations. Solved biquadratic equation independently by resolving into two quadratic factors. 1903 attacked solving quintic equation and failed. Beginning of 1904 began to investigate the series $1, \frac{1}{2}, \frac{1}{3}$ and calculated Euler's constant in a regular series, and also to 15 places of decimals, and also Bernoulli's numbers without knowing that these existed. Developed summation of series and subsequently learnt that this was only integral calculus. Then learnt what was differentiation and integration."

COLLEGE CAREER

At the end of 1904, at the age of 16, he passed the Madras university Matriculation and joined Kumbakonam College in the Intermediate class, going through the first year course. Taking, however, undue interest in mathematics to the sacrifice of other subjects, he failed to secure promotion to the second year course. In 1905 he was afflicted with an illness, but, notwithstanding this, he in that year investigated much of what is known as "hypergeometric series". At the end of 1905 and during 1906 he investigated the relation of many integrals and series, learning only subsequently that these investigations of his related to what are called Elliptic Functions. In 1906 he joined Pachaiyappa's College and for three months read in the first year course of that College. Unhappily, at this stage, having fallen seriously ill again, he was obliged to discontinue his College studies. In December, 1907, however, he appeared privately for the First Arts Examination and had the distinction of failing in *all* the subjects, doubtless, as the result of his illness. In 1908 he developed "continued fractions" and investigated "divergent series." In the following year he underwent a severe surgical operation and was completely prostrated for nearly twelve months. In 1910 he developed relations among Elliptical Modular equations. In 1911 he went in search of mathematicians in the hope of their appreciating his work, and got in touch, *inter alia*, with Mr. V. Ramaswami Iyer, Professor Ross, of the Christian College, Professor Middlemast, of the Presidency College, and last, not least, Mr. R. Ramachandra Rao, the Collector of Nellore, and now Dewan Bahadur and Secretary to Government. This last gentleman paid his boarding fees in Madras for

CHAPTER 6. RAMANUJAN RETURNS TO INDIA

nearly a year, and throughout up to the present date has been his kind and faithful friend.

AS A PORT TRUST CLERK

At this stage there began Mr. Ramanujan's connection with the Port Trust which has led to the recognition and the facilities of which he has made wonderful use. Supported by Mr. E.W. Middlemast and by Mr. S. Narayana Iyer, M.A., now Chief Accountant, Madras Port Trust, and himself a fine mathematician, he applied for a small billet in the Trust's Accounts Department. The former gentleman wrote of him:— "He is a young man of quite exceptional capacity in mathematics, and especially in work relating to numbers." The latter wrote of him as "a mathematical genius." Thereupon he found himself appointed as accounts clerk on Rs. 25 per mensem from the 1st of March, 1912. Eight months later Mr. C.L.T. Griffith, then Professor of Engineering in Madras Engineering College, wrote to Sir Francis Spring as follows:—

"You have in your office as an accountant on Rs. 25 a young man named S. Ramanujan, who is a most remarkable mathematician. He may be a very poor accountant, but I hope you will see that he is kept happily employed until something can be done to make use of his extraordinary gifts. I am writing to one of the leading mathematical professors at Home about him and sending copies of some of Ramanujan's papers and results. Our Mathematics Professor here says that very few people could follow or criticise the work. It is, of course, far beyond my scope, but I happen to know who is at work in the same line at Home, and I hope to get instructions as to what this fellow ought to do. If there is any real genius in him, he will have to be provided with money for books and with leisure, but until I hear from Home, I don't feel sure that it is worth while spending much time or money on him."

FIRST STEPS TO FAME

Dr. Bourne, F.R.S., at that time Director of Public Instruction, Madras, was next communicated with. Up to that date no intimation seems to have reached Dr. Bourne from any of the schools in which Ramanujan had attempted to study. So it may truly, perhaps be said that had his discovery depended on the Madras Educational Department, the poor struggling student might never have got a chance of offering to the world what he has since given to it and the first mathematicians of the age have been glad to publish. Meanwhile Mr. Griffith had enlisted the interest of Mr. W. Graham, I.C.S., Accountant-General, Madras, and a good mathematician—indeed a very high "wrangler"—since, alas! killed in command of his battery at the French front—whose letter reached Sir Francis Spring under cover of a note from Mr. Griffith. Mr. Griffith's letter and its inclosure from Mr. Graham ran as follows:—

Mr. Griffith's letter

"I enclose the letter re Ramanujan which Graham has been kind enough to send me. He says the man has brains, but may be of the calculating boy type: but, like the other mathematicians here modestly disclaims the power of criticizing Ramanujan's work.

Ramanujan seems to have seen Mr. Middlemast already, and he, also appears to be unable to criticise.

I think I was right in writing to Professor Hill in London, and we must wait for his opinion, which I hope to get during January."

Mr. Graham's letter

"Ramanujan came to see me today (27-11-1912). He seems to have done a great deal of work in one particular branch of calculus, and from the way he has done it he must have considerable mathematical aptitude. He has read no mathematics at all except calculus, apparently, and it is possible his brains are akin to those of the calculating boy."

"I should say however that, if he had the training, mathematics of a certain kind—algebra, Differential Equations, Hydrodynamics, and Calculus, i.e., pure theory and development of algebraic functions—juggling with symbols as it has been called—would have become very easy to him."

"Whether he has the stuff of great mathematicians or not I do not know. He gave me the impression of having brains. His original work is an interesting development of work already done—but interesting only to the purist. However I am not the best qualified to judge. Middlemast's opinion would be of value."

VIEWS OF BRITISH MATHEMATICIANS

The next thing of any importance to happen was the arrival from England of two successive letters, both dated December, 1912, from Mr. M.J.M. Hill, of University College, London University. These letters, which are perhaps rather technical for reproduction here, gave useful advice, chiefly in the direction of insisting on rigid proofs, step by step, instead of conclusions being jumped at which, however correct, had been reached intuitively or sub-consciously, without any intentional or deliberate working out of the intermediate stages. The receipt of these letters was a great assistance to Mr. S. Narayana Iyer, under whom Mr. Ramanujan was carrying on his humble clerical accounts duties, in insisting on the filling up of the requisite, intermediate steps by way of rigid proof instead of flying to conclusions, however, correct they might prove to be. Mr. Hill recommended the use, particularly, of Bromwich's Theory of Infinite Series, published by Macmillan. This letter of Professor Hill's reached Sir Francis Spring with the following note from Mr. Griffith and was passed on to Mr. S. Narayana Iyer. Mr. Ramanujan's constant friend and mathematical adviser:—

"I enclose Professor Hill's last letter re Ramanujan. Will you read it and hand it on to him? I have written to Mr. Ramachandra Rao, Collector of

Nellore, and asked him to buy the book mentioned and give it to Ramanujan. The fact that Professor Hill is prepared to consider Ramanujan's Elliptic Functions, even though the proofs may not be of that logical completeness that is demanded nowadays shows that it is possible that 'the intuitive results' may be of interest. All the same Ramanujan ought to do his utmost to make his proofs complete."

SIR FRANCIS SPRING'S ACTION

At this stage a letter, dated 8th February, 1913, was received from Professor G.H. Hardy, of Trinity College, and Cayley Lecturer in Mathematics at the University of Cambridge, fully appreciative of the specimen of Mr. Ramanujan's work which had been sent Home to him by Mr. S. Narayana Iyer, but insisting on rigid proofs of the theorems. At the same time Professor Hardy wrote to Mr. Mallet, of the India Office, to prevail upon Ramanujan to go over to Cambridge. Mr. Mallet communicated with Mr. Arthur Davies, then Secretary to the Students' Advisory Board. Mr. Davies wrote in his turn to the Madras University. On February 13th Dr. Gilbert Walker, F.R.S., happened to be at Madras Harbour in connection with observatory matters, and Sir Francis Spring took the opportunity to get Mr. Narayana Iyer to produce some of the work. This was pronounced on most favourably by Dr. Walker, who advised that it should be sent to Professor Hardy, who would probably be found to be able to judge of it as its true value, and also promised, at the instance of Mr. S. Narayana Iyer, to write to the University of Madras, recommending Ramanujan to a research scholarship. The next thing to happen was the receipt of a letter from Mr. Hanumantha Rao, asking Mr. Narayana Iyer to be present at a meeting of the Board of Studies of the University to decide what it was possible to do to help Mr. Ramanujan. Next, as stated in the note already published, the University of Madras offered Mr. Ramanujan a Rs.75 scholarship to aid him in continuing his studies, and this was accepted on the 12th April, 1913, whereupon the Port Trust gave him 2 years' leave, with the option of returning on the cessation of the scholarship. At this stage there came a letter from Mr. Arthur Davies to say that he was conveying to Mr. Mallet, of the India Office, the news of the grant of the scholarship and that he hoped some good might come of it from Cambridge. Mr. Littlehailes next came on the scene, and Mr. Narayana Iyer took Ramanujan to see this gentleman (now Director of Public Instruction, Madras, and a well-known mathematician). On the 5th February, 1914, Sir Francis Spring wrote as follows to His Excellency the Governor's Private Secretary:—

"If I understand right, His Excellency has the Educational portfolio. So I am anxious to interest him in a matter which I presume will come before him within the next few days—a matter which under the circumstances is, I believe, very urgent. It relates to the affairs of a clerk of my office named S. Ramanujan who, as I think, His Excellency has already heard from me, is pronounced by very high mathematical authorities to be a mathematician of a new and high, if not transcendental, order of genius."

"A few months ago the Madras University gave S. Ramanujan a scholarship to enable him to fill certain gaps in his education which operated to prevent his conveying his conceptions to the outside world. Meanwhile during the last 8 or 9 months various mathematicians in the first rank in Cambridge, Simla, and Madras have had before them selections from his work and have pronounced upon them in terms of the very highest eulogy."

"Just now, as probably His Excellency is aware, a Mr. Neville, who, I think, is a Senior Wrangler and a Fellow of Trinity, Cambridge, has been in Madras giving a series of lectures on certain phases of the Higher Mathematics to Honours students and others interested. Under a mandate from Cambridge he has interested himself greatly in Ramanujan, and there is every reason to hope that he may be persuaded to go to Cambridge for a year or two so that, under expert guidance, not only may the fruits of his genius be given to the world, but also, we may hope, his own fame, future usefulness and personal prosperity may be secured—matters probably quite impossible if he remained in a backwater like Madras for the rest of his life."

"I now come to the point where His Excellency may perhaps be able to interfere with advantage. Last evening I learnt from Mr. Littlehailes and others that the University Syndicate had decided, subject to sanction of Government, to set aside a sum of Rs.10,000 in order to secure Ramanujan's visit to England for a couple of years. Messrs. Littlehailes and Neville begged me to intercede with His Excellency with a view to the speedy confirmation of this action of the University Syndicate. But I wish to make it quite clear that I write under no mandate from the Syndicate but merely as a private individual interested in my own employee, Ramanujan, as well as in Mathematics.

"Mr. Arthur Davies will doubtless arrange for the voyage to England and that Ramanujan's orthodoxy may be maintained unimpaired. Mr. Neville assures me that he will meet him on his arrival in London and conduct him personally to Cambridge, and that when there he will interest himself personally in his welfare, generally and in all matters of Brahman orthodoxy, so that he may return to India without any loss of the esteem of his caste men."

"I myself am very far from being Mathematician enough to express adequately what has been said to me by several who are fully qualified to express an opinion on the subject of the potential value to Science of the new line of thought in which Ramanujan's investigations lie. I am assured, however, by those who ought to know what I am talking about, that they may conceivably be epoch-making and as such well worthy of financial support at the hands of the Madras University."

"Needless to say, Professor Hardy and other high mathematicians may be trusted to give Ramanujan the fullest credit in the scientific world for his work. My reason for saying anything so obvious as this is that I am told that certain of his Indian friends have been suggesting to him that all that the scientists of England desire is to steal his ideas—obviously an utterly impossible suggestion with men of their class."

A MADRAS UNIVERSITY SCHOLARSHIP

The result was that His Excellency Lord Pentland expressed cordial sympathy with the proposal that the University should provide Ramanujan with the means of continuing his studies at Cambridge and expressing a desire to do all that was possible to assist in the matter.

About the beginning of March, 1914, Mr. Ramanujan was prevailed upon by Sir Francis Spring to accept a scholarship of Rs.10,000 from the Madras University in order to continue his studies in the more favourable Cambridge atmosphere, and with the help of Mr. Arthur Davies, in caste matters and so on, he embarked from Madras on the 17th March in the ss. Navasa for England, and in May, 1914, he went into residence in Trinity College, Cambridge, in order to read with Mr. Hardy, Mr. Littlewood, and Mr. E.H. Neville. The last gentleman, a Fellow of Trinity College, Cambridge, had been in Madras as a special lecturer in Mathematics during the cold weather of 1914 and was most kind in helping Mr. Ramanujan on his arrival in England. Mr. Hardy wrote:—

"The first I knew of him was about fifteen months ago. He wrote to me explaining who he was, and sent a large number of Mathematical theorems which he had proved. There were a great many very remarkable results. His theorems were all in pure mathematics, particularly, in the theory of numbers and the theory of elliptic functions. While many of them were quite new, others had been anticipated by writers of whom he had never heard, and of whose work he was quite innocent. That is the wonderful thing; he discovered for himself a great number of things which the leading mathematicians of the last hundred years, such as Cauchy and Jacobi, had added to the knowledge of schoolmen."

CAREER AT CAMBRIDGE

Meanwhile the scholarship term was coming to an end, and then arose a question of the continuance of his studies in Cambridge: whereupon on receipt of a letter from Mr. Ramanujan, Sir Francis Spring wrote as follows to the Registrar of Madras University:—

"The accompanying papers refer to Mr. S. Ramanujan, in whom I am interested if for no other reason than that he was discovered while working as a clerk in my office. The papers have reached me from my Office Manager, Mr. S. Narayana Iyer, M.A., L.T., who himself a good mathematician, was chiefly interested in bringing Mr. Ramanujan into notice."

"In the interests of India's reputation, as the birthplace of some good mathematicians of which Mr. Ramanujan is, I am assured, one of the most remarkable, I hope it may be possible for the University to ensure the uninterrupted continuance of his studies, but chiefly of his original research work, at Cambridge, by promising him the enjoyment of his present scholarship up to the end of March, 1918.

"The Syndicate will, I have no doubt, give the matter their best consideration." The reply was as follows:—

"In reply to your letter, dated December 15, 1915, I have the honour to state that the University of Madras has requested sanction for the allocation of funds for the extension of the scholarship to Mr. S. Ramanujan for another year, i.e., to the end of March 1917. Your letter and enclosures will be placed before the Syndicate for orders. The Syndicate has been given to understand that it is quite possible that Mr. Ramanujan may become independent of a scholarship before the expiry of this extension. If this proves not to be the case a further extension can be considered at a later date. In the meantime it will be difficult for the Syndicate to consider the question of providing for a return of Mr. Ramanujan to India next year, in the absence of any representation on the point from him or from those connected with his work at Cambridge, Possibly Mr. Narayana Iyer may be able to furnish the Syndicate with more detailed information on the matter." This was replied to as follows:—

"With reference to the concluding portion of letter No. 7923, dated the 16th December, 1915, from the University Registrar, I attach an extract from Mr. S. Ramanujan's letter, dated 11th November, 1915."

"2. He wants to arrive in Madras in April, 1916 and to stay with his family during May and June, returning to England in July. If his scholarship is to end in March, 1917, that leaves him only 8 or 9 months more for study at Cambridge. If he is to compete for the Fellowship mentioned by Mr. Hardy in October, 1916, and if he is successful the continuance to him of the Madras' University scholarship will not be necessary."

"3. If the question of providing for his return to India next year can be considered only after a formal representation from him I will write to him to put in a formal application through the University authorities there. But I doubt if there is enough time for this formal representation."

S. Narayana Aiyar,
20th December, 1915

Forwarded to the Registrar of Madras University in reply to his letter No. 7928, dated the 18th December 1915."

Francis J.E. Spring.
20th December, 1915

FELLOWSHIP OF ROYAL SOCIETY

Later, on the 17th January, 1916, the Syndicate of Madras University, with sanction of Government decided to continue the scholarship for another year, and on the 18th October, 1916, the Port Trust gave him another year's leave. Early in 1918 the wonderful news was received that Mr. S. Ramanujan had been elected a Fellow of the Royal Society, the highest honour attainable in Science amongst the English race, thereby only confirming the correctness of his backers' diagnosis. Later, towards the end of 1918, it was announced that he had been elected a Fellow of Trinity College, Cambridge, whereupon Sir Francis Spring made over his Madras University Fellow cap and gown to be offered by Mr. Narayana Iyer to Mr. Ramanujan in the pretty certain event of his being made a Fellow of Madras University on his return to India—feeling highly honoured, he says, in so doing. Full details of the University Scholarships granted to Mr. Ramanujan will be found in the Registrar's note which we published on March 30th.

GREATEST GENIUS SINCE NEWTON

Mr. Ramanujan arrived in India in s.s. Nagoya last week, and it remains to be seen what his future is to be. Unhappily he is reported as not being in the best of health, but his many friends and admirers cannot but hope that his return to his Motherland, to his parents, to the friends of his early manhood, and to those who in the small Madras scientific world admire his work—even if unable to assimilate much of it—may have the effect of restoring him completely to health and of the world being given further instalments of the results of his wonderful genius. Somebody indeed wrote of him from Cambridge that since Newton there has been nobody his equal—than which, needless to say, there can be no higher eulogy.

Commentary. Much of the preceding article repeats what has already been recorded on earlier pages, but we have, nevertheless, thought it worth including without abbreviation. Several statements are incorrect, such as the year of Ramanujan's birth and J.F. Graham's initials. We do not think that it is necessary to give a complete list. It is revealed in the letter of 15 February 1921 from Sir Francis Spring to S.N. Aiyar that they are the authors of this article.

New India   **18 July 1919**

Mr. Ramanujam, F.R.S.

Our representative who called on this distinguished Mathematician recently reports that his condition is very far from satisfactory. He was to have gone to the Coimbatore district for a change, but although it is stated that he is on the way towards recovery, friends who knew him in England express pained surprise at the extent of the change in his health. The trouble with his lungs, Mr. Ramanujam told our representative, began in England chiefly as a result of the absence of proper arrangements for food, and after two years of vain effort in sanatoria and nursing homes, he returned to India on the advice of his doctors, who feared that further stay in England might react dangerously on his life. Mr. Ramanujam has been in bed ever since his return, and being ordered complete rest, all thought of continuing his work has to be put aside till he gets back some of his former vitality. It is extremely doubtful, he said, in view of his difficulties in England, whether he would return to that country. The more important consideration, however, is that he should be enabled to resume his great work, and we hope the Government will see that he has the best of medical attention and all possible care.

Report regarding Mr. Ramanujam, F.R.S.

Mr. Sitapati Ayyar, L.M.&S. of the King Institute, Guindy, is Mr. Ramanujam's medical attendant. He was treating Mr. Ramanujam for about a month and a half and he noticed a slight improvement in his condition, Ramanujam having gained 4 or 5 lbs in weight, though he was generally confined to bed. He seldom cared to carry out the doctor's instructions as regards either medicine or diet and he is now staying at Kodumudi in the Trichinopoly district.

It seems it was not possible to diagnose his disease accurately, that his lungs were found to be quite healthy and that tuberculosis of some organ (which nobody can say) had set in and that the open air treatment was the best for him.

Dr. Kesava Pai, B.A., M.D., of the King Institute is also reported to have visited Mr. Ramanujam more than once.

Who treats him at Kodumudi is not known.

Home (Education) Dept.

Ramanujam's arrangements were made by me. At first he was very weak—so weak he could not walk unaided from the train to my motor. I put him up first in some rooms in a bungalow in Cathedral Road; and when he picked up strength, I removed him to Venkata Vilas. He gradually picked up strength and even worked for a day or two. He did not then require medical treatment, but at my instance Dr. Seethapathy of the Kings Institute was attending.

When my wife followed me up to Ootacamund a month after I came here, she told me that Ramanujam was getting low fever every day. I then wrote to Drs. Kesava Pai and Bhujanga Rao to look him up. He has been very bad. But he is so disgusted with the excess medical treatment he got in England that he declines to submit himself to any here. I believe that it is tuberculosis but some tests are negative. The last I heard of him was on the eve of his going to Kodumudi—he was well enough to be moved.

The D.M.& S.O., Coimbatore or some other medical man from Erode may be asked to visit him casually and report on the present state of his health.

R.R.R.

Commentary. The first of the two reports above on Ramanujan's health is undated, but the date 30 July 1919 has been inserted by hand between the typed reports found in the Tamil Nadu Archives. The second is dated 31 July 1919.

Guindy is a section in the very southern part of Madras, and is the location of the official residence, Raj Bhavan, of the governor of Madras. Kodumudi is a small town on the southern bank of the Cauvery River, less than 20 kilometers from Ramanujan's birthplace, Erode. Kodumudi is known for its famous Magudeswara Temple. Ramanujan stayed in Kodumudi about two months.

The initials R.R.R. doubtless denote R. Ramachandra Rao. The Seethapathy mentioned is presumably the same person as the Sitapati previously mentioned. The initials L.M.&S. may stand for Licentiate in Medicine and Surgery.

After the two reports above is a governmental office memo, "H.M. Yes, please. As Secretary has already been kind enough to take much interest in this mathematical genius, I hope he will continue to do so and from time to time keep Government informed of his condition. What was the excess medical treatment referred to. I gather that he is rather an intractable patient. D. 1–8–19."

The following medical report is untitled and bears a government stamp with the date 1 August 1919.

Medical Report

The excess medical treatment as follows.

2. When he first got ill, Prof Hardy F.R.S. consulted a Harvey Street specialist. The first diagnosis was cancer and Prof Hardy wrote to me for instructions through the Secretary of State. We have been since exchanging cables and letters. Luckily this diagnosis proved inaccurate and he was sent to a sanatorium for consumptives. He got well after a long time; but when weak he walked out and injured his knee. All along he was under treatment.

3. The best medical opinion seems to be that it is a case of tuberculosis. But Ramanujan himself thinks that it is a case of malnutrition. He is a strict vegetarian.

4. I am writing to the D.M. & S.O., Coimbatore.

Commentary. Harvey Street should be Harley Street, a London street famous for its medical specialists.

Ramanujan severely injured his knee when, severely depressed, he threw himself on the track of an approaching train in London. Miraculously, his

P.A. Pires to District Medical Officer, Coimbatore 30 August 1919

From
> Mr. P.A. Pires,
>> Civil Apothecary,
>>> Erode.

To
> The District Medical & Sanitary Officer, Coimbatore

Dear Sir,

I again visited Kodumudi to day. Mr. Ramanujam was unable to come down to me and I was asked to go up to his room upstairs to see him. I found him seated in his bed looking pale and weaker and thinner than when I last saw him on the 10th instant. His face is slightly puffy and there are dark rings under his eyes. He looked tired too. He keeps mostly to his room and rarely comes down now. He has been getting fever often of an irregular type and yesterday he had fever all day. There is no cough and no night-sweats. His appetite is still fair and his sleep good. He receives no visitors and does not appear to be cheerful as before. His arms and hands look very thin and he feels pain in the bones when he exerts himself. He is still unable to clench his right hand but he can close his left fist without effort.

His finger nails are paler than before; he is not short of breath while conversing; he has no cough; his gait is still stooping. He has no visitors and scarcely sees anybody.

On the whole I find that his general health has deteriorated much since I last saw him.

I beg to be informed whether any further report is needed. If so required, I shall visit the patient as per further direction.

I beg to remain etc.,
P.A. Pires,
Civil Apothecary.
Lt–Col; I.M.S.,
District Medical & Sanitation Officer

Commentary. The city of Coimbatore is located near the west border of Tamil Nadu, about 60 kilometers southeast of Ootacamund. In 1921, Coimbatore had a population of 65,788, but it has expanded rapidly to its present population of about 1,100,000.

CHAPTER 6. RAMANUJAN RETURNS TO INDIA

S. N. Aiyar to S. Ramanujan   23 October 1919

From

THE CHIEF ACCOUNTANT,
MADRAS PORT TRUST.

To

<u>Mr S. Ramanujam B.A. F.R.S.</u>
<u>17 Sarangapanikoilsannidhi Street</u>
Kumbakonam.

<u>Memo</u>

Mr. S. Ramanujam is informed that there is a sum of Rs 103-12-8 standing to the credit of his Provident Fund account with the Port Trust. This sum will be paid to him by a cheque in the Bank of Madras on his furnishing a stamped receipt for the amount.

S.N.

Commentary. The files from the Madras Port Trust also contain the following memo providing a description of the origin of the funds paid to Ramanujan.

Entertained 1st March 1912
Leave without pay for 5 years
on 1st May 1913
Services terminated on 30th April 1918
Total No. of years of service = 6 years

Subscription	Rs 66-14-5
Interest on Subscription	7- 3-9
Trust contribution ($\frac{6}{15}$ Rs of Rs 66-14-5)	26-12-2
Interest on Trust contribution ($\frac{6}{15}$ Rs of Rs 7-3-9)	2-14-4
Total Rs.	103-12-8

The Provident Fund is a retirement benefit fund for employees of any organization in India. The amount of Subscription is the total that the individual (Ramanujan) contributed to this retirement plan. As indicated, the Madras Port Trust paid the management's contribution to the retirement benefits of an employee. *Entertained* is a curious word, and one might think it should be *Entered*, but the same usage occurs in S. Narayana Aiyar's letter of 10 March 1921 and in the earlier memo from the Madras Port Trust in 1912.

Only two blocks from Ramanujan's home in Kumbakonam lies the famous and beautiful Sarangapani Temple, a prominent place of worship for the Iyangar subcaste of Brahmins. The name of the street on which Ramanujan lived, Sarangapanikoilsannidhi, means road leading to the main entrance (sannidhi) of the temple (koil) for the God Sarangapani.

Sarangapani Temple

Francis Dewsbury to G.H. Hardy 22 December 1919

(at Kodaikanal)

SENATE HOUSE,
MADRAS

Dear Mr. Hardy,

Will you please let me know whether Mr. Ramanujan has paid what he owes to you and Dr. Russell. I can get his reply from him, and I am arranging to pay for him if he has not done so. He is still in very bad health and difficult to deal with, living up country with his family. Mr. Ramachandra Rao is doing what he can for him, but Mr. Ramanujan himself will not consent to live in a suitable environment under proper treatment. It is a great pity.

Yours sincerely,
Francis Dewsbury

Commentary. Located in the mountains of southwest Tamil Nadu near the border with Kerala, and blessed with a cool climate, Kodaikanal is a well-known resort town and "hill station"; in particular, it was a popular summer retreat for the British, and it remains an attractive holiday spot today.

The identity of Dr Russell is not known, since there are several medical practitioners of that name listed in the *Medical Register*. A possible identification is given in [**234**, p. 82].

S. Ramanujan to Madras Port Trust 12 January 1920

Chetpet

To
  The Chief Accountant
  Madras Port Trust

Sir,

With reference to your memo no. 7728A of 23-10-19 I request you to be good enough to send by the bearer of this letter the sum (Rs 103-12-8) of Rupees one hundred and three, annas twelve and pies eight being my Provident Fund contribution and subscription.

I have the honour to be,
Sir,
Your most obedient servant
S. Ramanujan

Commentary. Ramanujan had recently moved from Kumbakonam to Madras when this letter was written.

At that time, the Indian rupee was divided into 16 annas, and each anna was worth 12 pies (singular, pie) for a total of 192 pies per rupee. After Indian independence in 1947, the monetary system was decimalized so that 1 rupee now is equivalent to 100 paise (singular, paisa). The designation pie is British. The word "paisa" in Hindi means "money."

S.N. Aiyar to S. Ramanujan 15(?) January 1920

From
  THE CHIEF ACCOUNTANT
  MADRAS PORT TRUST

To
  Mr. S. Ramanujan B.A. F.R.S.
  'Kudsia' Harrington Road.
  Chetpet. Madras

<u>Memo</u>

With reference to his letter dated 12-1-1920, a cheque for Rs 103-12-8 being the amount due to him in final settlement of his Provident Fund account with the Madras Port Trust, is forwarded herewith. He is requested to acknowledge the receipt of the cheque.

SN

Commentary. The office memo reproduced above is a duplicate of a letter sent to Ramanujan.

S. Ramanujan to G.H. Hardy   **12 January 1920**

University of Madras

I am extremely sorry for not writing you a single letter up to now ... I discovered very interesting functions recently which I call "Mock" ϑ-functions. Unlike the "False" ϑ-functions (studied partially by Prof. Rogers in his interesting paper) they enter into mathematics as beautifully as the ordinary ϑ-functions. I am sending you with this letter some examples ... .

If we consider a ϑ-function in the transformed Eulerian form e.g.

(A) $\quad 1 + \dfrac{q}{(1-q)^2} + \dfrac{q^4}{(1-q)^2(1-q^2)^2} + \dfrac{q^9}{(1-q)^2(1-q^2)^2(1-q^3)^2} + \cdots$

(B) $\quad 1 + \dfrac{q}{1-q} + \dfrac{q^4}{(1-q)(1-q^2)} + \dfrac{q^9}{(1-q)(1-q^2)(1-q^3)} + \cdots$

and determine the nature of the singularities at the points $q = 1, q^2 = 1, q^3 = 1, q^4 = 1, q^5 = 1, \ldots$ we know how beautifully the asymptotic form of the function can be expressed in a very neat and closed exponential form. For instance when $q = e^{-t}$ and $t \to 0$

$$(A) = \sqrt{\dfrac{t}{2\pi}} \exp\left(\dfrac{\pi^2}{6t} - \dfrac{t}{24}\right) * + o(1) \dagger$$

$$(B) = \dfrac{\exp\left(\dfrac{\pi^2}{15t} - \dfrac{t}{60}\right)}{\sqrt{\dfrac{5-\sqrt{5}}{2}}} * + o(1) \dagger$$

and similar results at other singularities.* It is not necessary that there should be only one term like this. There may be many terms but <u>the number of terms must be finite.</u> † Also $o(1)$ may turn out to be $O(1)$. That is all. For instance when $q \to 1$ the function

$$\dfrac{1}{\left\{(1-q)(1-q^2)(1-q^3)\cdots\right\}^{\tfrac{1}{120}}}$$

is equivalent to the sum of five terms like (*) together with $O(1)$ instead of $o(1)$.

If we take a number of functions like (A) and (B) it is only in a limited number of cases the terms close as above; but in the majority of cases they never close as above. For instance when $q = e^{-t}$ and $t \to 0$

(C) $\quad 1 + \dfrac{q}{(1-q)^2} + \dfrac{q^3}{(1-q)^2(1-q^2)^2} + \dfrac{q^6}{(1-q)^2(1-q^2)^2(1-q^3)^2} + \cdots$

$\quad = \sqrt{\dfrac{t}{2\sqrt{5}}} \exp\left(\dfrac{\pi^2}{5t} + a_1 t + a_2 t^2 + \cdots + O(a_k t^k)\right)$

where $a_1 = \frac{1}{8\sqrt{5}}$, and so on. The function (C) is a simple example of a function behaving in an unclosed form at the singularities.

*The coefficient (of) $1/t$ in the index of e happens to be $\frac{\pi^2}{5}$ in this particular case. It may be some other transcendental numbers in other cases.

† The coefficients of $t, t^2, \ldots$ happen to be $\frac{1}{8\sqrt{5}}, \ldots$ in this case. In other cases they may turn out to be some other algebraic numbers.

Now a very interesting question arises. Is the converse of the statements concerning the forms (A) and (B) true? That is to say Suppose there is a function in the Eulerian form and suppose that all or an infinity of points $q = e^{2i\pi m/n}$ are exponential singularities and also suppose that at these points the asymptotic form of the function closes as neatly as in the cases of (A) and (B). The question is:—is the function taken the sum of two functions one of which is an ordinary ϑ function and the other a (trivial) function which is $O(1)$ at all the points $e^{2i\pi m/n}$? The answer is it is not necessarily so. When it is not so I call the function Mock ϑ-function. I have not proved rigorously that it is not necessarily so. But I have constructed a number of examples in which it is inconceivable to construct a ϑ-function to cut out the singularities of the original function. Also I have shown if it is necessarily so then it leads to the following assertion:—viz. it is possible to construct two power series in x namely $\sum_0^\infty a_n x^n$ and $\sum b_n x^n$ both of which have essential singularities on the unit circle, are convergent when $|x| < 1$, and tend to finite limits at every point $x = e^{2i\pi r/s}$ and that at the same time the limit of $\sum_0^\infty a_n x^n$ at the point $x = e^{-2i\pi r/s}$ is equal to the limit of $\sum_0^\infty b_n x^n$ at the point $x = e^{-2i\pi r/s}$.

This assertion seems to be untrue. Any how we shall go to the examples and see how far our assertions are true.

I have proved that if

$$f(q) = 1 + \frac{q}{(1+q)^2} + \frac{q^4}{(1+q)^2(1+q^2)^2} + \cdots$$

then

$$f(q) + (1-q)(1-q^3)(1-q^5)\cdots(1 - 2q + 2q^4 - 2q^9 + \cdots) = O(1)$$

at all the points $q = -1, q^3 = -1, q^5 = -1, q^7 = -1, \ldots$, and at the same time

$$f(q) - (1-q)(1-q^3)(1-q^5)\cdots(1 - 2q + 2q^4 - \cdots) = O(1)$$

at all the points $q^2 = -1, q^4 = -1, q^6 = -1, \ldots$. Also obviously $f(q) = O(1)$ at all the points $q = 1, q^3 = 1, q^5 = 1, \ldots$. And so $f(q)$ is a Mock ϑ function. When $q = -e^{-t}$ and $t \to 0$

$$f(q) + \sqrt{\frac{\pi}{t}} \exp\left(\frac{\pi^2}{24t} - \frac{t}{24}\right) \to 4.$$

The coefficient of q^n in $f(q)$ is

$$(-1)^{n-1} \frac{\exp\left(\pi\sqrt{\frac{n}{6} - \frac{1}{144}}\right)}{2\sqrt{n - \frac{1}{24}}} + O\left(\frac{\exp\left(\frac{\pi}{2}\sqrt{\frac{n}{6} - \frac{1}{144}}\right)}{\sqrt{n - \frac{1}{24}}}\right)$$

It is inconceivable that a single ϑ function could be found to cut out the singularities of $f(q)$.

Mock ϑ-functions

$$\phi(q) = 1 + \frac{q}{1+q^2} + \frac{q^4}{(1+q^2)(1+q^4)} + \cdots$$

$$\psi(q) = \frac{q}{1-q} + \frac{q^4}{(1-q)(1-q^3)} + \frac{q^9}{(1-q)(1-q^3)(1-q^5)} + \cdots$$

$$\chi(q) = 1 + \frac{q}{1-q+q^2} + \frac{q^4}{(1-q+q^2)(1-q^2+q^4)} + \cdots$$

These are related to $f(q)$ as shown below.

$$2\phi(-q) - f(q) = f(q) + 4\psi(-q)$$

$$= \frac{1 - 2q + 2q^4 - 2q^9 + \cdots}{(1+q)(1+q^2)(1+q^3)\cdots}$$

$$4\chi(q) - f(q) = \frac{(1 - 2q^3 + 2q^{12} - \cdots)^2}{(1-q)(1-q^2)(1-q^3)\cdots}$$

These are of the 3rd order.

Mock ϑ-functions (of 5th order)

$$f(q) = 1 + \frac{q}{1+q} + \frac{q^4}{(1+q)(1+q^2)} + \frac{q^9}{(1+q)(1+q^2)(1+q^3)} + \cdots$$

$$\phi(q) = 1 + q(1+q) + q^4(1+q)(1+q^3) + q^9(1+q)(1+q^3)(1+q^5) + \cdots$$

$$\psi(q) = q + q^3(1+q) + q^6(1+q)(1+q^2) + q^{10}(1+q)(1+q^2)(1+q^3) + \cdots$$

$$\chi(q) = 1 + \frac{q}{1-q^2} + \frac{q^2}{(1-q^3)(1-q^4)} + \frac{q^3}{(1-q^4)(1-q^5)(1-q^6)} + \cdots$$

$$= 1 + \left\{\frac{q}{1-q} + \frac{q^3}{(1-q^2)(1-q^3)} + \frac{q^5}{(1-q^3)(1-q^4)(1-q^5)} + \cdots\right\}$$

$$F(q) = 1 + \frac{q^2}{1-q} + \frac{q^8}{(1-q)(1-q^3)} + \cdots$$

$$\phi(-q) + \chi(q) = 2F(q).$$

CHAPTER 6. RAMANUJAN RETURNS TO INDIA

$$f(-q) + 2F(q^2) - 2 = \phi(-q^2) + \psi(-q)$$

$$= 2\phi(-q^2) - f(q) = \frac{1 - 2q + 2q^4 - 2q^9 + \cdots}{(1-q)(1-q^4)(1-q^6)(1-q^9)\cdots}$$

$$\psi(q) - F(q^2) + 1 = q \frac{1 + q^2 + q^6 + q^{12} + \cdots}{(1-q^8)(1-q^{12})(1-q^{28})\cdots}$$

Mock ϑ-functions (of 5th order)

$$f(q) = 1 + \frac{q^2}{1+q} + \frac{q^6}{(1+q)(1+q^2)} + \frac{q^{12}}{(1+q)(1+q^2)(1+q^3)} + \cdots$$

$$\phi(q) = q + q^4(1+q) + q^9(1+q)(1+q^3) + \cdots$$

$$\psi(q) = 1 + q(1+q) + q^3(1+q)(1+q^2) + q^6(1+q)(1+q^2)(1+q^3) + \cdots$$

$$\chi(q) = \frac{1}{1-q} + \frac{q}{(1-q^2)(1-q^3)} + \frac{q^2}{(1-q^3)(1-q^4)(1-q^5)}$$
$$+ \frac{q^3}{(1-q^4)(1-q^5)(1-q^6)(1-q^7)} + \cdots$$

$$F(q) = \frac{1}{1-q} + \frac{q^4}{(1-q)(1-q^3)} + \frac{q^{12}}{(1-q)(1-q^3)(1-q^5)} + \cdots$$

have got similar relations as above.

Mock ϑ-functions (of 7th order)

(i) $\quad 1 + \dfrac{q}{1-q^2} + \dfrac{q^4}{(1-q^3)(1-q^4)} + \dfrac{q^9}{(1-q^4)(1-q^5)(1-q^6)} + \cdots$

(ii) $\quad \dfrac{q}{1-q} + \dfrac{q^4}{(1-q^2)(1-q^3)} + \dfrac{q^9}{(1-q^3)(1-q^4)(1-q^5)} + \cdots$

(iii) $\quad \dfrac{1}{1-q} + \dfrac{q^2}{(1-q^2)(1-q^3)} + \dfrac{q^6}{(1-q^3)(1-q^4)(1-q^5)} + \cdots$

These are not related to each other.

Ever yours sincerely
S. Ramanujan

Commentary. The first paragraph and a sampling of formulas from the sequel of this letter can be found in Ramanujan's *Collected Papers* [226, pp. xxxi–xxxii, 354–355]. However, the complete text beginning with the second paragraph is photocopied with the "lost notebook" [227, pp. 127–131]. We have been unable to locate an original copy of the first paragraph of the letter, and it is highly probable that a portion of the letter following the first paragraph is missing. For almost 60 years, the only information available about

Ramanujan's work in 1919–1920 was contained in his last letter to Hardy. Now we know that the results Ramanujan described in this letter are part of a much longer collection, called the "lost notebook" [227].

In the 1930s, G.N. Watson [289], [290] proved most of the assertions in the last letter, and A. Selberg [254], [255, pp. 22–34] examined the behavior of the seventh order mock theta functions near the unit circle. After this work, the asymptotic formula for the coefficients of $f(q)$ was the only remaining unproved assertion in the letter. This formula was proved by L. Dragonette [76] and refined by G.E. Andrews [4].

While this work seemed to answer the immediate questions raised by the letter, there still remained subtle questions raised by Watson. These questions were made even more significant once the lost notebook was examined. Basing his study on several of Andrews' papers, D. Hickerson [132] proved the Mock Theta Conjectures (as they had come to be called).

In an important paper [67], H. Cohen studied the contrast of mock theta functions with modular forms. However, a complete understanding of mock theta functions must be reserved for the distant future. For a survey of work done on mock theta functions, see Andrews' paper [10].

Francis Dewsbury to G.H. Hardy   **15 January 1920**

SENATE HOUSE
MADRAS

Dear Mr. Hardy,

In continuation of my letter of the 22nd inst., I now write to add that Mr. Ramanujan reports that he has no cheque book for use with his Cambridge bankers but that he has written to them and to you as to his debts and as to subscriptions to mathematical societies. I trust that this will be sufficient to put matters right. Mr. Ramanujan gives his address as Kudsia, Harrington Road, Chetpet, Madras. This office can always forward communications to him if necessary.

Yours sincerely,
Francis Dewsbury

Commentary. Ramanujan was a member of the London Mathematical Society. Elected on 6 December 1917, he never signed the Members' Book, so that it is doubtful whether he attended any meetings of the Society after that date.

Through the cooperation of T.V. Rangaswami and K. Srinivasa Rao, we have obtained copies of some of the diets prescribed for Ramanujan in the last two months of his life. Some diets are given in Tamil; others are written in English. Below are translations of the diets for 26, 27 March 1920.

26 March 1920	Friday
7 a.m.	corn flakes, 4 ounces
	lime juice, 4 ounces
11 a.m.	yogurt, 8 ounces
2 p.m.	yogurt, 6 ounces
6 p.m.	yogurt, 4 ounces
	lime juice, 8 ounces
10 p.m.	kamala pazam, 5
total	yogurt, 18 ounces

27 March 1920	Saturday
8 a.m.	portions(?), milk, 3 ounces
	lime juice, 4 ounces
9 a.m.	kamala pazam, 1
10 a.m.	rasam sadam, 3 ounces
1 p.m.	yogurt, 8 ounces
4 p.m.	murukku, 4
	lime juice, 4 ounces
10 p.m.	kamala, 4
total	rice, 3 ounces, yogurt, 8 ounces

Commentary. Kamala pazam is a sweet orange, similar to a tangerine. Rasam sadam is a thin lentil soup with rice. Murukku is a very common crunchy light snack. The amount of food in these diets is meager indeed.

On the page opposite that of the two menus, the names Hari Narayana, Hari Govinda, Hari Gopala, and Hari Vasudeva are repeatedly written. These are names for the God Vishnu. It is common to write repeatedly the name of a Hindu God as a form of mental discipline or purification of the mind. We do not know the identity of the writer.

Chapter 7
After his Death

Article from *Madras Mail*   **27 April 1920**

S. Ramanujan—Some Reminiscences
by
P. Adinarayana Chetty, Barrister-at-Law

A fair complexion, a slightly pitted face, dreamy eyes with an absent-minded look about them, the average height and a body slightly inclined to be stout—such was Ramanujan, the great mathematician. His rooms in the outer court of Trinity College, Cambridge, were as unimposing as their occupant. An irregular shape, exposed on two sides, painted in some dirty tint of red, but with one shelf on which were some volumes of mathematical books of the sort not generally used by students, a German dictionary, a devotional work or two in Tamil and plenty of blank paper scattered on the table on which a faded hat was thrown—such was the sanctuary to which every Indian visitor to Cambridge during the years 1914 to 1918 turned his steps. Indeed visitors came from all parts of the World for such is fame. Even during the short period I was privileged to stay with him, I had many an occasion to hear from his visitors who were evidently taken by surprise of the unimposing figure of the great genius and the scantily supplied shelves. "Are you the great mathematician?" Naturally, the very prime object of the query used to shrink further into his chair on such occasions. But after the visitor left, Ramanujan used to ask me, "Can you suggest the appropriate reply to the question and describe the dramatic pose to be assumed when giving that reply?" And we used to have a hearty laugh. To speak the truth, I myself was first struck by the absence of learned treatises in his room. As regards the pencilled sheets of white paper, my wife asked him one day, "May I take a sheet of paper from the table?" He replied—"Take all the white paper you want; but only any pencilled or scribbled sheets, please leave alone". It shows he dreamt (if I may use that word) his mathematics or evolved it out of the inexhaustible sources of a wonderful brain.

We were staying in Oxford in the summer of 1917. When we proposed to the Indian friends with whom we were staying whether we should write to Ramanujan to look out for lodgings for us at Cambridge which we proposed to visit, our host said: "You may write him, but you will not get any reply. But depend on it, what you want to be done, would be done". That man was not evidently given to much correspondence. We ventured to Cambridge without previously writing to Ramanujan. We invaded his room. It was about 5 p.m. Ramanujan was absent from his room—a rather unusual thing with him. My wife remained in the room and I went out to arrange for our lodging during our stay at the place. Meanwhile Ramanujan returned, profusely apologised to my wife for his absence and disregarding her protestation that we were vegetarians and attributing it to a desire not to give him any trouble, Ramanujan ordered an elaborate dinner at the excellent Mess of the Cadet Corps which was stationed at the College then. My wife did not protest more strongly because she thought that Ramanujan, like most Indian students who go to England had no objection to eat meat. In a few minutes, however, I came back, and after a brief talk it transpired that all three of us were strict vegetarians, and the dinner order was cancelled. Such was our first acquaintance with Ramanujan. During the days we stayed in Cambridge we learned to love the simple soul, and we stayed in his room from 8 a.m. to 11 p.m. (the latest hour allowed for strangers to remain within the college precincts). And the time went pleasantly with Indian curries and talk of the old folk at home. But his health was even then giving way.

One day some of his friends prevailed on him to go to London to consult a specialist there. And he went saying he would be away for a week after securing permission from Dr Hardy for our stay in his rooms, during his absence. We afterwards learnt, though we had not taken advantage of it, that he had asked the grocer to send up all our requirements and to charge to his account. At London he suspected that the people were trying to give him some meat food (in liquid form) as they thought not unnaturally that his ill-health was due to bad feeding as a result of his insisting on cooking his own food without knowing how to do it. He promptly vomited what they gave him and next morning before the household got up he was seen wending his way to the railway station. He went to Liverpool Street Station. That was the day when the worst Zeppelin raid took place. Thousands were hurrying away from the station, on the platform of which shattered corpses were lying and whole trains were reduced to splinters. People were hurrying from the station for all they were worth. The absent-minded genius never noticed that he was a solitary figure struggling against that vast human stream. The raid was then actually in progress. Fortunately, however, a porter pushed him unceremoniously back saying: "Do you want to be killed?" It was only then that he realised what he was doing. He came by a later train with a peculiar satisfaction on his face at his having outwitted his friends.

None of his friends failed to note that he was unusually active during the time we were staying with him. He took fairly long walks. He accompanied us even to Music Halls. One day we went to the botanical gardens. His favourite pastime that day was to call the plants by some familiar Indian names and derive a sort of child-like delight from it. We went on the river often. But it was no easy matter, however. He could not row and what was worse he would not keep steady for a minute. Our friend from Oxford, Mr Sastri, had to manipulate the pole. We would give him no assistance of any kind and Ramanujan would keep moving and shifting every second. One day after elaborately cautioning Ramanujan, Sastri promised to take us for a longer time on the river. But when we came to the Lock, at a critical movement, Ramanujan's promise was broken and down went Sastri into the river and the three of us got a good ducking. We hurried home, changed or dried our clothes before the fire and went out again. But not to the river!

Afterwards his illness gradually became worse. I wrote some letters but got only a couple of notes from him in reply. He had to go to a sanitarium. But there too not satisfied with the diet they gave him and rebelling against the doctor's discipline he ran away more than once.

The next time I met him was at the Madras Central Station. A group of friends were on the platform. Although it was a warm day in March (1919) he had his heavy woolen comforter and his top coat. I gave him my arm and helped him out of the carriage. He could not then recognise me. He was shaking a bit and had a sort of dazed look in his eyes. But once in the car, he recognised me and when we came to my (friend's) house and had changed his dress, he resumed his former pleasantness of look and talked of the old times. The doctor had ordered complete rest. But one day when he was slightly better, I caught him reading and reminded him of the promise he had given to the doctor. With an apologetic smile to which I had grown so used, he said: "That is a book which I picked up at Heffer's on the last day of my stay at Cambridge. It is by a continental author and is the first real attempt to trace out the missing books of Euclid. Do allow me to read it. I assure you I feel better already. Moreover, you see I must do some work or other to justify the allowance I draw from the University."

He had kind friends and attention and he steadily began to improve. But it was patent to his friends that his system was undermined. He was advised to change to some other place. He went to Kodumudi and told me afterwards how he liked the broad expanse of the Cauvery and the quiet life on the banks of the river. The last time I met him was at his native place—Kumbakonam. It was a small or rather poor residence for the genius whose contribution to mathematical knowledge is of world-wide fame. He was reclining on a bed spread by the floor which was not over clean. He had on his side a half-finished green cucumber.

"Ramanujan you are at it again. The doctor has told you not to eat this kind of stuff". With a glint in his look such as is seen in that of a child caught stealing jam, he said apologetically, "But, I must eat something or other". I advised him into a better and ventilated house. He said that his mother had just gone out to look for a house and pressed me to wait until she came back. That was the last occasion when I met him.

Today all that remains of him is his fame. I am not gifted enough to appreciate his great intellect. But as an Indian, his memory will ever be sacred to me. The picture of the red-painted walls of his room, the pencilled sheets flying about, the smile with which he would greet us every morning and the child-like guilty face with which he would meet my chiding look whenever I caught him eat something prohibited by the doctor—these will remain in my memory as the stories of the absent-minded predecessor of his in the same college—the great Newton and his rooms in the beautiful Fountain Court of the famous college. Today the same court is there. There are no doubt flowers, fine spring flowers, round the fountain the same as before. The old clock on the magnificent entrance tower of Henry, the face of which we use to scan in the lingering summer twilight as we watched for the latest minute when we must leave Ramanujan's rooms for our lodgings; all these are there; but the man who brought fame to the land of his birth in an important department of human knowledge is not there today.

Commentary. The article appeared in the *Madras Mail* on the day after Ramanujan's death. Little is known about the author. He is mentioned briefly on page 92 of volume 1 of P.K. Srinivasan's books [**263**], but we have been unable to find more about him.

By the 'outer court of Trinity College,' the author means Whewhell's Court, which is on the other side of Trinity Street from the main college buildings.

A slightly different story concerning Ramanujan's return to Cambridge from London during the bombing of Liverpool Street Station is given by Dr C.D. Deshmukh on page 79 of S.R. Ranganathan's book [**229**].

The chief means of relaxation on the river Cam was, then as now, by punt, a vessel propelled by a long pole manipulated by someone standing on a small platform at the rear of the craft. The motion could be assisted—or, as in Ramanujan's case, disturbed—by other members of the party using paddles on either side. The Mr Sastri from Oxford, who is mentioned as being in charge of the pole, is probably P.P.S. Sastri, who was a member of Balliol College, Oxford, and graduated in 1919 in Oriental Languages.

On page 182 of volume 1 of [**263**] it is stated that Ramanujan arrived at Madras Central Station, not in March, but on 2 April. Heffer's is the bookshop in Cambridge at which Ramanujan purchased the book on Euclid. Professor Wilbur R. Knorr has suggested that this book may have been Michel Chasles's *Les trois livres des Porismes d'Euclide* (Paris, 1860). It is likely that Ramanujan had acquired a smattering of French, since he makes occasional references to French authors in his papers, and one of his papers written together with Hardy [**123**] is in that language.

An account of Ramanujan's stay in Kodumudi will be found on pages 319–321 of Kanigel's book [**142**].

GOMETRA, THE HOUSE IN MADRAS WHERE RAMANUJAN DIED

The Fountain Court mentioned at the end of the article is Trinity Great Court. The Henry mentioned in connection with the gate tower is Henry VIII, who in 1546 founded the college, the largest in Cambridge, by consolidating and extending two earlier foundations.

J.J. Hensman to G.H. Hardy 29 April 1920

From THE MANAGER IN CHARGE, OFFICE OF
THE REGISTRAR, UNIVERSITY OF MADRAS
To G.H. Hardy, Esq., M.A., F.R.S.,

Dear Sir,
By direction of the Syndicate, I write to communicate to you, with feelings of deep regret, the sad news of the death of Mr. S. Ramanujam, F.R.S., which took place on the morning of the 26th April.

Yours faithfully,
John J. Hensman
Manager-in-charge.

S. Lakshmi Narasimhan to G.H. Hardy 29 April 1920

Temporary Address
Gometra
Harrington Road
Chetpet
Madras

Dear Sir,
I am extremely very sorry to tell you that my brother Mr S. Ramanujan F.R.S. died on 26th inst. at an early age of 32.

All his mss that were in his trunk were handed over on the day of his death to Mr. Ramachandra Row's son in law, since the former is at Nilgris. Not only those but also the journals, magazines — all he possessed except the books which he had, were taken away by them. He has brought a dozen books from England and a week before his death he bought another one and he has sent for Hutton's Mathematical Tables. All the books and all the suits and all things are entirely under the control of his wife here. Mr. Ramachandra Row's son in law told me that he wrote a letter to the latter telling him that he would make arrangements with the Madras Government to send a monthly instalment of Rs.—(amount I don't know) to his wife. I am very sorry that why [sic] such an arrangement was not made for my family.

Sir, I have a lame grandmother who was greatly responsible for bringing my brother up and educating him. As soon as my brother's departure to England, my father has lost his two eyes and thus he remained a blind man for his stay in England for 5 years and after my brother's arrival here one eye is restored to him after medical treatment. His age is 65. Besides these, I have got a corpulent mother who resembles my brother in all his physical features.

I have an younger brother reading in the high school. I have no sisters. Coming to myself, I am a young man of 18, appeared this year for the Intermediate Examination and I did not tell you my studies are not interrupted this year but then while he was living, I did not care for this and being wrapped myself in ecstasy I remained with him for the past one year and served him cordially.

I have no uncles or cousins to protect me. We have no property, as you might have known very well. I have a great desire to study and I wish 'to neglect worldly ends, all dedicated to closeness and the bettering of my mind' (Tempest). I have no taste, sorry to say, for Mathematics. I like to read Shakespeare, Wordsworth, Tennyson and wish to travel in the fairy land 'half flying half on foot'. I do not know how to feed them. Therefore, I humbly request you to write to the Madras University to give a monthly allowance to us. I have been told that my brother is entitled to get a sum from the Cambridge University.

In short, I am a young man. One day passes with great difficulty. I do not know how to protect them. Unless any arrangement is made to support us, we have to go a-begging from door to door. I entrust the whole case into your hands. It is your bounden duty to protect us.

Please kindly reply me at your earliest convenience.

my permanent address
17, Sarengapani
  Sannidhi St.
    Kumbakonam
      S. India

Yours sincerely,
S. Lakshmi Narasimhan

Commentary. Srinivasa Lakshmi Narasimhan was born in Kumbakonam on 18 April 1898, and his younger brother Srinivasa Tirunarayanan on 19 April 1905 there. Both brothers obtained employment in the Post Office in Madras.

Nilgris is the hilly Nilgiris District west of Madras near the west coast. Therein lies Ootacamund (known as Ooty), the chief summer resort of South India and summer home of the governor of Madras.

Charles Hutton, F.R.S. (1737–1823) was Professor at the Military Academy, Woolwich, and the author of several mathematical works. Among these was a volume of Mathematical Tables which ran through several editions in the early 19th century. The tables, mainly to seven decimal places, were of standard functions, logarithms (ordinary and hyperbolic), and the trigonometric ratios. It is difficult to understand how they could have been of interest to Ramanujan.

The quotation from *The Tempest* is spoken by Prospero in Act I, Scene 2. The provenance of the following quotation, 'half flying half on foot', has not been determined; it does not occur in the works of the three authors mentioned earlier in the sentence. Nor have we found the origin of the writer's earlier phrase 'wrapped myself in ecstasy'. He was clearly anxious to impress Hardy with his knowledge of English literature.

MR RAMANUJAN'S ACCOUNTS

Expenditures	£		
Nov. 28, 1917, Dr Mutlin	3	13	11
Dec. 1, 1917, Dr L. Ram	14	0	0
Dec. 24, 1917, Dr L. Ram	12	18	2
Jan. 3, 1918, Matlock San.	15	18	1
Mar. 21, 1918, Dr Kincaid	63	15	0
Mar. 21, 1918, Capt. Wynn	21	0	0
Apr. 30, 1918, Dr L. Humphrey	2	2	0
Apr. 30, 1918, Dr Wingate	25	9	0
July 3, 1918, Dr Kincaid	87	5	10
Nov. 2, 1918, Home Hosp. Ass.	50	8	0
Dec. 5, 1918, Mr Neville	2	15	0
Dec. 8, 1918, Mr Phillips	21	0	0
Dec. 9, 1918, Home Hosp. Ass.	29	18	2
Jan. 13, 1919, Mr Phillips	20	18	8
Mar. 17, 1919, Dr Kincaid	38	18	8
Mar. 17, 1919, Nursing Hostel	30	6	2
Mar. 25, 1919, Mr Hardy	54	0	0
Aug. 23, 1919, Dr Bolton	<u>70</u>	<u>7</u>	<u>0</u>
	564	13	8

MR RAMANUJAN'S ACCOUNTS

Receipts	£		
Dec. 17, 1917, Mr Ramanujan	100	0	0
Mar. 20, 1918, Mr Hardy	50	0	0
July 4, 1918, Mr Ramanujan	100	0	0
Oct. 22, 1918, Mr Ramanujan	100	0	0
Feb. 12, 1919, Mr Ramanujan	100	0	0
Aug. 23, 1919, Mr Ramanujan	100	0	0
Aug. 23, 1919, Mr Hardy	<u>14</u>	<u>13</u>	<u>8</u>
	564	13	8

Commentary. These accounts were sent to the University of Madras by Hardy after Ramanujan's death. The names in the expenditure list are mainly those of the medical practitioners who treated Ramanujan. The first is Dr. Chowry-Muthu (not Mutlin), the proprietor of the sanatorium at Hill Grove. Dr. Ram and Dr. Kincaid practised in Matlock House; see A.S. Ram's long letter of 23 June 1918 to Ramanujan. Drs. W.H. Wynn, L. Humphrey, and W.W. Wingate were specialists who examined Ramanujan on various occasions, and Mr. S.M. Phillips was the proprietor of the nursing home in Colinette Road in Putney. Additional information about these gentlemen can be found in [233] and [234]. Dr. Bolton has not been identified; there were several doctors of that name at the time. The sums of money are in pounds, shillings, and pence, the totals being £564.68 in decimal currency.

E.M. Clark to G.H. Hardy   7 May 1920

BARCLAYS BANK LIMITED
CAMBRIDGE.

My dear Hardy,

S. RAMANUJAN

The document you sent us only requires a 2d stamp stuck on it to make it a perfectly valid one. Therefore strictly speaking we ought to pay over to you Ramanujan's balance which today amounts to £343: 4: 8: As I gather from your letter that the bills you have to settle only amount to about £40, you probably would rather not have so much money paid over to you. If you like therefore the Bank will pay you the exact sum you require and then write and tell Ramanujan what they have done and ask his instructions as to the disposal of the balance. Do you think it would be a good thing if we suggested to him that he should authorize you to draw on his account? After all he will have considerable sums to be paid to him by Trinity College before his Fellowship expires and in default of further instructions from him these will simply roll up his banking account here. If you wish us to write to Ramanujan please let me know his address in India. Is the University of Madras sufficient?

Yours ever,
E. Mellish Clark

Professor G.H. Hardy
    New College,
        Oxford.

Commentary. Edward Mellish Clark was the Local Director of Barclays Bank and had been an undergraduate at Trinity College. He graduated BA in 1896 and MA in 1900. As a graduate of some standing, he probably had dining rights at his old College, which is probably why he was familiar with Hardy. Hardy had been appointed Savilian professor of geometry at Oxford during the previous year and was now resident in New College, where he had been elected to a Fellowship.

Komalatammal to the Government of Madras   **14 May 1920**

From
    Srimati Komalatammal
        (mother of late Mr S. Ramanujan, F.R.S.)
            c/o
        M.R.Ry. P.V. Seshu Aiyar Avl.
        18 Pycroft's Road
            Triplicane

To
    The Secretary
        Educational Department
            Government of Madras

Sir,

    I am very sorry to inform you that my son Mr. S. Ramanujan F.R.S. passed away on 26 April '20 and that he leaves behind an old grandmother, aged and infirm parents and two brothers reading at school and his wife to bemoan the loss and the whole family is left in a distressed and destitute condition. He did not live long enough to make any provision and after his death our condition socially also has become much worse. So long as he was alive nobody raised any questions of caste or religion though he was an England–returned gentleman. The latter fact we are made to feel religiously for no one will officiate as Purohits at our religious ceremonies and socially we are ostracised because we associated with an England returned–gentleman.

    Even when it was proposed that he should be sent to England, I anticipated these difficulties and was not at first willing to send him there. But when the government and the Syndicate were so pleased to grant him a liberal

KOMALATAMMAL, RAMANUJAN'S MOTHER

scholarship and when my wellwishers appealed to me by saying that my son was a great mathematician and that he would bring credit to India and that I should not stand in his way of his being sent to England, I readily yielded and I only pray that the government will not give up the family especially when it is reduced to distressed and destitute condition and when Mr. S. Ramanujan was asked as regards the provisions, he wanted us to appeal to the Syndicate and government for an allowance.

I therefore pray that the government will be pleased to sanction an adequate allowance for his aged parents and scholarships to enable his brothers to continue their studies.

I beg to remain,
Sir,
Your most obedient servant,
Komalatammal

Commentary. The abbreviation Avl. after P.V. Seshu Aiyar's name denotes the Tamil word, Avargal, which can be translated as "respected sir."

A Purohit's function is similar to that of a priest.

E.M. Clark to G.H. Hardy   **17 May 1920**

BARCLAYS BANK LIMITED,
CAMBRIDGE.

My dear Hardy,

S. RAMANUJAN

I return the standing order form filled in on the assumption that the payments due on March 25th, 1920 has not been made. Please sign the form as I have pencilled the names.

I also enclose for your signature a cheque which please fill up for the exact amount that Ramanujan owes you and pay it into your banking account in the ordinary way.

As soon as this cheque has been passed through we will write to Ramanujan, if you will let me have his address, and tell him what we have done. I quite agree that as there will be so few transactions on the account the bank must either allow interest on current account or place the whole sum on deposit. Possibly the former course will be the simpler.

Yours ever,
E. Mellish Clark

Professor Hardy,
    New College,
        Oxford.

E.M. Clark to G.H. Hardy   **19 May 1920**

BARCLAYS BANK LIMITED,
CAMBRIDGE.

My dear Hardy,

S. RAMANUJAN

Thanks for the standing order to the Royal Society which we will pay.

I herewith return as requested the statement you sent us.

As soon as the cheque you have drawn has been presented for payment we will write to Ramanujan and tell him what we have done.

Yours ever,
E. Mellish Clark

P. S. From what I have heard of Cazalet I expect you are right and that he will beat Huband; it should however be a good match.

Professor Hardy,
    New College,
        Oxford.

Commentary. Clearly, the news of Ramanujan's death had not yet reached Barclays Bank by 19 May.

The postscript refers to the Inter-University Real Tennis Singles, which were played on 9 July 1920, when Victor Alexander Cazalet (1896–1943) of Christ Church, Oxford, beat George Darley Huband of Trinity College, Cambridge, by three sets to two. They were the 'first strings' in their respective university teams. Both Cazalet and Huband came to their universities after service in World War I, in which each won the Military Cross. Cazalet read Modern History and was awarded 'blues' for real tennis, lawn tennis, and racquets. He was elected M.P. for Chippenham in 1924, and served in World War II as liaison officer with the Polish forces. He died in mysterious circumstances in an air accident near Gibraltar, while he was accompanying General Sikorski, the Polish Prime Minister and Commander in Chief, back to London. Huband left after only one year at Cambridge and went into shipping. He became joint general manager of the Cunard Company in Canada, where he died. Hardy had an interest in games of every kind, and in his younger days was an accomplished real tennis player.

S. Janaki Ammal to Sir Freeman Freeman-Thomas   20 May 1920

To
> His Excellency the Right Honourable
> Sir Freeman Freeman-Thomas
> Baron Willingdon of Ratton,
> G.C.S.I., G.C.I.E., G.B.E.
> Governor of Madras in Council

The humble memorial of Janakiammal
> Widow of the late Mr. S. Ramanujan,
> F.R.S. (Mathematical Scholar of
> the University of Madras), now
> resident at 6, New Street,
> Mambalam, Saidapet.

May it please Your Excellency,

Your Excellency's humble memorialist believes that Your Excellency is aware of the death of her husband, Mr. S. Ramanujan, which took place in Madras on April 26, 1920; and she most respectfully requests that in view of the circumstances stated below Your Excellency's Government may be graciously pleased to sanction the grant of an adequate allowance for her future maintenance.

2. Mr. Ramanujan is said to have shown considerable promise as a mathematician at a very early age. Owing to the extreme poverty of his parents he was unable to obtain a complete University education, and was obliged to seek employment as a clerk soon after his school career, though he never ceased to devote all his spare time to the study of Mathematics. In April 1913, the University of Madras, in recognition of his reported exceptional

S. Janaki Ammal, Ramanujan's wife

abilities, and with the sanction of Government, appointed him a Research Scholar on a scholarship of Rs.75 per mensem, thus enabling him to give his undivided attention to his researches, which by this time had attracted the favourable notice of other eminent mathematicians in India and abroad. In February of the following year, however, Government sanctioned another proposal of the University that Mr. Ramanujan "should be enabled by the grant of a scholarship, to proceed to England so as to continue his studies and researches in Mathematics at the University of Cambridge." The scholarship sanctioned was of the value of £250 per annum. Mr. Ramanujan was at first unwilling to go to England on social and religious grounds, but eventually yielded to the view put to him by his friends that by doing so he might be able to advance the progress of science and raise the prestige of India one more step, as well as improve his own financial position. Accordingly, Mr. Ramanujan left India in March 1914, and after spending five years in England, returned to India in April 1919.

3. Your Excellency's humble memorialist begs to quote the following extracts from a report furnished in 1916 by Prof. G.H. Hardy, F.R.S., of the Trinity College, Cambridge, to this University of Madras on "Mr. S. Ramanujan's Mathematical Work in England."

"It is already safe to say that Mr. Ramanujan has justified abundantly all the hopes that were based upon his work in India, and has shown that he possesses powers as remarkable in their way as those of any living mathematician. His work is only the more valuable because his abilities and methods are of so unusual a kind, and so unlike those of a European Mathematician trained in the orthodox school."

"I have said enough, I hope, to give some idea of its (Mr. Ramanujan's work) astonishing individuality and power. India has produced many talented mathematicians in recent years, a number of whom have come to Cambridge and attained high academical distinction. They will be the first to recognise that Mr. Ramanujan's work is of a different category. In him India now possesses a pure mathematician of the first order, whose achievements suggest the brightest hopes for its scientific future."

Perhaps a more eloquent and authoritative testimony to the value and importance of his contribution to the Science of Mathematics is afforded by his election as a Fellow of the Royal Society which followed in 1918.

4. Your Excellency's humble memorialist begs to take this opportunity to express her gratitude for all that has been done to enable her late husband to devote himself to his mathematical researches untrammelled by the cares and anxieties of poverty, but at the same time ventures respectfully to submit that in proceeding to England he took a step which eventually proved fatal to his life; and that, at any rate, if he had not left India, it is improbable that he would have died at the early age of 31. Mr. Ramanujan belonged to the highly conservative Sri Vaishnava Brahmin community, and the social traditions, beliefs and prejudices of that community to which he clung with some tenacity before he went to England, peculiarly unfitted him for life in the wholly different conditions of that country. It has been said for instance, that he was the only Indian in England to have thought it necessary to cook his own food, and he has been known to take stale food rather than take the food touched by other hands. This caste exclusiveness was one of the main contributory causes of the results that his health in England was soon undermined that he was easily attacked by, and suffered from, two successive epidemics of pneumonic influenza, and that during the latter part of the period of his stay in England, he was a constant inmate of the hospitals. The War perhaps intervened to prevent his repatriation in proper time and possible restoration to health, so that when he returned to India in April 1919, he returned thoroughly shattered in health and with the seeds of tuberculosis sown in his system to which he succumbed a year later.

5. The death of Mr. Ramanujan has rendered Your Excellency's humble memorialist thoroughly destitute. Owing to his prolonged illness he had not been able to save anything, nor is there any likelihood of the memorialist being maintained by her parents-in-law, who are themselves poor. Your Excellency's humble memorialist is now only about 22 years of age and the

prospect of the greater part of a lifetime yet to be lived, with no means of livelihood, and with the social and religious disabilities attaching to the widow of an England-returned gentleman adding bitterness to the poverty, obliges her to make this earnest appeal to the generosity of Your Excellency's Government. Prof. G.H. Hardy wrote some time ago in a private letter: "It is inconceivable that India is going to allow its one and only F.R.S. to starve!" Your Excellency's memorialist craves permission to express the humble hope that Your Excellency's Government may be willing to do something to relieve the distress and despair of the widow of that solitary Indian F.R.S. who endeavoured successfully and died in his endeavour to increase the sum of human knowledge.

The memorialist begs to remain
Your Excellency's most obedient and humble servant,
S. Janaki

Commentary. This letter and a brief note dated 5 August 1921 are the only letters from Mrs. Ramanujan that we have been able to locate. The location of her residence evidently was near the boundary of two sections of south Madras, Mambalam and Saidapet. In 1920, Saidapet perhaps was outside the municipal boundary of Madras. Both Mambalam and Saidapet are north of Guindy, the home of the King Institute providing the medical report of Ramanujan on 30 July 1919.

Since Mrs. Ramanujan did not know English, it is clear that the letter must have been written by a friend or adviser.

Freeman Freeman-Thomas (1866–1941), who later became the first Marquis of Willingdon, was Governor of Madras from 1919 to 1924. He served as Vice-Roy and Governor-General of India from 1931 to 1936.

E.M. Clark to G.H. Hardy   **25 May 1920**

BARCLAYS BANK LIMITED,
LOCAL HEAD OFFICE,
CAMBRIDGE.

My dear Hardy,

S. RAMANUJAN

His death complicates matters to this extent, namely, that it automatically cancels his instructions to us to pay the balance of his account to you. Consequently no one can draw on the balance of his account until either Probate of his Will has been exhibited to us or someone has taken out Letters of Administration.

Have you any idea whether he made a will and if so whether it would be in this country or in India?

I suppose Trinity owe him the proportion of his Fellowship up to the date of his death.

I am writing as you suggest to the Registrar of the University of Madras.

Yours ever,
E. Mellish Clark

Professor G.H. Hardy
    New College,
        Oxford.

G.H. Hardy to Sir J.J. Thomson   May 1920

I have just heard from India that Ramanujan is dead. It is a great shock to me: for I thought, before he went back, he had begun to turn the corner.

Could not Trinity do something to commemorate him permanently in a small way? After all he was a most extraordinary genius, of whom even Trinity may justly be proud. I am only just beginning to hear from abroad how greatly everyone has been impressed by his work.

One possibility (and perhaps as good as any) would be to help to bring out an edition of his published work, with such additions as can be made from the MSS he has left. Of course Madras may be proposing to do this: I don't know what funding they have available for such a purpose. And in that case some other form of memorial might be more appropriate. In any case, would you mention the question to the Council? I feel very clear that <u>some</u> action of some kind would be both reasonably and politically graceful.

Yours sincerely,
G.H. Hardy.

It would be worth while for you to ask Littlewood what he thinks.

Commentary. The beginning of the letter, which is written from New College, Oxford, is missing. The original (Add. 7654, H29) is in the University Library, Cambridge. In due course, Ramanujan's *Collected Papers* [226] were published with funding from Trinity, Madras, and the Royal Society.

G.H. Hardy to Francis Dewsbury   26 May 1920

New College,
Oxford

Dear Sir

It was a great shock and surprise to me to hear of Mr. Ramanujan's death. When he left England the general opinion was that, while still very ill, he had turned the corner towards recovery; he had even gained over a stone in weight (at one period he had wasted away almost to nothing). And the last letter I had from him (about 2 months ago) was quite cheerful and full of mathematics.

There is a substantial balance in his account at Cambridge (some £343, if I remember right), you may remember that there were a few unpaid debts (some £30 odd). He sent me, not a cheque, but a form transferring his whole account to me; and I arranged with the Bank to settle the debts and place the balance back to his credit on deposit. (He had no doubt forgotten that there would be substantial payments from Trinity—apart from these the balance would only barely have sufficed to settle the bills—so this action was not so absurd as it seems). In any case I understand that his death invalidates the whole transaction.

In any case there will be a substantial sum for his heirs, about which the bank will write to you—it will presumably be increased by the portion of a year's Fellowship up to his death.

However, these are rather sordid details. There will of course be universal regret among European mathematicians at such a tragedy—his work was just beginning to be known abroad, especially in Germany. Prof. Landau of Göttingen, Prof. Schur of Berlin were both interested and impressed by it.

Is it possible that Madras would consider the question of publishing the papers in a collected form? There should be some permanent memorial of so remarkable a genius: and this memorial would certainly be the most appropriate form. It might be possible to get some financial assistance in Cambridge.

For my part, it is difficult for me to say what I owe to Ramanujan—his originality has been a constant source of suggestion to me ever since I knew him, and his death is one of the worst blows I have ever had. There was nothing perhaps more remarkable in him than his persistent modesty. Few Indians would not have had their heads a little bit turned by the praise he had received, but he was too big a man.

I have two letters from relatives which I will forward to you shortly.

I am,

Yours very faithfully,
G.H. Hardy

Commentary. Edmund Landau (1877–1938) was a distinguished number theorist and analyst, who was the author of more than 250 papers and seven important books. In 1909 he succeeded H. Minkowski as ordinary professor at the University of Göttingen, but was forced to resign his chair in 1933 because of the political situation. For further details see the obituary by G.H. Hardy and H. Heilbronn [113]. A discussion of the work of Landau and Ramanujan on the distribution of numbers that are the sum of two squares is given in our commentary after Ramanujan's first letter to Hardy.

In the letter, as printed on page 78 of Volume 1 of [263], "Cohus" should be replaced by "Schur." This is clearly a misprint, and we are indebted to Professor W. Schwarz of Frankfurt for suggesting that it is the distinguished algebraist Issai Schur (1875-1941) who is intended. Hardy's handwriting was usually very clear, but in his letter of 23 July 1940 (see later) to A.F. Scholfield, the first three letters of the latter's name could be misread as "Coh." Schur did important work on algebra and analysis. Like Landau, and for the same reasons, he retired prematurely in 1939 from his chair in the University of Berlin. Subsequently, he emigrated to Israel. In 1917, Schur [251], [252, pp. 117–136] independently discovered and proved what are now called the Rogers–Ramanujan identities, which are briefly discussed in our commentary after Ramanujan's second letter to Hardy.

One of the letters mentioned in the last sentence of Hardy's letter above was, doubtless, the one from Ramanujan's brother S. Lakshmi Narasimhan.

John J. Hensman to the Education Department 26 May 1920

From
>John J. Hensman, Esq., B.A.,
>>Manager-in-charge, office of
>>the *Registrar, University of Madras*,

To
>The Secretary to Government,
>>Home (Education) Department.

Sir,

I have the honour to acknowledge receipt of the Government memorandum No. 977-B-1 of the 21st instant and to inform you that the memorandum and the petitions received in this office from the mother and the wife of the late Mr. S. Ramanujan F.R.S., will be laid before the Syndicate for consideration at its next meeting to be held in the latter part of July, 1920, and the decision of that body thereon will thereafter be communicated to Government.

>I have the honour to be,
>>Sir,
>>Your most obedient servant,
>>>John Hensman
>>>Manager-in-charge.

Komalatammal to Educational Department 19 June 1920

From
Komalatammal
>c/o Mr. P.V. Seshu Aiyar
>>18, Pycrofts Road
>>>Triplicane

To
>The Secretary
>>Educational Department
>>>Government of India

I humbly beg to tender the few following lines to your kind perusal.

CHAPTER 7. AFTER HIS DEATH

On Monday 17th May Mr. R. Srinivasa Iyengar brother to my daughter in law came to Mr. P.V. Seshu Aiyar's house and I also went over to the latter's house where he expressed her (daughter in law's) wish that she refused to perform the funeral ceremonies. Upon this, on Tuesday 18 May I with my son went to the Choultry at Royapuram (133, Mannar Swamy St) and remained there for 3 days and my son, S. Lakshminarasimhan has performed the whole thing. I got Rs.150 on mortgage of my bangles. I need not tell you that God has ordained that I should become an enemy to him (S. Ramanujan) before his death. Thus my daughter-in-law, it was, that kept the accounts and had not given anything to me except Rs.60 sent to Kumbakonam for the expenses of the family.

All brass vessels (except a few which I have) of my family, and silver vessels, all chairs, beds, quilts, my son's dresses, his books—all these have been moved to Mambalam. Why should not I get at least 2 chairs for my sons? What is the use of suits to her? Why should not my other son's wear? Everything I have told to Mr. P.V. Seshu Aiyar. I am in desperate straits. My son is going to join the College Kumbakonam. Another son is reading in the town high school Kumbakonam. My pecuniary troubles on one side and the grief for the loss of my son on the other side grieve me most.

Therefore I humbly request you to send Rs150 from the amount due to my son for the month of April. Since I have spent it on funeral ceremonies and I request you to grant two scholarships to two of my sons and an amount for the expenses of my family.

I beg to remain,
Sir,
Your most obedient,
Most humble servant,
Komalatammal

Commentary. A Choultry is normally a place to stay for people on a pilgrimage. Conditions at a Choultry are usually austere, but food and sleeping accommodations are provided. A Choultry could also be a place where people stay during a religious ceremony such as a funeral, as is the case here. Funeral ceremonies often last thirteen days, with a large feast held on the thirteenth day. Therefore, those coming from afar would stay in a Choultry. Royapuram is a northern section of Madras, while Ramanujan and his relatives lived in southern Madras several miles away. Thus, it was necessary for them also to stay at a Choultry. Royapuram was known as a place where funerals for poorer people were conducted. But, as indicated in the letter, the cost of a funeral was still high. In particular, those attending the ceremony would be provided with food during their stay.

As Komalatammal had no knowledge of English, this and other letters from her on later pages will have been written by others on her behalf.

Komalatammal to the Government of Madras   **26 July 1920**

From
    Komalatammal
      (mother of S. Ramanujan F.R.S.
        Late of Trinity College, Cambridge)
    17, Sarangapani Sannadhi Street
        Kumbakonam

To
    The Government of Madras
      Ootacamund.

Most Respected Sir,

I Komalatammal, mother of S. Ramanujan late of Trinity College, most humbly pray to let you know, that my family is extremely poor, and consequent to the death of my son Ramanujan, we have been reduced to the most abject condition. We educated him, though we were in very humble circumstances and did our best to bring him up to a good position. At the desire of the government, I even sent him to England. After he returned from England we were receiving every month nearly Rs270: He was the sole prop of the family and after his death, we are in an utterably miserable state of condition.

The funeral rites and ceremonies that ought to have been performed were performed by Ramanujan's brother Lakshmi Narasimhan. The debt nearly runs up to Rs300. The amount due to Ramanujan for the month of April has not been given to us, and I will most humbly request your government to pay the same to us.

Ramanujan has got two brothers. One is S. Lakshmi Narasimhan, reading in the Senior Intermediate Class of the Kumbakonam College. The other is S. Tiru Narayanan, reading in the Vth Form of the Town High School in Kumbakonam. I entrust these boys to the care of your benign government. I will most humbly request that your government will be pleased to grant them scholarships, and kindly look to their future welfare. My husband is an old man, bereft of eyesight. I have a very old mother. We are suffering very greatly for want of timely food even. We have neither lands nor money.

I would most humbly entreat your government to consider our family as yours and bestow on us every month some allowance for our maintenance. Ramanujan has left some amount of money and furniture in England. I would request you to send them to me at your earliest convenience. We solely depend upon your benign government for my maintenance. Mr Ramanujam, while he was in England, was not able to support us, but the government gave us Rs60. Even then it was a little difficult for us to manage. We are five members in our family, and we solely trust in you for our future maintenance and living.

We most humbly pray that the monthly allowance may be sent to S. Lakshmi Narasimhan, 17, Sarangapani Sannadhi Street, Kumbakonam.

I beg to remain,
Most Respected Sir,
Your most obedient servant,
Komalatammal

Commentary. The name at the top of this letter is actually Komalammal, which is a common spelling or version of Komalatammal.

R. Littlehailes to the Government of Madras   28 July 1920

Station Ootacamund
From The Hon'ble Mr. R. Littlehailes, M.A.
    Director of Public Instruction
To The Secretary to the Government of Madras, Department

Sir,

I have the honour to submit after enquires the following report with reference to Government Memorandum No. 977 B–1 dated 21st May 1920, endorsement No. 1024 B–1 dated 26th May 1920, and Memorandum No. 1233 B–1 dated 28th June 1920 regarding the family of the late Mr. S. Ramanujam.

2. Mr. Ramanujam left behind him a father, a mother, a grand-mother, two younger brothers and a wife.

3. The father appears over 60 years of age. He is infirm in body, his eye-sight is very poor and he is not now doing any work although when he was in good health he acted as a selling agent to a cloth merchant.

4. His mother is about 50 years of age, is of good health and has no special occupation. She looks after household duties and occasionally helps friends and relatives to purchase clothes and ornaments on commission. She is able to do little or nothing of this work at present as she has become aged. I understand that the mother and the widow are not on friendly terms with each other.

5. His grand-mother is aged about 70 and is a leper.

6. The elder of his brothers S. Lakshminarasimhan was a student of the Government College, Kumbakonam for the past two years reading in the Intermediate class with group I for his special subject. He has just been readmitted into the college. The Principal reports that his performance in the classes were always poor as also in examinations. His tutors do not speak well of him; his main defect is want of steady application to his studies. He pays too much attention to his toilet and dress and not sufficient to study and games. He has just failed in the Intermediate examination and the boy

states that his failure is due to the frequent disturbances during the whole of last year in his studies since he had to attend on his late brother.

7. The younger brother, S. Thirunarayanan, is at present reading in the V form of the Town High School, Kumbakonam. His attainments are reported to be in general above the average and his conduct to be very good.

8. Steps are being taken to persuade his widow to enter into the widows' Home Triplicane and be trained as a teacher. She is hesitating as to whether she should take this step and her mother and sister are persuading her not to take it since the latter consider that such a course would be irreligious and unorthodox.

9. The case of the late Mr. Ramanujam was considered by the Syndicate at its meetings on the 19th and 20th July. There is a balance of about £ 300 standing to the account of Mr. Ramanujam with Barclay's Bank Ltd., Cambridge, but they cannot pay over this balance until probate of his will has been exhibited to them or some one has taken out letters of Administration. This sum of £ 300 goes to the widow.

10. The Syndicate decided that it had no legal power to make provision for the widow or other members of the family of the late Mr. Ramanujam. It decided however to arrange if possible for the purchase of the copyright of the works of Ramanujam in consideration of the grant of an annuity to his widow amounting to Rs.20 per mensem. Sanction of Government will no doubt be sought in due course for this proposal.

11. The Syndicate further decided to recommend that the case of Ramanujam's family is a fit case for generous treatment at the hands of Government.

12. I recommend that the younger brother be granted a scholarship equal in value to the ordinary scholarships awarded together with the concession that he be not required to pay fees until he completes his school and college course provided that he obtains promotion and passes all school, college and University examinations in due course. Failure to pass these examinations or to obtain promotion should cause these concessions automatically to cease unless any very special reasons are adduced in explanation of failure.

13. I recommend further that a similar concession be granted to the elder boy who is now reading for a second year in the senior Intermediate class at the Kumbakonam college.

14. I consider that for further aid the family of Ramanujam should rely upon public and private beneficence.

> I have the honour to be,
> Sir,
> Your most obedient servant,
> R. Littlehailes
> Director of Public Instruction.

K. Ramunni Menon to Education Department   28 July 1920

From

    The Ag. Registrar,
        University of Madras

To

    The Secretary to Government,
        Home (Education) Department.

Sir,

With reference to Government Memoranda Nos. 977.B–1 and 1024.B–1, dated 21st and 26th May 1920, forwarding for remarks petitions from the mother and the widow of the late Mr. S. Ramanujam for pecuniary assistance, I have the honour, by direction, to inform you that the Syndicate regrets that it has no power to make any pecuniary grant to the parents of the deceased but considers that the case is a fit one for generous treatment by the Government.

2. I have further to inform you that the Syndicate has resolved that it would be prepared, subject to the sanction of the Senate and the Government, to purchase from the widow of Mr. Ramanujam the copyright of Ramanujam's works in consideration of an annuity, at the rate of Rs20/– per month, to be paid to her for life on her transferring all her rights in Ramanujam's manuscripts and the copyright in his publications unconditionally and handing over the manuscripts to the University.

I have the honour to be,
Sir,
Your most obedient servant,
K. Ramunni Menon
Ag. Registrar.

Government Memo   3 September 1920

The government have under consideration the question of the grant of a compassionate allowance to the family of the late Mr. S. Ramanujam, F.R.S. They have decided on the grant to his parents of an annuity of Rs. 50 p.m. payable to the father, which, on his death, might be reduced to Rs. 25 p.m. for the mother only, unless her brothers are still under the age of 21, in which case each should be given in addition to the Rs. 25 for the mother Rs. 10 p.m. until they reach the age of 21. They also propose to grant a separate annuity of Rs. 50 p.m. to his widow.

Commentary. The writer incorrectly refers to Komalatammal's brothers instead of her sons.

The Tamil Nadu Archives contain several letters and notes pertaining to pensions for Ramanujan's family. Efforts were made to increase the proposed annuities, but these efforts failed, and considerable acrimony among certain members of the government developed. The awarding of scholarships to Ramanujan's brothers did not generate controversy. One of the letters in the Archives indicates that "Ramanujan's trip to England has alienated the sympathy of the neighbors [of Ramanujan's family]."

Komalatammal to the Secretary to the Government, Madras
29 September 1920

To,
       The Secretary to the Government,
          Madras.

Humble petition of Komalathammal, mother of the mathematician S. Ramanujam, 17, Sarangapany Swami Sannadhi, Kumbakonam.

Honored and Respected Sir,

1. I beg to acknowledge the receipt of your kind order no 1072 dated 17 September 1920 (Education), and bring the following to your kind consideration and favour.

2. Ramanujam was dearly brought up by me up to his twenty fifth year with a heavy heart, obeying the will of the government. I consented his sailing overseas, though such a procedure caused, in addition to a long and painful separation for five years, the severance from me of many of my relatives and friends who held strong religious convictions. With a heart heavier still, I had to bear the separation. Soon after his return, to show us his kind face and give us his loving words, he was cut away from us by an unnatural and precocious death.

3. Ever during his life, he respected us, loved us, eagerly watched our smiles, and gave us many balmy words in times of distress. To him, we owed our food and clothes. To him only, we owed our joy. To him alone, and to no one else, we owed our life even, by his strong affections, and never ceasing care for us. In short, he had been everything to us, and we in turn owed everything to him, so much, that his sudden death aimed at us a blow to our body and mind, severe and unrelenting in its stroke, as it was beyond the hope of human recovery.

4. Let me beg to draw tears from all feeling and tender hearts, that pass over these lines of mine. I will not live long, to see my other sons well up in life. I leave them entirely to the care of my benign government, to support them, educate them, and award them good positions in life. I have to feed at present an old mother, an old husband stark blind, and the two boys. Up to the time of Ramanujam's death, his wife also remained with us as members of one undivided family, and we were all enjoying in Ramanujam's bounty and his alone. Ever since his death, apart from the mental anxieties, our family had been in indescribable straits, famishing at times from want

of proper food and clothes, and being pressed by debts. These debts were incurred by me to meet the expenses of Ramanujam's funeral, which was performed by my younger son. I thought, I could repay them, soon as the Senate sanctioned to me the deposit in their hands (I have not as yet received the same).

5. When death lay in wait demanding his dear life, he gasped a few words to me, that he bequeathed to <u>me</u> all he has and that I might get all help and protection from our benign Government and University. In accordance with those sacred words that died away with his last moments, I applied for relief to the University, who told me they had referred the matter to the decision of the government. I heard that they sanctioned without my knowledge an annuity of rupees twenty per mensem to Ramanujam's widow, intending thereby to purchase Ramanujam's manuscripts. I did not find cause to think that the tenth part at least of the responsibility of birth, or nourishment, or education, or his career, of mathematician S. Ramanujam stood or fell, with the woman who was his wife. It was <u>we</u>, his father and mother, who produced him, fondled him, nourished him, educated him, laughed with him, sighed with him stood by him in all places at all times, felt keenly the pang of separation from him, and were ever ready to give up our lives for him. So it was, that he left us all he had. It was said, that the wife <u>alone</u> is to enjoy the price of the manuscripts, viz. the annuity. This, we beg to state, had been completely against the sacred intentions of Ramanujan, and strangely infringing the elementary rules of justice and humanity.

6. The pay of Ramanujan for the last April is with the University. There is also a deposit in the Barclay bank at Cambridge. We most humbly pray, the Government will be graciously pleased, to order the Syndicate to revise the decision about the annuity as to award it to us, and to make arrangements, for giving me the sum that is with the University here, and the sum that is in the bank at Great Britain.

7. Expecting anxiously your kind favour, that may give me immediate reply,

I most humbly beg to remain,
Honored and Respected Sir,
Your obedient servant,
Komalatammal

Commentary. At the beginning of this letter, Komalathammal is another variation of Komalatammal.

R. Srinivasa Iyangar to The Registrar, University of Madras

1 November 1920

From
    R. Srinivasa Iyangar,
      (Attorney of Mrs. Ramanujan),
        c/o Financial Dept., Secretariat, Poona

To
    The Registrar,
      University of Madras.

Sir,

I beg to state that since writing my letter of the 25th October regarding the purchase of Mr. Ramanujan's works by the University, certain facts have been communicated to me which render that letter unnecessary. I therefore beg respectfully to withdraw it and apologise for having written it.

I also beg on Mrs. Ramanujan's behalf the terms of the resolutions of the Syndicate and of the Senate on the subject.

              Your most obedient servant,
              R. Srinivasa Iyangar

Commentary. Poona is a city with a population approaching 3,000,000, located about 120 kilometers southeast of Bombay.

In a previous letter, which is not printed here, R.S. Iyangar had argued that the value of Ramanujan's manuscripts was more than that appraised by the government and that Janaki should be more substantially compensated in one large payment.

Although not mentioned in the letter, R.S. Iyangar was Janaki's brother, and was Assistant Commissioner of Income Tax in Bombay. Janaki stayed with her brother for about eight years until 1931 before returning to Madras. During this time she learned tailoring, which was her lifelong occupation. She taught tailoring at home.

R. Ramachandra Rao to G.H. Hardy

3 December 1920

Madras

Dear Sir,

Many thanks for your letter of the 21 Oct.

Ramanujan's m.s.s. whatever they are with me and will be handed over intact to the University of Madras who I understand is already in correspondence with you regarding methods of publication.

You want me to send some details from the Indian point of view. I am not sure what you want exactly. I send you by book today
(1) a newspaper cutting from the Madras Mail 27 April 1920 by T.A. Chettiar
(2) an article by the same gentleman in Every Man's Review June 1920

(3), (4) articles by Prof. P.V. Seshu Aiyar and myself in the *Journal of the Indian Mathematical Society* June 1920.

I trust that these four documents will furnish you the information you require. If not, please specify clearly what nature of information you want and I shall try to supplement.

The post nowadays is very irregular; so I shall be obliged if you will please drop a line of acknowledgment.

Yours truly
R. Ramachandra Row

Commentary. For help in writing accounts of Ramanujan's life and work, Hardy wrote R. Ramachandra Rao (Row) for biographical information. Hardy's obituary of Ramanujan was originally published in the *Proceedings of the London Mathematical Society* [98] and is reproduced at the beginning of Ramanujan's *Collected Papers* [226] and in Hardy's *Collected Papers* [112, pp. 702–720]. A shorter version appeared in the *Proceedings of the Royal Society*.

The articles by P.V. Seshu Aiyar and R. Ramachandra Rao are [1] and [243], respectively.

Sir Francis Spring to S.N. Aiyar   **15 February 1921**

Victoria 8392

8, Chester Terrace
Eaton Square
London S.W.

Dear Narayana Aiyar

I wrote suggesting to Professor Hardy that he and I should write a chapter about our poor friend Ramanujan for the information of Mathematicians throughout the World. He has now replied welcoming the suggestion and promising to write the mathematical part if I would write the biographical part. All I have to go on is the article which you helped me to write that appeared in the Madras Times of April 5th 1919. In many respects this is redundant and in others inadequate and it would help me very much and would be a fitting tribute to our friend's great genius if you, and if possible also Mr R. Ramachandra Rao, could at once send me any additional notes that occur to you both.

I say "at once" because Mr Hardy wants the Notice to appear in the London Mathematical Society's Proceedings Vol. 19. He says "about March or April". So if you can put a few notes together quickly, there may be time, even if we have to print them off in an Appendix.

There is some question about printing all Ramanujan's papers. But did not the Madras University buy the copyright of them by an annuity to his mother?

A few notes, which surely Mr Ramachandra Rao could furnish of Ramanujan's last few months of life after his return to Madras would be interesting.

Also any notes you can give in regard to his preservation of his Brahminical "Orthodoxy" in matters of food, avoidance of "defilement" and so on.

I would like very much to get the life records of this wonderful genius as complete as practicable. They are certain to be of high interest to the entire scientific world.

My very kind regards to old friends. I hope you keep well and your family circle also.

Yours sincerely
Francis J.E. Spring

Sir Francis Spring to G.H. Hardy   18 February 1921

8, CHESTER TERRACE,
EATON SQUARE
LONDON S.W.

Dear Mr. Hardy

On sitting down to try and compose the part of S. Ramanujan's obituary notice which you suggest I should undertake I find that I know nothing else at all than is contained in the newspaper cuttings sent herewith, gummed onto a copy of a Report of your own, furnished to me by the Registrar, Madras University.

The 5 or 6 columns of the "Madras Times" (very badly printed!) of April 5th 1919 (written by me in concert with Mr. S. Narayana Aiyar) appear to me now to be rather an unliterary melange of memoir and mathematics, and I am afraid to trust it lest I should go wrong with references to the latter—being nowadays no mathematician.

It occurs to me that the obituary notice will be likely to be better worthy of its subject if, with the aid of the cuttings—also please see other cuttings at end of pamphlet—you were to write the whole notice. Does this idea commend itself to you?

When I wrote the Madras Times notice I was working some 12 or 14 hours a day in and out of doors on some very heavy harbour construction work and the Note is a patchwork of other notes flung at me and undigested.

It is in the hope that, with this aid of such facts as are contained in the cuttings, you will do all. Rather than I was two years ago to make a blend of the available information less unworthy of its subject than I could make, that I venture to pass them on to you. But of course if you prefer it, and send the notes back, I will do my best to write something readable and thereby perhaps save you some trouble.

What do you wish in the matter?

Yours sincerely
Francis J.E. Spring

Commentary. Indeed, Sir Francis Spring did not become a coauthor; Hardy was the sole author. As indicated earlier, the article did appear in the *Proceedings of the London Mathematical Society* [**98**].

S.N. Aiyar to Sir Francis Spring   **10 March 1921**

Harbour, Madras

Dear Sir,
    I enclose copies of two notices about our late Ramanujan which give all details known to me. I have nothing more to add. There are one or two misstatements in these notices. In the article by Professor Seshu Aiyar he says that the letter to Mr. Hardy was written at his instance. This is not a fact. I suggested to Ramanujan to write to Professor Hardy and I even dictated the letter to him. In Mr. Ramachandra Rao's notice he says he introduced Ramanujan to you and you gave the appointment. You came to know of him some months after he was entertained in the Port Trust Office.
    Ramanujan's parents were in my house about November 1920 and Ramanujan's father got ill and died in my house on 28th November 1920.
    With kind regards,

Yours sincerely,
S.N.

Commentary. The following two letters bring to a close the discussions of the annuities to be paid to Mrs. Ramanujan and to Ramanujan's family.

The pension paid to Mrs. Ramanujan evidently was not sufficiently increased to compensate for inflation, and she lived for many years in poverty. In a letter dated 18 March 1994 to one of us (BCB), Professor S. Chandrasekhar informed us, "With regard to my conversations with Prime minister Jawaharlal Nehru and Professor M.S. Thacker, Director general of the Council of Scientific and Industrial Research: I had dinner with Nehru in September 1961 and mentioned to him the ridiculous amount of Mrs. Ramanujan's stipend (which at that time was, I believe 100Rs). He asked me to talk to Thacker, which I did. Thacker assured me that the stipend would be increased to 500 Rs. I later understood it was. I hope I was informed correctly."

Born in April of 1900, Mrs. Janakiammal Ramanujan quietly passed away early in the morning of 13 April 1994. K. Srinivasa Rao [**241**] immediately wrote an obituary notice which appeared in several Indian newspapers. In this summary of her life, he mentions that the government pension had risen to about 500 Rs. per month. Furthermore, the Madras Port Trust gave her

JANAKI AMMAL, MRS. RAMANUJAN, IN 1984

a pension for the remainder of her life beginning in 1962, the 75th anniversary of Ramanujan's birth. Since then, the governments of Tamil Nadu, Andhra Pradesh, and West Bengal; the Indian National Science Academy; the Hinduja Foundation; and Trinity College, Cambridge, also had given her monthly pensions.

K. Ramunni Menon to the Secretary of the Government

24 March 1921

From
    The Registrar,
        University of Madras.
To
    The Secretary to the Government
        Home (Education) Department.
Sir,

With reference to paragraph 3 of G.O.No. 1072, dated 17th September 1920, confirming the offer of the Syndicate to purchase the copyright of late Mr. S. Ramanujam's works by payment to his widow of an annuity for life amounting to Rs20 per mensem, I have the honour, by direction of the Syndicate, to inform the Government that the Senate at its meeting held on the 15th October 1920 approved of the purchase and the payment of the annuity from the Vacation Lectures Account from which the scholarship was paid to Mr. Ramanujam. I am now directed to request that the Government will be pleased to accord sanction for the payment of the annuity from that Account

MRS. RAMANUJAN WITH ADOPTED SON, W. NARAYANAN

with effect from the 15th October 1920, the date of the resolution of the Senate.

The favour of very early orders on this communication is requested.

> I have the honour to be,
> Sir,
> Your most obedient servant,
> K. Ramunni Menon.
> Ag. Registrar.

Commentary. Kankoth Ramunni Menon (1872–1949) received a B.A. from the University of Madras. He was admitted to Christ's College, Cambridge, and obtained First Class Honours in both parts of the Natural Sciences Tripos. After obtaining his M.A., he returned to India to become, in 1910, Professor of Zoology at the Presidency College, Madras. He served as Vice-Chancellor of the University of Madras from 1928 to 1934, and was knighted in 1933.

H. Sharp to the Secretary of the Government of Madras   4 May 1921

From
    H. Sharp, Esq., C.S.I., C.I.E.,
        Secretary to the Government of India,
To
    The Secretary to the Government of Madras,
        Home (Education) Department.
Sir,

With reference to the correspondence ending with your letter no. 132, dated the 26th January 1921, I am directed to forward herewith a copy of a despatch from the Secretary of State, no.22–Financial, dated the 24th February 1921, sanctioning the proposals made in your letter no.1073, dated the 17th September 1920, for the grant of a compassionate allowance to certain members of the family of the late Mr. S. Ramanujam, F.R.S.

2. The attention of the local Government is invited to the remarks made in paragraph 3 of the Secretary of State's despatch.

I have the honour to be,
Sir,
Your most obedient servant,
H. Sharp.
Secretary to the Government of India.

Commentary. Henry Sharp, a graduate of New College, Oxford, entered the Indian Civil Service in 1894, and was appointed Joint Secretary to the Government of India, in the Department of Education, in 1910. He was knighted in 1922.

Chapter 8
Ramanujan's Papers and Manuscripts

K. Ananda Rau to G.H. Hardy   26 May 1921

> Presidency College
> Madras (India)

Dear Prof. Hardy,

You will have no doubt received a copy of the Minutes of the Meeting of the Syndicate of the Madras University relating to the publication of Ramanujan's works. Mr Seshu Iyer and I were asked to sift and arrange, as far as possible, his manuscripts. Most of them we are able to recognise as manuscripts of his published work; but there are some which we are not able to piece together. Mr R. Ramachandra Rao told me that you had written to him some months ago that Ramanujan was working on a certain topic in his last days, and possibly there may be some record of this work left. If you will please tell us the nature of this investigation, we may find it easier to sift the papers. The whole of the manuscripts will of course be finally sent to you in accordance with the Resolution of the Syndicate. You will have noticed also in the Minutes that the Syndicate has asked Mr Seshu Iyer and me to arrange for the preparation of a transcript of Ramanujan's note book, with a view to having it incorporated as an Appendix to the Memorial Volume. I do not know if this will serve any useful purpose; I fear it may look a little incongruous by the side of his mature work. But there are some here, who think that the Note Book may contain valuable algorithms providing starting points for future investigations.

I read with much interest and profit your Inaugural Address to the Oxford University, in which you give such a beautiful account of Waring's Problem and its developments. I should very much like to read the detailed investigation, which is however not available here. I have been subscribing for the *Mathematische Zeitschrift* since recently, and I have received the number containing your and Littlewood's second paper on "Partitio Numerorum". The first paper on the subject is, I understand from your reference, published in the Göttinger Nachrichten. I shall be glad to have a copy of it, if there is one you can spare me. Indeed the dearth of good journals is so great here that I shall feel thankful to receive any reprints of your papers you can conveniently send me.

I enclose herewith a short note on a geometrical subject written by R. Vythynathaswamy, a research student of this University. He wishes it to be communicated to the London Mathematical Society. I am not able to form any opinion about its merits. If it is considered worthy of being communicated, please do so.

I owe you an apology for not having kept my promise to have published at an early date the Tauberian Theorems concerning derivatives (Generalisations of Theorems in your and Littlewood's paper "Contributions to the Arithmetic Theory of Series") to which you have been good enough to refer in one or two places. I sent a copy of my paper containing these theorems and their applications to Dirichlet's series, to Riesz more than six months ago (I addressed him to the University of Stockholm) with the request for permission and suggestions for publication. I have not yet heard from him in reply. I shall be glad if you can help me in expediting the publication by mentioning the matter to Riesz at any suitable opportunity.

With kind regards,
Yours sincerely,
K. Ananda Rau.

Commentary. In the final months of his life, Ramanujan developed the theory of mock theta–functions, as outlined in his letter of 12 January 1920 to Hardy. Ramanujan never published any of his work on mock theta–functions, but most of his discoveries can be found in the 'lost notebook' [227].

The paper by R. Vythynathaswamy was communicated by title to the London Mathematical Society, but was not published in their *Proceedings*.

Hardy and Littlewood published seven papers in their famous series of papers on 'Partitio Numerorum'. The first two, [118], [105, pp. 405–426] and [119], [105, pp. 427–440], had been published at the time Ananda Rau wrote his letter of 26 May 1921. The papers have the Roman numerals I–VI, VIII attached to them. For a description of the unpublished paper, VII, see [105, pp. 379–380], where all seven papers are reprinted.

The paper of Hardy and Littlewood on series to which he refers is [114], [109, pp. 428–495].

Born in Madras in 1893 and dying in Bombay in 1966, Krishnaswami Ananda Rau was a relative of R. Ramachandra Rao; Ramachandra Rao's nephew R. Krishna Rao was the cousin of Ananda Rau's mother. Educated at Hindu High School and Presidency College in Madras, where he obtained a first class honours degree in Mathematics, Ananda Rau sailed to England in August 1914, only five months after Ramanujan departed for Cambridge. Ananda Rau entered King's College, Cambridge, and obtained first class honours in both parts of the Mathematical Tripos. He then did research in mathematics under the guidance of G.H. Hardy, and was awarded one of the two Smith's Prizes in 1918. Much of the content of that prize-winning essay appeared in the paper mentioned in the close of his letter. For unknown reasons, this paper [3] was not published until 1932.

After returning to India shortly after Ramanujan, in July 1919, Ananda Rau became Professor of Mathematics at Presidency College, Madras, where he stayed throughout his career; for some time he served as Principal of the College. Ananda Rau had a reputation as an excellent teacher. Among his students were S. Minakshisundaram, K. Chandrasekharan, C.T. Rajagopal, and S. Chandrasekhar. At that time, only students of mathematics could attend lectures in mathematics. Being a student of physics, Chandrasekhar had to obtain special permission to attend Ananda Rau's lectures in mathematics and therefore to miss the lectures in physics.

Ananda Rau was a prominent Indian mathematician who worked on summability of series, in particular, of Dirichlet series. He also did research on functions of a complex variable and problems concerning theta-functions and the representation of a number as a sum of squares.

He was a tall man, always immaculately dressed, and was one of Ramanujan's closest friends in Cambridge. In the photograph published as a frontispiece to the "lost notebook" [227], it is possible that he is the man standing two places to the right of Ramanujan on the occasion when the latter was awarded his B.A. degree (see photo, p. 135).

See obituary notices by C.T. Rajagopal [193], [194] for further information.

S. Janaki Ammal to the Secretary of the Government   5 August 1921

Triplicane

To

The Secretary in the Government
Home (Education) Dept.
Fort Saint George
Madras.

Sir,

I had presented a memorial for the grant of a pension as widow of the late Professor Ramanujan and am awaiting the orders of the Government thereon.

I had given my address as "Care of Mr S. Iyengar, Mambalam." I have now changed my residence and request you to address any communication to me to my new address which is 29 South Tank Square. Triplicane.

I remain
Yours very faithfully
S. Janaki Ammal.

R.H. Fowler to G.H. Hardy 9 June 1922

Trinity College
Cambridge

Dear Hardy,

I raised the Ramanujan works business at the Council today with the result that Trinity have decided "to guarantee £100 on the condition that total guarantees of £500 are obtained".

It is my own opinion not for repetition that if pressed they would go further. Perhaps you will let me know what luck you have from other sources as soon as you can.

J.E.L. raised the point that he hoped you would include as much of Ramanujan's unpublished stuff as possible. No doubt he will discuss this with you himself.

I don't think T.C.C. want this idea to drop, so that if you are short of guarantees let me know. But they would want full details of your exact proposals to go any further.

Yours,
R.H.F.

Commentary. Ralph Howard Fowler (1889–1944) was a Fellow of Trinity, who in 1932 became Plummer Professor of Applied Mathematics at Cambridge. He was elected to the Royal Society in 1925 and was knighted in 1942. J.E.L. is, of course, Littlewood, and T.C.C. denotes Trinity College, Cambridge.

J.H. Jeans to G.H. Hardy 15 June 1922

The Royal Society
Burlington House
London, W.1

Dear Hardy,

Your letter of the 10th of June, with reference to the issue of a volume of Ramanujan's memoirs, was considered by Council to-day, and I have pleasure in informing you that the Council are prepared to guarantee to the extent of £200, subject to the terms of your letter.

Yours sincerely
J.H. Jeans
Secretary R.S.

Francis Dewsbury to G.H. Hardy 13 July 1922

SENATE HOUSE

From
THE REGISTRAR, UNIVERSITY OF MADRAS.

To
G.H. Hardy Esq., M.A.
New College, Oxford.

Dear sir,

Your letter of the 6th June 1922 on the matter of the publication of the proposed volume of the Memoirs of the late Mr. S. Ramanujam, F.R.S., has been considered by the Syndicate, and I have been directed to inform you that the University of Madras is prepared to guarantee the sum of £300 required for the guarantee fund. I shall address you later with respect to this publication.

Yours faithfully,
Francis Dewsbury,
Registrar.

Commentary. The Watson bequest in Trinity College Library contains numerous letters concerned with printing Ramanujan's papers and the possibility of including his unpublished manuscripts in a second volume. It was eventually decided that it would be too expensive to do the latter; the possibility of doing so at a later stage was left open. In January 1927, the Cambridge University Press estimated the cost of printing as £460–£500, plus binding at not more than two shillings (£0.1) a copy, and £20 for advertising. The volume of *Collected Papers* [**226**] appeared in 1927. The account submitted on 16 December 1927 was for £499.17.6 (£499.875) including corrections. The costs were amply covered by the guarantees, totalling £600, from the University of Madras, the Royal Society, and Trinity College, and a contribution of £100 from the Syndics of the Press to cover binding and advertising. By 23 August 1929, 282 copies had been sold and, from the royalties, the three sponsoring bodies had been reimbursed in proportion to their grants.

Francis Dewsbury to G.H. Hardy 2 August 1923

UNIVERSITY OF MADRAS.

From
FRANCIS DEWSBURY, ESQ., B.A., LL.B.,
Registrar, University of Madras.

To
G.H. Hardy Esq., M.A., F.R.S.

Sir,

With reference to this office letter No. 2542, dated 22nd March 1921, on the proposed memorial volume of the late Mr. S. Ramanujam's work, I am now in a position to send you enclosed herewith the biographical sketch (which was to have been written by Diwan Bahadur R. Ramachandra Rao, C.S.I., but for which in its present form not only he but Mr. P.V. Seshu Aiyar, B.A., L.T., Presidency College, Madras, and the Rev. Dr. E.M. Macphail, D.D., Vice–Chancellor of this University are also responsible). I hope to be able to send you by next mail or very soon thereafter 4 bound manuscript books containing a transcript of Mr. Ramanujam's notebook which is to appear as an appendix to the memorial volume. In notebook 4 will also be found at the end transcripts of three quarterly reports which Mr. Ramanujam made upon his work to the Board of Studies in Mathematics in this University while he was in receipt of a special research scholarship here. It has been thought possible that some parts of these reports might be suitable for publication.

In this connection I have to ask you whether you have any information as to a second notebook containing some of Mr. Ramanujam's original work done and written up by him therein before his departure to Europe. Mr. P.V. Seshu Aiyar states that he remembers this notebook being in Mr. Ramanujam's possession. Mr. K. Ananda Rau, who studied Mathematics at Cambridge while Mr. Ramanujam was there, is said to have seen this notebook in Mr. Ramanujam's possession in Cambridge. If this notebook is in existence, I have to ask you to extract from it such of Mr. Ramanujam's original work as you may consider suitable for publication in the memorial volume and to send the notebook for preservation with the other notebook in the library of this University.

I have the honour to be,
Sir
Your most obedient servant,
Francis Dewsbury
Registrar.

Notes on the biographical sketch of
Mr. S. Ramanujam, F.R.S.

1. The biographical note may fall into two parts, though not indicated as such. The first part may give a sketch of his life, somewhat on the lines contained herein. The second part may contain a critical account of his works, explaining his special powers and tracing the growth and development of these powers—in short that part may deal exhaustively with his mathematics. Of course, Mr. Hardy alone can supply this part.

2. Page 4. The second notebook is not at present in the Madras University Library. It must be decided where it is to be kept permanently and arrangements made accordingly. Of course, the second notebook is bound to be more important.

3. Page 9. The original reports are now in the Registrar's office files. Probably, it may be better to have them removed to the University Library. If that is not possible, at least a copy of the reports must be preserved in the Library and a copy sent to Mr. Hardy so that they may be edited for inclusion in the collected works as appendix.

4. Page 11. About Ramanujam entitled 'Ramanujam's work in England', Mr. Hardy's report may be included in toto in the memorial volume.

Commentary. The biographical sketch of Ramanujan by R. Ramachandra Rao and P.V. Seshu Aiyar was originally published in the *Journal of the Indian Mathematical Society* [1], [243] and an amalgam appeared in Ramanujan's *Collected Papers* [226].

Earle Monteith Macphail (1861–1937) was an Edinburgh graduate on whom his University conferred the honorary degree of Doctor of Divinity in 1922. He was ordained a missionary of the United Free Church of Scotland in 1890, and was appointed Principal of the Madras Christian College in 1921. He served as Vice-Chancellor of the University of Madras from 1923 to 1925.

Dewsbury appears to have placed Ramanujan's notebooks in the wrong order. When Ramanujan returned to India in 1919, he took his second and third notebooks with him. The first notebook was left with Hardy, and it was this notebook that Hardy used when writing his paper [99], [107, pp. 505–516] on Ramanujan's contributions to the theory of hypergeometric series. Evidently, shortly after receiving this letter, Hardy sent the first notebook to the University of Madras.

Original plans called for the publication of Ramanujan's quarterly reports along with his collected papers. However, this proposal was eventually dropped. Mr T.A. Satagopan at the University of Madras Library made a handwritten copy of the reports on 51 foolscap pages. Evidently, the four bound manuscript books to which Dewsbury refers contain Mr. Satagopan's transcriptions of Ramanujan's second notebook and quarterly reports. G.N. Watson subsequently produced two handwritten copies of the second notebook and the reports for his personal use. Both copies of the reports, as well as one of Watson's copies of the second notebook, can now be found in Trinity College Library, Cambridge. The other can be found in the library of the Mathematical Institute in Oxford. Unfortunately, the library at the University of Madras has lost Ramanujan's original quarterly reports. Ramanujan's three notebooks are preserved at the same library. The reports have never been published. However, a complete description and discussion of the reports has been published by Berndt [31], [32]; a briefer account can be found in [30].

Francis Dewsbury to G.H. Hardy 6 August 1923

UNIVERSITY OF MADRAS

From   FRANCIS DEWSBURY, Esq., B.A., LL.B.,
        Registrar, University of Madras.

To

    G.H. Hardy Esq., M.A., F.R.S.

Sir,

In continuation of my letter No. 6796, dated 2nd August 1923, I have now seen your letter dated 19th November 1921 to Mr. K. Ananda Rao, in which you state that you have one of the late Mr. S. Ramanujam's notebooks with you. I shall be glad if you will return the original notebook when you have extracted from it the selected matter for printing in the Memorial Volume.

I am informed by Mr. P.V. Seshu Aiyar, one of the authors of the biographical sketch, that there is a mistake of fact in it. The sketch states that Mr. Ramanujam went to school first at Erode. This is wrong. He first went to school at Kumbakonam. Will you kindly make the necessary alteration in the manuscript.

                I have the honour to be,
                         Sir
                Your most obedient servant,
                        Francis Dewsbury
                            Registrar.

Francis Dewsbury to G.H. Hardy 30 August 1923

UNIVERSITY OF MADRAS

From

    FRANCIS DEWSBURY, Esq., B.A., LL.B.,
        Registrar, University of Madras.

To

    G.H. Hardy Esq., M.A., F.R.S.

Sir,

With reference to correspondence ending with my letter No. 6938 dated the 6th instant, I have the honour to advise despatch to-day to your address per registered and insured parcel post of the four manuscript note-books referred to in my letter No. 6796 of the 2nd idem.

I also forward a packet of miscellaneous papers which have not been copied. It is left to you to decide whether any or all of them should find a place in the proposed memorial volume. Kindly preserve them for ultimate return to this office.

                I have the honour to be,
                         Sir,
                Your most obedient servant,
                        Francis Dewsbury
                            REGISTRAR.

Commentary. It seems very likely that the "packet of miscellaneous papers which have not been copied" contained the "lost notebook" [227]. If this indeed is the case, then it appears certain that Hardy did not return the miscellaneous papers to the University of Madras. At some later time, probably after Watson's interest in Ramanujan's work had waned, Hardy evidently relinquished the lost notebook to Watson who had it in his possession at his death in 1965. At Rankin's suggestion, Mrs. Watson donated her late husband's papers, including material related to Ramanujan, to the library of Trinity College, Cambridge. There the lost notebook remained until G.E. Andrews rediscovered it in the spring of 1976. For further details, see Rankin's obituary article on Watson [230], Andrews' paper [7], and Andrews' monograph [8].

G.H. Hardy to B.M. Wilson   **10 June 1925**

Dear Wilson,

Thanks for the revised MS, which I think will do now. I can have it printed in the records almost at once.

<u>Ramanujan</u>. All letters now recovered and sent herewith. What is wanted now is, I think,

(I) a type-written statement (about 6 copies) of what <u>we</u> propose: this to be sent to

  (1) Press

  (2) Royal Society

  (3) Trinity

  (4) Madras

  (5) Mr Seshu Aiyar (I think this is the man. I get muddled between the various editors).

We recommend (roughly):

(1) that all work of Ramanujan already printed, together with the appropriate notices of him, should be printed as soon as it can be collected. This volume to be called 'Collected Mathematical Memoirs.'

  length: 400–450 pages.

Estimated cost (bearable by guarantors) not over £500. Binding, advertising, etc. to be paid by Press.

(2) that the note books etc. should be printed later (if finance permits) as a second volume (possibly with a different title).

(3) The first £100 from sales to go to Press (or whatever is necessary to meet <u>their</u> out of pocket expenses). All later proceeds from sales to go proportionally to guarantors, until they are paid off. In (unlikely) event of a profit, profits to be divided equally between Press and guarantors (for them, proportionally).

(4) We want a definite undertaking from each guarantor that such part of the guarantee as may be required to meet the printing bills will be paid when the bills are due.

B. M. Wilson

(II) There should be a separate letter to each guarantor. My own view is that we might ask Trinity College and the Royal Society to promise the whole of their guarantees at an early stage, Madras to pay up only what is actually needed for Vol. 1, keeping the rest in reserve for Vol. 2.

Mr. Aiyar should have a separate personal letter asking for his approval as editor.

(III) When the answers are received, no doubt the Press will want a formal agreement. It is probable that they will want to make this with me leaving me to recover from the guarantors. If so I do not object.

This is merely a rough sketch. Letters should be kept as short as is consistent with clearness.

If the first volume can be arranged for on these lines, it should not take more than a year or so to get out.

Incur any reasonable expense for typing and I will be responsible for it.

Yours sincerely,
G.H. Hardy.

Commentary. Bertram Martin Wilson (1896–1935) entered Trinity College, Cambridge in 1916. He obtained First Class Honours in both parts of the Mathematical Tripos in 1917 and 1919, respectively. Being unfit for military service, he was able to remain in Cambridge throughout the war. He was appointed to a post in Liverpool University in 1920, where he remained until he went to Dundee in 1933 as Professor of Mathematics in University

College, in the University of St. Andrews. He died there after holding his appointment for barely eighteen months. He was one of the three editors of Ramanujan's *Collected Papers* [226], and bore the brunt of the administrative and editorial work involved. With G.N. Watson, he undertook in 1929 the task of editing Ramanujan's notebooks in a form suitable for publication. They worked independently, with Wilson taking Chapters I–XIV of the second notebook and Watson the later chapters. Wilson did not publish any of this work, and after his death it was passed on to Watson. Apart from this, he was the author of eight papers on analysis and number theory during his early years. However, his heavy load of teaching and examining, in addition to his work on Ramanujan's notebooks and papers, limited his output in later years. G.H. Hardy, in a letter of 30 September 1921 to the Swedish mathematician Mittag-Leffler, assessed him as 'quite a good, though not at all original, mathematician,' but this judgement, made at a very early stage in Wilson's career, is probably too severe. For obituaries of Wilson, see [127] and [268].

P.V. Seshu Aiyar to B.M. Wilson   **8 July 1925**

Government College,
Kumbakonam
Madras Presidency, India

Dear Mr. Wilson,

I thank you very much for your kind letter of 11th June '25.

I have been wondering till now as to what has become of Prof. Hardy's proposal to edit the late Mr. S. Ramanujam's works and I am glad to find from your letter that the editing and printing of the works is to be taken up soon.

I accept in entirety all your proposals regarding the printing etc., of the two volumes and shall be glad to do my portion of the work with pleasure. If you want me to move the Madras university authorities to place any amount at your disposal, I shall be glad to do so.

No kind of formality need be observed between us, as business is our main point and you need not therefore feel or make any kind of apology for opening the correspondence between us so very abruptly.

I am of an optimistic temperament and I feel sure that we will be able to bring out both the volumes without much undue delay.

Kindly note my change of address. I was Professor of Applied Mathematics in the Presidency College, Madras till last year when I was transferred to this College as its Principal.

With kind regards,

Yours sincerely,
P.V. Seshuiyer

To
> B.M. Wilson Esq., M.A.,
> Liverpool, England.

Commentary. Wilson's letter of 11 June 1925 has been lost.

Ramanujan's papers [226], in fact, appeared in one volume, not two volumes.

The Principal of a college corresponds to the President of an American university.

B.M. Wilson to P.V. Seshu Aiyar   **18 August 1925**

> Trinity College,
> Cambridge

Dear Professor Aiyar,

Thank you for your letter of the 8th July, and in particular for your ready acceptance of the steps already taken by Professor Hardy and myself towards the publication of Mr S. Ramanujan's collected works. Both Professor Hardy and I have been in Cambridge for some time this summer, and appreciable progress has now been made both in concluding arrangements with the University Press and in getting ready the contents of the first volume. A certain amount of avoidable overlapping was found to exist as between the biographical sketch of Mr. S. Ramanujan prepared by Mr. R. Ramachandra Rao and yourself and Professor Hardy's obituary notice of him (which is also to be included in the volume): a number of verbal alterations have therefore been made to the sketch which we trust that you will sanction for the avoidance of repetition. Most of these changes are, I may add, quite trivial. In addition I have now read through all the memoirs, notes, questions and solutions contributed by Ramanujan to the *Journal of the Indian Mathematical Society*, so that these are now substantially ready for press.

In the course of this week a number of points have arisen about which we wish to ask for your help. These I detail below as follows:—

1. So far as re-printing Ramanujan's contributions to the *Journal of the Indian Mathematical Society* are concerned, the Press will be able to do this from the volumes themselves. But when it comes to the correction of proofs it would be extremely convenient if off-prints were available. Would you then be so good as to ascertain from the Society whether offprints of some or all of Ramanujan's contributions to its Journal still exist, and, if they do, have as complete a set as possible sent <u>to me</u>. I should of course be willing to return them as soon as I had finished the correction of proofs.

2. <u>Note-books</u>. Professor Hardy has in his possession the four manuscript note-books sent to him by the Registrar of the University of Madras. These are, I understand, copies of Ramanujan's own note-books. He also had earlier one original note-book, but returned this to the University of Madras

at their request. He now finds, on examining the copies more closely that they appear to omit some things which he remembers to have seen in the original volume. Would it then, in view of all this, be possible to recover from the University of Madras the note-book which he returned so that it may be collated with the ones we have. If this is possible would you kindly direct that the note-book be sent straight <u>to Professor Hardy</u>.

3. <u>Guarantee</u>. I notice that in your letter you kindly offer to use your influence with the authorities of the University of Madras if it is found necessary to have stronger financial support in order to complete publication. I think that it is now quite clear that our present guarantees are more than sufficient to ensure publication of the first volume. But unless that sells rather unexpectedly rapidly it is very likely that more money will be required to guarantee the second volume; if this proves to be the case we shall certainly be glad of your good offices with the University of Madras.

4. <u>Frontispiece to volume</u>. It is obviously fitting that the volume of collected memoirs should have, as frontispiece, a photograph of Ramanujan. The best photograph of which we know is the one reproduced in Vol. XI of the *Journal of the Indian Math. Soc.*, showing head and shoulders and with academic dress. I have failed, after enquiry at all the Cambridge photographers, to find any trace of this photograph, or, rather, of the group of which it formed part, and so am forced to conclude either that they have missed it or that it was not taken in Cambridge at all. Would you therefore kindly find out for us from where the photograph from which that illustration was reproduced was obtained? If it was in India would you then have <u>the negative</u> if possible (rather than a print or reproduction) sent direct to the Cambridge University Press here, — telling them who it is? If, on the other hand, it was obtained from England would you let me know the name and address of the photographer, and I will obtain the negative for the Press.

These four points cover all of importance that has arisen between Professor Hardy and myself as calling for consultation with you. Thanking you once again,

    I remain,
      Yours Truly,
        B.M. Wilson

If you happen to know Professor N.M. Shah you might be so good as to give him my very kind regards when next you meet him. He and I were contemporaries at Trinity and took the Tripos together.
    B.M.W.

Commentary. This letter was obtained by P.K. Srinivasan from Dr. P.S. Srinivasan, the youngest son of P.V. Seshu Aiyar, and was printed in a centenary tribute to Ramanujan by the Association of Mathematics Teachers of India [**17**].

Wilson's letter indicates that as of 18 August 1925 the publication of the notebooks was still intended. Evidently, financial conditions were not sufficiently ameliorated to allow for the publication of the notebooks in 1927.

No frontispiece of Ramanujan appears in his *Collected Papers* [**226**], and so the sought negative apparently was not found.

Born on 14 July 1894, Nagindas Maneklal Shah received First Class Honours in Part I (1917) and Part II (1919) of the Mathematical Tripos, as did B.M. Wilson. He received his BA in 1919 and MA in 1923 from Cambridge University. After returning to India, Shah held various teaching posts in the state of Gujarat. Shah and Wilson coauthored a paper [**259**] that was published in 1919.

P.V. Seshu Aiyar to B.M. Wilson   **22 April 1926**

RAO BAHADUR P. V. SESHU IYER,   KUMBAKONAM
PRINCIPAL, GOVERNMENT COLLEGE.

Dear Mr. Wilson,

I thank you very much for your kind letter of 28/3/26. I am sorry I don't remember having received any letter of the kind you say, you wrote to me in last summer and I do not know how much inconvenience the non-reply to it has caused you.

The original of the portrait reproduced in the *Journal of the Indian Mathematical Society* is not with me nor with the Society. It was returned to Ramanujan's family, i.e. his brothers and mother and I am writing for it to his brother and as soon as I get it, I shall send it on to you. If for any reason I am not able to get it, I can certainly send you the block prepared for the Society or the reproduction.

I am glad to hear that you are proceeding with the proof reading. Soon the manuscript copy of his note book which is being prepared will be sent to you or to Mr Hardy.

Yours truly
P.V. Seshu Iyer

Commentary. Aiyar, Aiyer, and Iyer are interchangeable spellings.

Wilson's letter of 28 March 1926 is lost.

The last paragraph of P.V. Seshu Aiyar's letter is somewhat puzzling. According to the letter of 30 August 1923 from Francis Dewsbury to G.H. Hardy, a copy of the notebooks was mailed to Hardy on that same date. Did Seshu Aiyar not know that a transcription of the notebooks was sent almost three years earlier? Possibly Seshu Aiyar is referring to a photographic facsimile of the notebooks that was being prepared for publication. It is not known if such a copy was ever made or sent to Wilson or Hardy.

Komalatammal to G.H. Hardy   **25 August 1927**

18 Hanamanta Rayan
Koil Street
Triplicane
Madras

From,
   Komalatammal
     mother of late
     S. Ramanujan, F.R.S.
To
   G.H. Hardy Esq., M.A F.R.S.
Sir,

   You will be pleased to hear after seven years some news about the bereaved members of the illustrious mathematician S. Ramanujan, F.R.S.

   I have two other sons except S. Ramanujan. S. Lakshmi Narasimhan the older, who wrote a letter to you while Ramanujan was ill and who received a letter from you about Ramanujan's health, studied up to Intermediate after passing the matriculation examination and is now employed in the Triplicane Post Office Madras. S. Thirunarayanan the younger joined the Presidency College Madras and got his B.A. degree in March 1927.

   The Govt. of India, after my illustrious son's death, sanctioned an annuity of Rs 50/- or nearly £3/- per mensem to my husband, and a like amount to the unfortunate widow. I lost Ramanujan in April 1920 and my husband in Nov. 1920. Go. no. 633 of 4/5/21 Govt. of India (Education) Department also provided that I should get an annuity of Rs 25/- or nearly £1-10-0 per mensem, and my two sons should get Rs 10/- or 13 s. per mensem till they reach 21 years of age. The widow is getting an annuity of Rs 20/- or nearly 26 s per mensem from the Registrar and University of Madras for transferring rights to Ramanujan's mss. to the university. My own sons were getting Govt. scholarships with the concession that they were not required to pay fees.

   So from the year 1920, I am bringing up my two sons with the greatest difficulty. I sent my son to England in 1914 thinking that my family ever wedded to poverty would become rich, that my son would become famous. Like Achilles, he won everlasting fame; like Achilles, he died young.

   Now, I have written the Govt. of India to select my younger son S. Thirunarayanan, B.A. for the post of a probationery, superintendent of post offices in India. (*marginal note*: The post of a probationery superintendent's carries with it a salary of £12/- per mensem and that of an inspector is £6- per mensem.) I have also requested them to select my elder son S. Lakshmi Narasimhan for the post of an inspector of post offices in Madras. I have also mentioned in the application the various benefits conferred upon the bereaved members.

I have also requested Mr Littlehailes M.A. (who was in 1914 Director of Public Instruction and who is now educational Commissioner with the Govt. of India) to recommend my two sons, the unfortunate brothers of the first Indian Fellow of the Royal Society, to the Govt. of India.

I need not to tell you, Sir, that had my illustrious son, lived, my two sons would have held high positions in life. I have heard of your paternal care interest which you took in the health of my son. I also request you to extend your sympathy towards my sons.

I humbly request you to write to the Govt. of India to select my two sons for the posts I have mentioned and also to the Secretary of State for India London for giving them suitable appointments.

I hope you would kindly excuse the trouble that I am giving you. I hope to expect a reply from you soon.

I beg to remain
Sir,
Your most obedient
servant

My address.
  Komalatammal
  18 Hanumanta Royan Koil Street
    Triplicane Madras India
My son's address.
  S. Lakshmi Narasimhan
  18 Hanumanta Royan Koil Street
    Triplicane Madras India

Commentary. As previously mentioned, Ramanujan, in fact, was the second Indian to be elected as a Fellow of the Royal Society.

There is no evidence that Hardy wrote to the government of India on behalf of Ramanujan's brothers. He may have passed the letter to the Registrar of the University of Madras, as he had done before with Lakshmi Narasimhan's letter of 29 April 1920. See the last sentence of his letter of 26 May 1920 to Francis Dewsbury. However, both brothers did obtain positions in the Madras Post Office.

E.H. Neville to B.M. Wilson   **28 December 1927**

160 Castle Hill
Reading

Dear Wilson,

Rummaging just before Christmas, I came across this old <u>Nature</u> and remembered with regret that I promised at Oxford to send you either the copy

CHAPTER 8. RAMANUJAN'S PAPERS AND MANUSCRIPTS

itself or a reference so that you could read my letter on Ramanujan (p. 661). Perhaps if that second volume is ever completed, some of this might be reprinted. If only I could come across the bucket of cold water from the India Office, which I know I preserved carefully for some years!

I haven't another copy of this <u>Nature</u>, so please let me have this back some day within the next eighteen months: that is as much as I dare propose by way of a time limit.

Yours,

E.H. Neville.

Commentary. Enclosed with the letter was No 2673 of <u>Nature</u>, issued on 20 January 1921, which contains on pp. 661–662 Neville's letter on Ramanujan; this is reproduced in [**227**, pp. 413–414]. The 'bucket of cold water' doubtless refers to unhelpful letters received from the India Office.

T. Vijayaraghavan to B.M. Wilson   **4 January 1928**

142, Elgin Avenue,
London, W. 9.

Dear Mr Wilson,

Regarding the justification of the formal processes in Ramanujan's solution of a question (the justification is given on p. 348 of Ramanujan's Collected Papers) I now do not remember whether I justified Ramanujan's solution or simply established the criterion of convergency for expressions of the form

$$\sqrt{a_1 + \sqrt{a_2 + \sqrt{a_3 + \cdots}}},$$

from which the convergency of

$$\sqrt{\left[1 + 2\sqrt{1 + 3\sqrt{(1 + \cdots)}}\right]}$$

follows. To prove that the above expression has the value 3 some additional remarks are necessary, as will be evident from the following consideration: though

$$(1) \quad 2 = \sqrt{1 + \sqrt{9}} = \sqrt{1 + \sqrt{1 + \sqrt{64}}} = \sqrt{1 + \sqrt{1 + \sqrt{1 + \sqrt{6969}}}}$$
$$= \sqrt{1 + \sqrt{1 + \sqrt{\cdots + \sqrt{k_n}}}}$$

where $k_n \leq 3^{2^n}$, still it is not valid to deduce that

$$2 = \sqrt{1 + \sqrt{1 + \sqrt{1 + \cdots}}}.$$

Ramanujan's results are, however, correct, and the only proof I can think of is unconscionably long and tedious, and a rough sketch of it is as follows:

Let

$$x_{m,m} = \sqrt{1 + m + 1},$$
$$x_{m-1,m} = \sqrt{1 + mx_{m,m}},$$
$$x_{m-2,m} = \sqrt{1 + (m-1)x_{m-1,m}}$$
$$\cdots\cdots\cdots$$

(2)
$$x_{r,m} = \sqrt{1 + (r+1)x_{r+1,m}}$$
$$x_{1,m} = \sqrt{1 + 2x_{2,m}}.$$

We have to prove that when $m \to \infty$, $x_{1,m} \to 3$.

(As is pointed out by Ramanujan, (p. 323), it is evident that if $x_{m,m} = m + 2$, then for all $r < m$

(3) $$x_{r,m} = r + 2,$$

and $x_{1,m} = 3$. Since our $x_{m,m} = \sqrt{m+2}$, the convergency of the sequence $x_{1,m}$ becomes obvious. This is by the way.)

It is easy to verify that

$$x_{m,m}^2 = (m+1) + 1$$
$$x_{m-1,m}^2 > m^{3/2} + 1,$$
$$x_{m-2,m}^2 > (m-1)^{7/4} + 1,$$

(4) $$x_{r,m}^2 > (r+1)^{(2^{m-r+1}-1)/2^{m-r}} + 1.$$

It is not difficult to prove that if m is large (say, $m > 1000$) and r is small (say, $r < m/2$) then

(5) $$x_{r,m}^2 > (r+1)^2$$

so that

(6) $$x_{r,m} > r + 1.$$

At this stage I require a lemma:
<u>If</u>
$$0 < r + 2 - x_{r,m} = \delta < 2,$$
<u>then</u>
$$0 < r + 1 - x_{r-1,m} < \frac{\delta}{2}.$$

Since $x_{r,m} < r + 2$, it is clear that $x_{r-1,m} < r + 1$. Also

$$\begin{aligned} x_{r-1,m} &= \sqrt{1 + rx_{r,m}} \\ &= \sqrt{1 + r(r+2-\delta)} \\ &= \sqrt{(r+1-\frac{\delta}{2})^2 + \delta - \frac{\delta^2}{4}} \\ &> r + 1 - \frac{\delta}{2}, \end{aligned}$$

since $\delta - \frac{\delta^2}{4}$ is positive, δ being between 0 and 2. The lemma at once follows.

Now, formally to state the proof.:

Given $\epsilon > 0$, choose an r so large that

(7) $$2^{-r+1} < \epsilon,$$

and also find an M so large (say, $m \geq \max(1000, 2r)$) that for all $m > M$

(8) $$r + 2 > x_{r,m} > r + 1.$$

Now, repeatedly applying the lemma it is seen that

(9) $$3 > x_{1,m} > 3 - \epsilon.$$

My sincere congratulations for the fine edition of Ramanujan's collected papers which you have brought out.

Yours sincerely
T. Vijayaraghavan

P.S. Please don't bother to acknowledge this letter.
T.V.

Commentary. Tirukkanapuram Vijayaraghavan (1902–1955) spent the years 1926–1928 at Oxford as a research student of Hardy. He was an able mathematical analyst. After he returned to India, he became Professor of Mathematics at Andhra University. He resigned this post to become Director of the (at that time) privately endowed Ramanujan Institute in Madras. For obituaries of Vijayaraghavan, see [64] (reprinted in [144]) and [69].

We offer a few remarks about Vijayaraghavan's proof. (The numbers attached to certain displayed expressions are due to the authors.)

In (1), the last number on the right side should be 3969, instead of 6969.

In his parenthetical remark, Vijayaraghavan defines $x_{m,m} = m+2$, but for all $r < m$, the previous definition is being taken for $x_{r,m}$. The conclusion (3) then easily follows by induction on r.

We have not checked that (5) is valid for the stated values of r and m. However, such precise bounds for r and m are not important. After some elementary algebra, we deduce that (5) will follow from (4) if

$$2^{m-r} \log\left(\frac{(r+1)^2}{(r+1)^2 - 1}\right) \geq \log(r+1).$$

It is now clear that (5) is valid for m sufficiently large and r sufficiently small in comparison with m.

The first inequality of (8) follows from (2) and induction on r. The second inequality in (8) is simply (6).

Of course, the first inequality of (9) follows from (8). The second inequality in (9) arises from induction as follows: As in the lemma, set $\delta = r + 2 - x_{r,m}$. Note that $\delta < 1$. Apply the lemma and set $r + 1 - x_{r-1,m} = \delta_1$. Thus, $\delta_1 < \delta/2$. Proceeding by induction, we eventually find that

$$3 - x_{1,m} < 2^{1-r}\delta.$$

Since $\delta < 1$, the second inequality in (9) follows from (7).

For further work of Ramanujan on nested sequences of radicals, see Berndt's book [35, pp. 107-112].

G.H. Hardy to G.N. Watson   early 1928

Dear Watson,

Looking through my old collections of Ramanujan's letters, etc., etc., I find a few relevant notes. I've no desire to but in in your field, but there is one formula which goes further than anything in the letters and which I think I might write out for the Journal: Besides this I put down the other notes, which you can make any use you like of.

1. <u>The new formula</u>. This is

(A) $\quad \alpha^k \left\{ \frac{1}{2}\zeta(1 - 2k) + \frac{1^{2k-1}}{e^{2\alpha} - 1} + \cdots \right\} = (-\beta)^k \{\text{same function of } \beta\},$

if $\alpha\beta = \pi^2$ and k is an integer > 1. It gives your result when k is odd and $\alpha = \beta = \pi$, since then the two sides must vanish. I have 2 interesting proofs.

(1) <u>By Mellin's integral</u>,

$$e^{-\alpha} = \frac{1}{2\pi i} \int_{(x)} \Gamma(s)\alpha^{-s} ds \qquad (x > 0)$$

$$\frac{1}{e^{2\alpha} - 1} = \sum_1^\infty e^{-2n\alpha} = \frac{1}{2\pi i} \int \Gamma(s)(2\alpha)^{-s}\zeta(s)ds \qquad (x > 1)$$

CHAPTER 8. RAMANUJAN'S PAPERS AND MANUSCRIPTS

$$\psi(\alpha) = \sum_1^\infty \frac{n^{2k-1}}{e^{2n\alpha} - 1} = \frac{1}{2\pi i} \int \Gamma(s)(2\alpha)^{-s}\zeta(s)\zeta(s - 2k + 1)ds \qquad (x > 2k).$$

Move the contour to $x = -\delta$. There is no pole for $s = 1$ when $k > 1$ (if $k = 1$ there is and a correction is required). We get

$$\psi(\alpha) = -\frac{1}{2}\zeta(1 - 2k) + (2\alpha)^{-2k}\Gamma(2k)\zeta(2k) + \frac{1}{2\pi i}\int_{(-\delta)}.$$

Use the functional equations for Γ and ζ, and write

$$s = 2k - S$$

and you will find after reduction that

$$\frac{1}{2\pi i}\int_{(-\delta)} = \left(-\frac{\pi^2}{\alpha^2}\right)^k \psi(\beta),$$

whence the result.

(2) <u>By the 'Cauchy–Poisson' formula.</u> [and this, surely, must have been Ramanujan's method!]

The formula is

$$g(x) = \sqrt{\frac{2}{\pi}} \int_0^\infty f(t) \cos xt \, dt$$

$$\sqrt{A}\left(\frac{1}{2}f(0) + f(A) + f(2A) + \cdots\right) = \sqrt{B}\left(\frac{1}{2}g(0) + g(B) + g(2B) + \cdots\right)$$

$$(AB = 2\pi)$$

(but don't imagine this is going to give the result straight off — Ramanujan is cunning enough to make you expect that, but no!).
Here

$$f(x) = \frac{x^{2k-1}}{e^{2x} - 1}$$

$$g(x) = (-1)^{k-1}\sqrt{\frac{2}{\pi}}\left(\frac{d}{dx}\right)^{2k-1}\int_0^\infty \frac{\sin xt}{e^{2t} - 1}dt \qquad (x > 0)$$

$$g(0) = \sqrt{\frac{2}{\pi}}\int \frac{t^{2k-1}dt}{e^{2t} - 1} = \sqrt{\frac{2}{\pi}}2^{-2k}\Gamma(2k)\zeta(2k)$$

$$g(x) = (-1)^{k-1}\sqrt{\frac{\pi}{2}}\left(\frac{d}{dx}\right)^{2k-1}\left(\frac{1}{e^{\pi x} - 1} - \frac{1}{\pi x}\right)$$

$$= g_1(x) + g_2(x) \text{ (dividing in the obvious manner)}.$$

We have to calculate

$$g(B) + g(2B) + \cdots \text{ for } g_1 \text{ and } g_2.$$

(1) $$g_2(x) = (-1)^{k-1}\sqrt{\frac{\pi}{2}}\frac{1}{\pi}(-1)^{2k}\frac{\Gamma(2k)}{x^{2k}},$$

and this contributes a multiple of $\Gamma(2k)\zeta(2k)$ or of $\zeta(1-2k)$.

(2) $$g_1(x) = (-1)^{k-1}\sqrt{\frac{\pi}{2}}\left(\frac{d}{dx}\right)^{2k-1}\sum_1^\infty e^{-2nx}$$

$$\sum_1^\infty g_1(nB) = (-1)^k 2^{-\frac{1}{2}}\pi^{2k-\frac{1}{2}}\sum_1^\infty\sum_1^\infty \nu^{2k-1} e^{-\nu\mu\pi B}$$

$$= (-1)^k 2^{-\frac{1}{2}}\pi^{2k-\frac{1}{2}}\sum_1^\infty \frac{\nu^{2k-1}}{e^{\nu\pi B} - 1}.$$

This brings in $\psi(B)$: we have now all the material and the rest is a matter of not making a mistake.

Observe the Ramanujan touch: the question looks as if it ought to mean that $\dfrac{x^{2k-1}}{e^{2\alpha x}-1}$ is substantially its own Fourier transform. You try it on those lines and are discouraged to find that obviously it isn't: but the final formula is, so to say, what it would be if it were.

I cannot find (A) anywhere in the 'papers' or indeed anywhere but in my own notes: but I know it is Ramanujan all right.

This, I think, is well worth a Journal note.

Miscellaneous observations on other questions (these are merely what I find and you can utilise them or not as you please: it is certain that Mellin and Cauchy–Poisson are powerful aids for solution of many. Of course Mellin and the functional equation habitually lead to results equivalent to those got directly from C–P: there was a German who developed this systematically in the Mathematische Zeitschrift or Mathematische Annalen, I fancy.)

(1) The particular result for $\sum_1^\infty \dfrac{n^{4k-1}}{e^{2\pi n}-1}$ also comes directly from

$$P\int \frac{x^{4k+1}}{e^{2\pi x}-1}\frac{dx}{e^{2\pi ix}-1}.$$

(the 2 series of residues coincide and the integrals practically cancel.)

(2) p. xxvi, formula 8. I have also the lines of a proof of this by Mellin.

$$\sum_1^\infty \frac{k^{4n}}{(e^{k\pi}-e^{-k\pi})^2} = \frac{1}{2\pi i}\int \Gamma(s)(2\pi)^{-s}\zeta(s-1)\zeta(s-4n)ds \qquad (x > 4n+1).$$

Using the functional equation, and putting

$$4n + 1 - s = S$$

I get

$$\frac{1}{2\pi i}(2\pi)^{-4n-2}(2\pi)^{4n+1}\int (4n-S)(2\pi)^{-S}\Gamma(S)\zeta(S)\zeta(S+1-4n)dS.$$

Divide this into two pieces (corresponding to $4n$ and $-S$). One of these reproduces the original integral with its sign changed. Hence we get

$$\text{original} \int = \frac{n}{\pi} \frac{1}{2\pi i} \int (2\pi)^{-s} \Gamma(s) \zeta(s) \zeta(s - 4n + 1) ds$$

(plus residue corrections).

The integral here is now expressible as

$$\frac{n}{\pi} \left(\frac{1^{4n-1}}{e^{2\pi} - 1} + \cdots \right)$$

and we get the result.

(3) p. xxvi, formula 7. I have a proof of this also by Mellin (unfinished because here the other way is much simpler).

Here use (again Cauchy)

$$\sqrt{\frac{2}{\pi}} \int_0^\infty f(x) \sin xy \, dx = g(y)$$

$$\sqrt{\lambda} \sum_0^\infty (-1)^n f\left((n + \tfrac{1}{2})\lambda \right) = \sqrt{\mu} \sum_0^\infty (-1)^n g\left((n + \tfrac{1}{2})\mu \right),$$

$$\lambda\mu = 2\pi.$$

In this case Ramanujan's formula is a straightforward deduction. Taking

$$f(x) = e^{-A/x} x^{-3/2}$$

I find

$$g(x) = \sqrt{\frac{2}{A}} e^{-2\sqrt{Ax}} \sin \sqrt{2Ax}$$

and all goes straight. The Mellin proof uses

$$\text{left-hand side of formula} = \frac{n}{2\pi i} \int n^{-2s} \Gamma(s) L(\tfrac{3}{2} - s) ds$$

where

$$L(s) = 1^{-s} - 3^{-s} + 5^{-s} - \cdots.$$

The functional equation reduces this to

$$\frac{n}{2} \sqrt{\frac{\pi}{2}} \left(e^{-3\pi i/4} \phi\left(\frac{n^2 \pi}{2i} \right) + \text{ditto} - i \right)$$

where

$$\phi(A) = \frac{1}{2\pi i} \int A^{-s} \Gamma(s) \Gamma(s - \tfrac{1}{2}) L(s - \tfrac{1}{2}) ds,$$

and I think all that is required now is a straightforward calculation of $\phi(A)$ by residues.

Yours,
G.H.H.

G. N. Watson

No more for me, unless you run up against something very tricky you can't manage! But merely checking these old notes of mine has meant a day's work.

Ramanujan never used Mellin's integral in this sort of way, but of course he had an equivalent to use in his own (see my 'Carlson–Wigert' note in the Acta)

$$\left[\int_0^\infty w^{t-1}(a_0 - a_1 w + a_2 w^2 - \cdots) dw = \frac{\pi}{\sin \pi t} a_{-t}\right.$$

— in substance a sort of form of 'Mellin's inversion formula.']

Commentary. George Neville Watson (1886–1965) was, like J.E. Littlewood, educated at St Paul's School, London. In 1904 he entered Trinity College, Cambridge, and in 1907 was classed as Senior Wrangler. He obtained First Class Honours in Part II of the Tripos in the following year. In 1910 he was elected a Fellow of Trinity College, and held his Fellowship until 1916. However, in 1914 he went to University College, London as Assistant Lecturer, and was promoted to an Assistant Professorship in the following year. In 1918 he was appointed Professor of Mathematics at Birmingham University, and remained there until he retired in 1951.

Watson was an outstanding analyst and the author of more than 150 papers and three books. Of the latter, perhaps the best known is [**294**], affectionately known as W & W', an expanded version of the original textbook written by

E.T. Whittaker and published in 1902. Watson's monumental 800-page treatise [**274**] on Bessel functions appeared in 1922. Although now somewhat dated, it superseded all earlier books on the subject and, in an area where many different notations had been used, Watson's notations became standard. During the years between 1928 and the beginning of World War II, much of his work was concerned with problems connected with Ramanujan's work, and this was Watson's most prolific period. He wrote numerous papers on subjects such as singular moduli, mock theta functions, and partition functions. His well-known work on transforms also dates from this period.

Over the years from 1928 onwards, Hardy passed on to Watson all the Ramanujan manuscripts that remained in his possession, but, as remarked earlier, Watson never carefully examined Ramanujan's 'lost notebook' [**227**], perhaps because Hardy gave it to Watson toward the latter part of his interest in Ramanujan's work. Watson's last published paper on Ramanujan's work (apart from a table of the Ramanujan function $\tau(n)$, calculated as relaxation during the war years, and published in 1948) appeared in 1939.

From 1951 until his death in 1965, he was well-known to one of us (RAR), and retained his interest in Ramanujan. Watson had a very strong devotion to his alma mater Trinity College, and this, together with the fact that Ramanujan and Hardy had also been Fellows of Trinity, influenced his widow in her decision to deposit all Watson's Ramanujan manuscripts there. For further information about Watson, see [**230**] and [**295**].

The letter above is undated. However, it was probably written in early 1928, for the content of the first half of the letter is the subject of a paper by Hardy [**101**], [**107**, pp. 537–539] received by the editors on 15 May 1928. Although he could not find formula (A) in Ramanujan's papers when he wrote Watson, by the time Hardy prepared his paper for publication, he had indeed found it there. In fact, (A) is stated without proof in Ramanujan's paper [**221**, p. 269], [**226**, p. 190]. Also, (A) is Entry 13 of Chapter 14 in Ramanujan's second notebook [**225**], [**35**, p. 261]. The first published proof of (A) was not by Hardy but by M.B. Rao and M.V. Ayyar [**242**] in 1923. S.L. Malurkar's proof [**165**] also predated the two proofs by Hardy. For references to many of the proofs in the literature, see Berndt's book [**35**, p. 261]. Moreover, in Entry 21(i) of Chapter 14 of his second notebook [**225**], Ramanujan records a more general theorem that holds for every integer k. See Berndt's treatise [**35**, pp. 275–276] for this theorem and references to many proofs of it.

Pencilled against the first sentence of the letter on the left of the page is the remark (not reproduced in [**227**, p. 393]) 'The relic, I fancy, of a time some years ago when I began doing more or less what you are doing now: but I soon got discouraged.' In [**227**], for some reason, the fourth page of the letter is not reproduced. It will be found on pages 358 and 359 of [**235**].

The summation formula at the beginning of (3) is Poisson's summation formula for Fourier sine transforms [**267**, p. 66].

The page references under 'Miscellaneous observations' are to [**226**].

The 'Carlson–Wigert' paper by Hardy is [97], [107, pp. 610–622]. The formula akin to Mellin's inversion formula is discussed in the commentary on Hardy's letter of April 1930 to Watson.

G.N. Watson to B.M. Wilson 28 June 1929

THE UNIVERSITY,
UNIVERSITY ROAD,
EDGBASTON,
BIRMINGHAM.

Dear Wilson,

I never can remember the date of the June quarter-day (the others are easy enough), but it is near enough to make my Quarterly Report appropriate. In the last, I had done 24 of 39 problems in Ch. XVI. My present score is

XVI   24 out of 39
XVII 18 out of 18
XVIII 15 out of 24
XIX   14 out of 19
XX   7 out of 24.

I did not start on Ch. XX till about ten days ago & those which I have done are the first seven. Although I have nominally done only 71% out of Chs. XVI–XIX, those which I have written out are largely the sets which contain a good number of subdivisions. If each subdivision is counted as a separate problem, I have done 370 out of 469 in Chs. XVI–XIX, that is to say roughly 80%.

Chapters XX–XXI are the most fascinating of the lot. I am hoping to break the back of each of them before the inundation of O.&C. papers on about July 15.

Of course in Chs. XVIII & XIX (like Ch. XVI) many of those which I have not done are known results, & I want time to consider what to say about them & whether R. did them by classical methods or whether it seems more likely that he had his own devices.

I have written to G.H.H. and got his blessing. He is quite agreeable to (A) "R's notebooks by G.N.W. & B.M.W." (B) an introduction by G.H.H. giving a critical estimate of his work, based on more thorough knowledge than the Obituary. (C) reading and mss. & correcting any proofs on which we ask for his opinion.

He suggests 1 vol. of 600 pp. — though I don't know how it is to be compressed into that — & is optimistic about guarantees & finance generally if the work does not exceed this order of magnitude.

I am about to write to Vaidyanathaswamy to try to get the similar blessing of the Indian Committee.

My respect for R. has increased considerably in the last three months.

Livens was here for a couple of days last week; the first evening he drank the infant's health in the '91 Léoville and started a bottle of Tokay (presented by my father-in-law) & the second night we had a bottle of Asti. So you see that External Examiners in Mathematics are well treated here.

We are going up the Rhine & round the Black Forest in Aug–Sep parking the infant with his grandparents; that should provide the energy for attacking the heterogenous mass of material which follows Ch. XXI.

Yours sincerely,
G.N. Watson

I have retired from the J.M.B. to gain time for R.

Commentary. Watson and Wilson were invited by the University of Madras to become joint editors of the work contained in Ramanujan's notebooks. It appears from this and the following letter that they had an agreement to send each other quarterly reports on their work. No other reports, however, have survived. The chapters mentioned by Watson are from Ramanujan's second notebook, which is a revised, enlarged edition of the first. The second notebook contains 21 chapters of fairly organized material; Wilson was responsible for editing Chapters 1–14, while Watson assumed the task of editing Chapters 15–21. Evidently, Watson was also to edit the 100 pages of unorganized material at the end of the second notebook. After this letter was written, Watson finished proving most of the remaining formulas in Chapter 20 and a majority of the results in Chapter 21, which is considerably shorter than the foregoing chapters. He edited little from the aforementioned 100 pages at the end of the second notebook. Watson's efforts in editing the notebooks have fortunately been preserved in the library at Trinity College, except that a small portion has evidently been lost. Unfortunately, Wilson died in 1935 after incurring an infection during surgery. Watson's interest in completing the editing was waning at that time, and so their work was not completed.

With the help of the notes left by Wilson and Watson, Berndt [32], [35], [37] has completed the editing of the 21 chapters of organized material in the second notebook. These three volumes total 1226 pages, far exceeding the limit of 600 pages set by Hardy. Furthermore, the 100 pages at the end of the second notebook and the 33-page third notebook will be examined in two additional volumes.

For many years Watson was actively involved in the work of the Oxford and Cambridge Schools Examination Board (from which he was expecting later to receive examination papers for marking) and also of the Northern Universities Joint Matriculation Board (mentioned in his postscript).

It is of interest that, at this stage, Watson was hopeful that the work on the notebooks would be published, and to learn of Hardy's agreement to write the introduction. R. Vaidyanathaswamy (1894–1960) was at the time head of the newly formed Research Department of Mathematics of Madras University. He was a graduate of that university and had done research at St. Andrews University between 1922 and 1925. He was the author of more than 90 papers, mainly on geometry and algebra. For further information about Vaidyanathaswamy, see [**240**].

George Henry Livens (1886–1950) had been an undergraduate at Jesus College, Cambridge. He was 4th Wrangler in Part I of the Mathematical Tripos in 1909 (being bracketed equal with W.E.H. Berwick and three others) and obtained First Class Honours in Part II in the following year. After holding University posts in Sheffield and Manchester, he went in 1922 to University College, Cardiff, as Professor of Mathematics, and remained there until his death. An applied mathematician, he worked mainly on electromagnetic theory.

The infant mentioned is Watson's only child George, a graduate of Birmingham University.

G.H. Hardy to G.N. Watson   April 1930

Dear Watson,

Miss Stanley, who is staying here this week-end, and came to see me recently, brought me a large bundle of Ramanujan MSS, mostly fragmentary, but clear and neat.

Some of the notes are old ones, and refer to things plainly in the Papers or in his note books: but not all. There is one set, rougher than most, about the formula

$$\int w^{s-1}(a_0 - a_1 w + \cdots) dw = \frac{\pi}{\sin \pi t} a_{-t}$$

(a sort of alias of Mellin: see my 'Carlson–Wigert' paper in Acta 42), Fourier's $\int\int$, etc., etc. — all mixed up in a fine old stew. This I gave back to her, after a lecture, because I thought it might suggest something for her to work at. There are some other things I mean to talk to her about tomorrow. But there is a substantial chunk which I think I should be wise to send to you at once—by now you can probably 'place' most of them.

Of course I have a good many more which need sorting out.

Contents:

3 connected blocks (the most important being the suppressed part of 'H.C.N.'). These Miss S has attached slips to.

block of miscellaneous formulae: I have noted in pencil the sort of field of his work they hang on to.

You may expect some more after my further interview with her.

Yours
G.H. Hardy

N.B. I have a pupil working (inter alia) at

$$\sum c_n x^n = \frac{1}{(1-ax)(1-ax^2)\cdots}, \quad c_n \sim \sqrt{1-a}\frac{(kn)^{1/4}}{2n\sqrt{\pi}}e^{2\sqrt{kn}}$$

—so don't queer his pitch!

Commentary. The letter, presumably written from New College, Oxford, is undated but postmarked April 1930.

Gertrude Katherine Stanley (1898–1974) graduated from Westfield College, University of London, in 1917, and spent the next six years teaching in a girls' school. In 1925, she enrolled as a member of the Society of Oxford Home–Students. Hardy accepted her as a research student, and she obtained the B.Sc. degree in 1927; at that time in Oxford this degree was a postgraduate one corresponding to the present degree of M.Sc. She returned to Westfield College in 1931 and remained there until her retirement in 1964, being then head of the Mathematics Department. She was the author of five papers on analytic number theory published between the years 1927 and 1930. See [**233**] and [**232**].

In 1968, one of us (RAR) wrote to Miss Stanley to ask if she had any recollection of how the manuscripts mentioned in Hardy's letter above came into her possession. In her reply, she stated "I have only a vague recollection of Hardy's passing on a whole bundle of Ramanujan's papers and saying that some of them might suggest problems to me — also that he would like the papers back 'eventually'. The date 1930 means I was in Manchester, and I was at that time very intrigued by a problem suggested by Mordell. ... I *think* Hardy said he had a great many scraps of Ramanujan's work which came to him at Ramanujan's death and was trying to sort them."

The correspondence makes it clear that the bundle of manuscripts brought by Miss Stanley to Hardy had originally been given by him to her.

The variant of the Mellin transform displayed above is one of Ramanujan's fondest theorems. It is the primary focus of his quarterly reports, described by Berndt in [**32**, pp. 295–336], [**30**], and [**31**]. For a rigorous formulation and proof of Ramanujan's formula, see Hardy's book *Ramanujan* [**110**, pp. 189–191]. Ramanujan found numerous applications of this theorem, and many of these are offered in his quarterly reports [**32**]; see also [**110**, Chapter 11].

The 'Carlson–Wigert' paper [**97**], [**107**, pp. 610–622] is the same as that mentioned in the letter of Hardy to Watson in early 1928. The initials H.C.N. denote Ramanujan's article on highly composite numbers [**217**], [**226**, pp. 78–128]. Although the paper is long, it is not complete. The London Mathematical Society was in some financial difficulty at the time, and, to save expenses, part of the paper was omitted. The deleted sections have been published in [**227**, pp. 280–308]. J.-L. Nicolas [**173**], [**174**] has written short commentaries on this unpublished fragment and, together with G. Robin [**175**] has examined this fragment in complete detail.

Miss Stanley's bundle of papers is listed as Add.Ms.94.a.18 in Trinity College Library. For further details, see [233] and [235].

The pupil mentioned in the postscript is Edward Maitland Wright (1906–), who later collaborated with Hardy in the classic treatise [126], and is the author of many papers on the theory of numbers. Wright later became Professor of Mathematics and subsequently Principal of Aberdeen University; he was knighted in 1977.

G.N. Watson to B.M. Wilson 1 October 1930

THE UNIVERSITY,
EDGBASTON,
BIRMINGHAM.

Dear Wilson,

This quarterly Report is not going to be quite so meagre as the last two. It is true that I have not done much with the Note-books; but I have finished off the "unsolved problems" in the J.I.M.S., one of these being a joint effort with Vijayaraghavan. I have also written out three papers for the J.L.M.S., thus concluding the joint series with Preece. We decided to omit 18–22 on pp. xxix and 353 of the Collected papers, since these will fit more naturally into the Note-books; but 23 which deals with the singular modulus associated with 1353 is included; I was pleased at getting this out, because the bulk of the singular moduli in the Note-books can be obtained in the same way; I have not yet written out any of the latter properly, but have worked at many of them enough to make the formal writing out an easy matter. You will be interested to hear how R. got no. 23, particularly when you look at the length of the answer. I am absolutely convinced that he guessed it; I get out the others mentioned above by the same process of guessing.

I have also worked out about a quarter of Ch. XIV (which is really one of yours, but I got interested; apologies).

I fully agreed with what you said about our Registrarship, but the Powers that be decreed that we should have a man with actual administrative experience; so there was nothing doing.

Yours sincerely,

G.N. Watson

P.S. In all I have written out 9 papers this vac., 3 for J.I.M.S., 3 R. for J.L.M.S., one on windtunnels for R.S., one on Lebesgue's constant for Q.J. and one on hypergeometric functions for L.M.S. Not bad, though I says it as shouldn't.

Commentary. The postscript actually appears at the top of Watson's letter. At this time, Watson was remarkably active in publishing papers. Thus, over the years 1929–1931, 14 papers appeared, of which eleven were in the

Journal of the London Mathematical Society. Seven of these were under the heading 'Theorems stated by Ramanujan.' Of these, the three alluded to in the postscript are [282], [283], [284]. The joint paper with Vijayaraghavan is their solution of Problem 784 in the *Journal of the Indian Mathematical Society* [271]. Watson appears to have published only two, not three, papers in the *Journal of the Indian Mathematical Society*, namely [277] and [271].

Ramanujan calculated an amazing total of 102 singular moduli. Of these, 56 were also calculated by Weber [293]; the remainder were not previously calculated. A total of 46 were stated without proof by Ramanujan in his paper [210], [226, pp. 23–39]. In two papers [286] and [287], Watson established 24 singular moduli that appear in Ramanujan's paper [210] but not in his notebooks. Watson's proofs in the former paper depend upon "guessing," as indicated in the letter above, while the proofs in the latter paper arise from modular equations. In [286], Watson remarks that the 21 remaining singular moduli, which appear in both [210] and in Ramanujan's second notebook, will be proved in his account with Wilson of Ramanujan's notebooks. These proofs were never published and have apparently been lost. For an examination of the remaining singular moduli, see Berndt's book [41]

The last paragraph suggests, perhaps, that Wilson had been a candidate for the vacant Registrarship at Birmingham University, to which a Mr. Donald John Cameron was appointed in 1930. It would certainly have been convenient for their work on the notebooks if Wilson had been nearer at hand than Liverpool.

K. Ananda Rau to G.H. Hardy   **7 April 1936**

THE PRESIDENCY COLLEGE,
MADRAS.

Dear Prof. Hardy,

I received your letter of 24 Feb. Cayley's Elliptic Functions and Greenhill's Elliptic Functions have been in the Presidency College's Library and also in the Madras University Library for a long time, and it is not unreasonable to suppose that Ramanujan got his ideas on Elliptic Functions from these books. I am told that there has been a copy of Greenhill in the Library of the Government College at Kumbakonam, but I have not been able to verify this.

Ramanujan, as you probably know, spent a good part of his student days at Kumbakonam. While at Cambridge I remember Ramanujan having with him a private copy of Cayley, and he used to tell me that he found it useful.

I was looking at Ramanujan's early Notebook a few days ago at the Madras University Library, and I found Ramanujan had given a geometric construction for some problem in the oscillation of a pendulum. This was hardly a matter in which Ramanujan was primarily interested, and I can only suppose that he worked it out under the influence of Greenhill's book. There such discussions are given. Cayley does not treat of pendulums.

On the whole I think it may be taken for granted that Ramanujan had access to these books and used them.

In your letter you refer to a note by yourself and Prof. Littlewood on Lambert series. I shall be glad to have some details or if this note has been published, an offprint.

With all kind regards
Yours very sincerely
K. Ananda Rau

Commentary. Cayley's book [61] gives an extensive account of the theory of modular equations, and because Ramanujan recorded several hundred modular equations in his notebooks, it can be conjectured that Ramanujan first learned about modular equations from Cayley's treatise. However, Ramanujan's methods are primarily analytic and unlike those of Cayley. Greenhill's book [84] is frequently mentioned as a source of Ramanujan's knowledge about elliptic functions, because complex variables and double periodicity, in which Ramanujan showed little interest, are not mentioned until page 248. However, Ramanujan's development of the theory of elliptic functions shows no resemblance to that in Greenhill's text. One of us (BCB) has visited the library at the Government College in Kumbakonam and has verified that Greenhill's book is found there, although we have not been able to ascertain for certain that the book was present when Ramanujan was a student in Kumbakonam. In conclusion, although Ramanujan might have learned the rudiments of the theory of elliptic functions and modular equations from Cayley and/or Greenhill's texts, they seem to have provided little inspiration to Ramanujan's development of the subjects. In fact, Ramanujan's theory appears to have been influenced by no other author.

Ananda Rau's remarks on the pendulum are correct. Ramanujan considers the oscillation of a pendulum at three places in his second notebook, namely, in Section 24 of Chapter 18 and in Sections 7(iii) and 16(ii) in Chapter 19. Ramanujan's results are connected with modular equations of degrees 2, 3, and 5, respectively.

The paper of Hardy and Littlewood mentioned at the end of the letter is [122], [109, pp. 790–803], published in 1936.

G.H. Hardy to S. Chandrasekhar   19 February 1936

TRINITY COLLEGE,
CAMBRIDGE.

Dear Chandrasekhar,

I am going to give some lectures (here and at Harvard) on Ramanujan during the summer.

I want your reactions, as a sophisticated Indian, about the varying statements which have been made about Ramanujan and 'God'. See in particular the passages on pp. xviii (bottom) — xix (top), and p. xxxi of the two 'notices' in the Papers. Which view is the more likely to be right?

I take it that, in general, there is very little correlation, among Hindus, between 'observance' and 'belief'. And my own view is that, at bottom and to a first approximation, Ramanujan was (intellectually) as sound an infidel as B. Russell or Littlewood. But I cannot pretend to have any clear view about the matter.

One thing I am sure. Ramanujan was not in the least the 'inspired idiot' that some people seem to have thought him. On the contrary he was (except for a period when his mental equilibrium was definitely upset by illness) a very shrewd and sensible person: very individual, of course, and with a reasonable allowance of the minor eccentricities of genius, but fundamentally normal and sane.

I would be extremely grateful for any thing you can say. Of course, you did not know him; but there is a good deal you might find easier to understand than I can.

Yours very sincerely
G.H. Hardy

Commentary. The passages to which Hardy refers are from the articles on Ramanujan by P.V. Seshu Aiyar and R. Ramachandra Rao, and by himself in Ramanujan's *Collected Papers* [226].

The first states: Ramanujan had definite religious views. He had a special veneration for the Namakkal goddess. Fond of the *Puranas*, he used to attend popular lectures on the Great Epics of Ramayana and Mahabharata, and to enter into discussions with learned pundits. He believed in the existence of a Supreme Being and in the attainment of Godhead by men by proper methods of service and realisation of oneness with the Deity. He had settled convictions about the problem of life and after, and even the certain approach of death did not unsettle his faculties or spirits.

The second passage reads: He adhered, with a severity most unusual in Indians resident in England, to the religious observances of his caste; but his religion was a matter of observance and not of intellectual conviction, and I remember well his telling me (much to my surprise) that all religions seem to him more or less equally true.

Subrahmanyan Chandrasekhar is arguably the preeminent astrophysicist of the 20th century. Born in Lahore in 1910, he comes from a learned family that includes C.V. Raman, a Nobel laureate in physics, who made a fundamental discovery about the molecular scattering of light, now called the Raman effect. It is interesting that Chandrasekhar's grandfather, Ramanathan Chandrasekhar, received his high school education in Kumbakonam, and his grandmother Parvati Chandrasekhar's family village is Tiruvanakoil, which is very near Kumbakonam.

Chandrasekhar was 9 years old when Ramanujan died, but the success and accomplishments of his countryman, as related by his mother and uncle, C.V. Raman, had an indelible effect on Chandrasekhar. Absorbed and remarkably gifted in mathematics, he graduated at the age of 15 from The Hindu High School in Triplicane. At Presidency College in Madras, Chandrasekhar studied chemistry, physics, and mathematics, as his primary interests gradually focused on physics and applied mathematics. Upon receiving his B.A., he went to Cambridge, where he obtained a Ph.D. in astrophysics in 1933. Later in the same year, he was made a Fellow of Trinity College. In 1937, Chandrasekhar assumed a position at The University of Chicago, where he has remained throughout his career. He was elected to the Royal Society in 1944. The Nobel Prize was awarded to Chandrasekhar in 1983 for, among many other accomplishments, his well-known contribution to the study of white dwarfs and setting a limit (Chandrasekhar limit) on the mass of a star that could become a white dwarf. This discovery was the forerunner to our understanding of neutron stars and black holes.

The influence of Ramanujan on Chandrasekhar is detailed in an address [63] he gave at the Ramanujan Centenary Conference at the University of Illinois in June 1987 and in a book by K.C. Wali [273, pp. 260–265], an excellent personal and scientific biography of Chandrasekhar.

S. Chandrasekhar to G. H. Hardy   4 August 1937

THE UNIVERSITY OF CHICAGO
Yerkes Observatory
Williams Bay, Wis.

Dear Hardy,

I remember your telling me that when Ramanujan's *Collected Papers* were being edited, it was your original intention to include a portrait of Ramanujan, but eventually you had to abandon the plan as no good photograph was available.

So, when I was in India last summer, I made an effort to find a reasonably good photograph. I met Mrs. Ramanujan—his wife, who incidentally is having a rather difficult life. Some of her unscrupulous relatives having swindled her out of such financial resources as Ramanujan had left her—and it transpired that the only photograph of Ramanujan (other than the one where he is "cap and gown") available is his passport photograph taken prior to his leaving England.

I had a negative taken of his passport photograph (which is now in my possession) and had an enlargement professionally made. I am sending you by separate cover an enlarged photograph of Ramanujan, and as you will see it (is) a reasonably good one. I am not in a position to judge how 'true' it is, but I have on Mrs Ramanujan's authority that the rather worried look Ramanujan has in the picture was extremely frequent during his last year.

RAMANUJAN'S PICTURE IN THE HOME OF HIS WIFE

I do not know if you would want to include this photograph in your book on Ramanujan—I refer to your Harvard lectures—and I do not know either if the enclosed enlargement would be good enough for reproduction purposes. In any case I think a good professional must be able to use the original negative I have and if it should be needed, I shall be only too glad to loan it. I have some more spare copies of the enlargements of Ramanujan and if you should know others who may like to have a copy of Ramanujan's photograph perhaps you could ask them to write to me.

With best regards
Yours very sincerely
S. Chandrasekhar

P.S. In any case, I should be glad to know what you think of the photograph.

Commentary. Indeed, Ramanujan's passport photograph was reproduced by Hardy in his book on Ramanujan [110]. This cropped photograph can also be found in Berndt's book [32] and on the cover of the 'lost notebook' [227]. A fuller, less cropped version appears as the frontispiece in [11], [40], and in this book.

G.H. Hardy to S. Chandrasekhar   15 December 1937

TRINITY COLLEGE,
CAMBRIDGE.

Dear Chandrasekhar,

I ought of course to have replied long ago.

I was very glad indeed to have the photograph, which seems to me an extremely good one. He looks rather ill (and no doubt was very ill): but he looks all over the genius he was.

I certainly propose to include it in the book. A few days ago I took about 250 pages of typescript to the Press, to get an estimate of its length (I have, I think, the material now, not all in perfect form, for about 80% of the book). I took the opportunity of shewing the photo to Roberts, who quite agreed with me. He said it would be quite easy to reproduce it from the copy, and that the original negative would not be wanted. I suspect, however, that the prints from the negative will be better than any reproduction: and I should like to have a copy of my own (the one I have is sure to get knocked about a bit when being worked with). Also, I am sure, the College Library would like one for its collection of Fellows. I don't know whether they have one of the 'cap & gown' picture, but this is incomparably better.

With all kind regards
Yours very sincerely
G.H. Hardy

Commentary. This is in reply to the preceding letter from Chandrasekhar. The 'cap & gown' photograph (see Chapter 4, p. 137) is the one appearing as the frontispiece of [**227**] or opposite page 5 of [**263**]. Another photograph, in which Ramanujan is sitting on a chair, appears opposite page 1 of [**263**], and has been reproduced in various forms (see Chapter 4, p. 127). When Hardy gave one of the authors (RAR) in 1940 a copy of the photograph procured by Chandrasekhar, Hardy remarked that Ramanujan was in his dressing-gown and pyjamas when the photograph was taken. If Ramanujan had left the nursing home to have his photo taken, it is fairly certain that he would have put on a tie and dressed more formally. On the other hand, if the photograph was taken in the nursing home, he might well have not dressed for the occasion. The dressing gown in the picture is typical of those worn at that time. For more details on the history of this photograph, which as Hardy has indicated is incomparably better than any other existing photograph of Ramanujan, see Chandrasekhar's account at the beginning of [**32**].

Sydney Castle Roberts (1887–1966) was Secretary of the Cambridge University Press from 1922 to 1948 and Master of Pembroke College from 1948 to 1958; he was knighted in 1958.

D.H. Lehmer to G.H. Hardy   2 November 1937

Lehigh University
Bethlehem, Pa.

Professor G.H. Hardy
Trinity College
Cambridge University
Cambridge, England

Dear Professor Hardy:

I enclose for your amusement the first 62 (non-zero) terms of the H-R series for $p(14031)$ together with the corresponding remainders.

I have made considerable progress in studying the coefficients $A_k(n)$. These turn out to be closely related to generalized Kloosterman sums and are capable of being expressed as a single term. Thus it is a simple matter to carry the series to any number of terms. The only remaining difficulty lies in the computation of the huge exponential occurring in the first term. This difficulty I have reduced to some extent by the application of singular moduli.

The theoretical error after taking this many terms of the series does not exceed .26 in absolute value, while the actual error is about .00016.

The value thus obtained for $p(14031)$ is actually divisible by 11^4, in accord with Ramanujan's conjecture. It is not divisible by 11^5.

Professor G.N. Watson writes me to say that he has nearly completed the proof of a revised conjecture. The case of powers of 11 was still incomplete. There seems little doubt that Ramanujan was wrong in this case.

I am also writing you to warn you that you may hear sometime this winter from the Guggenheim Foundation to which I have applied for a travelling fellowship for next year. I have taken the liberty of mentioning your name in my application. However, past statistics show that the foundation has not been very partial to mathematics, so perhaps you will not hear from them after all.

Very truly yours
D.H. Lehmer

The Hardy–Ramanujan Series for $p(14031)$

T_k	k
92 85303 04759 09931 69434 85156 67127 75089 29160 56358	
46500 54568 28164 58088 38068 24656 61188 14283	1
33426 32345 44042 85709 15815 03179 90098 16010.80863	
−6 87664 77899 86064 18387	2
85518 80186 83621 41879 15224 38734 94037 51839.57288	
11205 95260 28340 75710 80319 22182 56728 12266.45638	3

	−378 97477 37584 89350 97660 06356.17860	4
	−52 05646 48432 11186.68338	6
	−30 68279 92999.62308	8
	52826 19485.36581	9
	98 25681.34285	11
	3 57552.44132	12
$r_n(k)$	33572.53285	13
	−1331.35005	16
−12.60694	−123.14679	18
−5.88991	6.71703	22
−0.69755	5.19236	23
1.18750	1.88505	24
0.00331	−1.18419	26
−0.4280	−0.43140	27
0.19801	0.62610	29
−0.03185	−0.22986	31
0.07876	0.11061	32
0.01090	−0.06786	33
−0.00613	−0.01703	36
−0.02130	−0.01517	39
−0.01350	0.00780	41
−0.00539	0.00811	44
−0.00995	−0.00456	46
0.00082	0.01077	47
−0.00023	−0.00105	48
0.00170	0.00193	52
0.00119	−0.00051	54
0.00155	0.00036	58
−0.00127	−0.00282	59
0.00016	0.00143	62

$\sum_{k=1}^{62} T_k = 92\,85303\,04759\,09931\,69434\,85156\,67127\,75089\,29160\,56358$
$46500\,54568\,28164\,58081\,50403\,46756\,75123\,95895$
$59113\,47418\,88383\,22063\,43272\,91599\,91345\,00745.00016$

$$T_k = \frac{\sqrt{12}}{\mu(24n-1)\sqrt{k}} A_k(n)(\mu - k)e^{\mu/k} \qquad (\mu = \frac{\pi}{6}\sqrt{24n-1})$$

$$p(n) = \sum_{k=1}^{N} T_k + r_n(N)$$

Commentary. Derrick Henry Lehmer (1905–1991) was a pioneer in the applications of computers to number theory. In 1927, while he was an undergraduate at the University of California at Berkeley, Lehmer designed and built an electromechanical sieve from bicycle chains and relays [**153**]. In 1933, he built from gears another sieve, which used a light ray and a photoelectric cell, and which was 100 times faster [154], [155]. In 1965, he built an even faster sieve (a million bits per second) from delay-lines and (later) shift registers [157]. These parallel machines were used to solve problems in number theory, such as the factoring and primality testing of integers, finding pseudo-squares, and counting solutions of Diophantine equations. His understanding of practical computing made him a strong advocate of parallelism, in strong contrast to J. von Neumann, who insisted that parallelism was unnecessary because serial computers were already fast enough [158]. For additional information about D.H. Lehmer, see [53].

Ramanujan's original conjectures on congruences satisfied by the partition function $p(n)$ are as follows:

(1) $\quad\quad$ If $24m \equiv 1 \pmod{5^n}$, then $p(m) \equiv 0 \pmod{5^n}$,

(2) $\quad\quad$ if $24m \equiv 1 \pmod{7^n}$, then $p(m) \equiv 0 \pmod{7^n}$,

(3) $\quad\quad$ if $24m \equiv 1 \pmod{11^n}$, then $p(m) \equiv 0 \pmod{11^n}$.

These remarkable conjectures were made upon closely examining a table of $p(n)$, $1 \leq n \leq 200$, compiled by MacMahon. H. Gupta [86] extended MacMahon's table up to $n = 300$. In particular, Gupta's table shows that

$$p(243) = 133978259344888,$$

a number not divisible by 7^3. Since

$$24 \cdot 243 \equiv 1 \pmod{7^3},$$

as first observed by S. Chowla, this contradicts Ramanujan's conjecture (2). To correct the conjecture, in the congruence for $p(m)$ in (2), replace 7^n by $7^{[(n+2)/2]}$. Ramanujan proved several special cases of his conjectures (1)–(3) but did not establish any of the three general conjectures. Conjecture (1) and the aforementioned corrected version of (2) were first proved by Watson [291] in 1937. Conjecture (3) was established by Atkin [18] in 1967. Lehmer's calculation gave firm evidence to the truth of Ramanujan's conjecture for powers of 11. Proofs of the Ramanujan congruences for powers of 5 and 7 can be found in Knopp's book [146].

Lehmer eventually published his study of $A_k(n)$ in [156]. In particular, he showed that these coefficients satisfy remarkable multiplicative properties. A. Selberg found a simpler formula for $A_k(n)$ which led to simpler proofs of the multiplicativity by Rademacher [188], [189, pp. 460–474].

S. Chandrasekhar to G.H. Hardy   28 December 1937

THE UNIVERSITY OF CHICAGO
Perkes Observatory
Williams Bay, Wis.

Dear Hardy,

Many thanks for your letter of Dec. 15. I am very glad to learn of your intention to include the photograph in your book.

I am enclosing the negative from which the enlargement I sent you was made. Also a rather good copy of the same size as the negative and four rather poor copies of the enlargement. I am afraid that my own efforts at copying from the enlargement proved less successful than I had hoped! I am however certain that a professional photographer can make even better enlargements from the negative than the one I sent you.

My impression is that the college has no portrait of Ramanujan in its collection of fellows — perhaps you would be so kind as to pass on one of the four copies I am sending.

With the season's greetings and with kind regards

Yours very sincerely
S. Chandrasekhar

G.H. Hardy to G.N. Watson

Trinity College
Cambridge

Dear Watson,

Another question you might answer on Thursday:

<u>Is</u> there a proof of

$$p(\cdots) \equiv 0(11^2)$$

in Ramanujan's MS? None seems ever to have been printed by anybody — Darling, Rogers, Mordell, Kreczmer are all silent on this point. The only evidence I can find is my statement in <u>Papers</u>, p. 343 (near bottom).

Yours
G.H. Hardy

Commentary. The letter is undated but was probably written in the late 1930s when Hardy was preparing his book [**110**]. As Watson may have informed him, Ramanujan had, in fact, proved that

$$p(121n - 5) \equiv 0(\mod 121).$$

This is to be found on page 166 of [**227**] in the '$p(n) - \tau(n)$' manuscript, which, presumably Hardy had passed to Watson at an earlier stage. On page

100 of [**110**], Hardy refers to a congruence of this form but does not give it explicitly. The first published proof of Ramanujan's congruence modulo 11^2 is by J. Lehner [**159**] in 1943.

The Thursday mentioned in the first sentence was, no doubt, the day when the London Mathematical Society met.

G.N. Watson to H. Heilbronn   **19 March 1939**

Department of Mathematics   THE UNIVERSITY
PROFESSOR G.N. WATSON, Sc.D. F.R.S.   EDGBASTON,
PROFESSOR R. PEIERLS, D.Sc.   BIRMINGHAM 15

Dear Heilbronn,

I have written out and enclose a list of the results that I was discussing with you on Friday. The eight statements headed definitions explain the notation used, and the five statements about Dirichlet series headed conjectures are the things which require proof. The notation used suggested the query at the foot of the page, but it is not impossible that the answer to the query is no.

Actually I think that Ramanujan uses only three of the five conjectures in his work on 121, but I thought it advisable to let you have the complete set. It is not impossible that the two missing functions in my query occur in some other mss. of Ramanujan, but I have not found them in a hasty search.

The paper by Mordell is <u>Proc. Camb. Phil. Soc.</u>, 19 (1920), 117-124.

With kind regards,
Yours sincerely,
G. N. Watson

Definitions:

$$P = 1 - 24 \left(\frac{x}{1-x} + \frac{2x^2}{1-x^2} + \frac{3x^3}{1-x^3} + \cdots \right)$$

$$Q = 1 + 240 \left(\frac{1^3 x}{1-x} + \frac{2^3 x^2}{1-x^2} + \frac{3^3 x^3}{1-x^3} + \cdots \right)$$

$$R = 1 - 504 \left(\frac{1^5 x}{1-x} + \frac{2^5 x^2}{1-x^2} + \frac{3^5 x^3}{1-x^3} + \cdots \right)$$

so that

$$Q^3 - R^2 = 1728 x \left\{ (1-x)(1-x^2) \cdots \right\}^{24}$$

[Definitions of $\tau_2(n), \tau_3(n), \tau_4(n), \tau_5(n), \tau_7(n)$]

$$\sum_{n=1}^{\infty} \tau_2(n)x^n = Qx\left\{(1-x)(1-x^2)(1-x^3)\cdots\right\}^{24}$$

$$\sum_{n=1}^{\infty} \tau_3(n)x^n = Rx\left\{(1-x)(1-x^2)(1-x^3)\cdots\right\}^{24}$$

$$\sum_{n=1}^{\infty} \tau_4(n)x^n = Q^2x\left\{(1-x)(1-x^2)(1-x^3)\cdots\right\}^{24}$$

$$\sum_{n=1}^{\infty} \tau_5(n)x^n = QRx\left\{(1-x)(1-x^2)(1-x^3)\cdots\right\}^{24}$$

$$\sum_{n=1}^{\infty} \tau_7(n)x^n = Q^2Rx\left\{(1-x)(1-x^2)(1-x^3)\cdots\right\}^{24}$$

Conjectures:

$$\sum_{n=1}^{\infty} \frac{\tau_2(n)}{n^s} = \prod \frac{1}{1-\tau_2(p)p^{-s}+p^{15-2s}}$$

$$\sum_{n=1}^{\infty} \frac{\tau_3(n)}{n^s} = \prod \frac{1}{1-\tau_3(p)p^{-s}+p^{17-2s}}$$

$$\sum_{n=1}^{\infty} \frac{\tau_4(n)}{n^s} = \prod \frac{1}{1-\tau_4(p)p^{-s}+p^{19-2s}}$$

$$\sum_{n=1}^{\infty} \frac{\tau_5(n)}{n^s} = \prod \frac{1}{1-\tau_5(p)p^{-s}+p^{21-2s}}$$

$$\sum_{n=1}^{\infty} \frac{\tau_7(n)}{n^s} = \prod \frac{1}{1-\tau_7(p)p^{-s}+p^{25-2s}}$$

Query. Do similar number theoretic functions $\tau_1(n)$, $\tau_6(n)$ exist with

$$\sum_{n=1}^{\infty} \tau_1(n)x^n = ?, \qquad \sum_{n=1}^{\infty} \tau_6(n)x^n = ?$$

and

$$\sum_{n=1}^{\infty} \frac{\tau_1(n)}{n^s} = \prod \frac{1}{1-\tau_1(p)p^{-s}+p^{13-2s}}$$

$$\sum_{n=1}^{\infty} \frac{\tau_6(n)}{n^s} = \prod \frac{1}{1-\tau_6(p)p^{-s}+p^{23-2s}}?$$

Commentary. Hans Arnold Heilbronn (1908–1975) was born in Berlin and studied at the universities of Berlin, Freiburg, and Göttingen. In 1930 he was appointed assistant to the distinguished number theorist Edmund Landau. After the Nazis came to power in 1933, he, like other Jewish scientists in Germany, lost his position and came to England. After holding various college and university posts and doing war service in the Pioneer Corps, he was appointed Professor of Mathematics at the University of Bristol in 1949. He resigned in 1963 and went to Canada, where, in 1964, he was appointed Professor at the University of Toronto, a post he held until his early death. He was elected Fellow of the Royal Society in 1951, and served as President of the London Mathematical Society from 1959 to 1961. Heilbronn did important work in the theory of numbers on several topics, and is perhaps best known for his work on the class number of imaginary quadratic fields. For further information, see [131].

The five Euler products listed are associated with the unique newforms of weights 16, 18, 20, 22, and 26 for the modular group. Watson seems to have thought that there may be two 'missing functions' of weights 14 and 24, respectively. In fact, there is no function for weight 14, as the space of cusp forms of this weight has dimension zero. On the other hand, for weight 24 the dimension of the space is 2, and there do exist two conjugate newforms, but their Fourier coefficients are algebraic integers in the quadratic field $Q(\sqrt{144169})$, as shown by E. Hecke.

It is rather astonishing that in March 1939 Watson should have believed that the five Euler products require proof, since, after Mordell's work [168], one would have thought he could himself have provided one. Moreover, he cannot have been aware of the fundamental work of Hecke [129], [130, pp. 644–671] on Euler products, which, appearing two years earlier, contains these results as particular cases.

G.H. Hardy to A.F. Scholfield 23 July 1940

TRINITY COLLEGE
CAMBRIDGE

Dear Scholfield

I have looked in all likely places, and can find no trace of the missing pages of the first letter, so I think we must assume that it is lost. This is very natural, since it was circulated to quite a number of people interested in Ramanujan's case.

On the other hand I have found two more (much shorter) letters which I think should be bound up with the others. The four were (so far as I can remember) all I ever had from Ramanujan before I met him here. E.H. Neville was in India and took charge of the later arrangements. The exact account of the letters will then be as follows. Nos. 1 and 2 are the letters printed in the <u>Papers,</u> pp. xxiii—xxix and 349–353, except for the lost beginning of No. 1 (printed on p. xxiii). As stated on page 349, one page of mathematical formulae was also lost. Nos. 3 and 4 have not been printed before, except for a fragment of No. 3, printed on p. xxix .

I have other letters, but they are much more fragmentary: these are, so to say, a complete set, on paper of uniform size, and easy to bind.

Yours sincerely

G.H. Hardy

Commentary. The dates of these four letters are 16 January 1913, 27 February 1913, 17 April 1913, and 22 January 1914. These four letters have, in fact, been bound, together with Hardy's covering letter to Scholfield, and are deposited in Cambridge University Library. The reference is Add. 7011 (C).

Alwyn Faber Scholfield (1884–1969) studied law at King's College, Cambridge. He was Keeper of the Records to the Government of India from 1913 to 1919, Librarian of Trinity College, Cambridge from 1919–1923, and University Librarian from 1923–1949.

R.A. Rankin to The Librarian, Trinity College   1 September 1965

Department of Mathemathics
University of Glasgow
Glasgow G12

The Librarian
Trinity College
Cambridge

Dear Sir,

I have been asked by the London Mathematical Society to write an obituary notice of the late Professor George Neville Watson, who was professor of mathematics at Birmingham University and was a past fellow of Trinity College. Professor Watson died on 2 February, 1965.

At the invitation of his widow, Mrs E.G. Watson of 46 Warwick New Road, Leamington Spa, I have been looking through his books and papers. Among these are the following manuscripts relating to the work of the Indian mathematician Srinivasa Ramanujan (1887–1920), who was also formerly a fellow of Trinity College:

A. One bound manuscript, approximately $1\frac{1}{2}$ in. thick. This is a copy of a notebook of Ramanujan made by T.A. Satagopan.

B1–4. Four bound manuscripts, each $\frac{3}{4}$ in. thick. This is a copy of another notebook of Ramanujan made by an unknown copyist.

GNW 1–3. Three green–backed files containing copies of Chapters XVI onwards of B. These copies were made by Professor Watson from B and contain in addition his proofs of Ramanujan's formulae. The work is incomplete and, particularly in GNW3, gaps have been left for proofs to be filled in at a later stage.

BMW. These are manuscripts belonging to the late Dr B.M. Wilson, of Liverpool University, and the first 7 items listed below are contained in University of Liverpool Examination Books. Dr Wilson was a former member of Trinity College.

CHAPTER 8. RAMANUJAN'S PAPERS AND MANUSCRIPTS

A. Account of Ramanujan and his work (notes for a lecture?).

B. List of corrections to Chapters I–X of B with references to published work of other mathematicians.

I Notes and proofs of results in Chapters II–VI of B.
II Ditto for Chs VI–IX.
III Ditto for Chs IX–X.
IV Ditto for Chs X–XI.
V Ditto for Ch. XII.
5 sheets clipped together on Ch. XIII.
20 further quarto sheets.

St Andrews University Examination Book (green quarto) containing excerpts from various chapters up to Ch.XI.

6 foolscap sheets giving a general account of Ramanujan's life and work.
19 foolscap sheets on Ch. VIII.

Mrs Watson feels, and I agree with her, that it would be appropriate that all this material should go to the Library of Trinity College, and I am writing to ask whether the College will accept these manuscripts. If they are accepted I shall arrange for them to be sent to you.

The following additional information may be of service to you.

From the enclosed offprint of an address given by Professor Watson to the London Mathematical Society in 1931, it is clear that A and B are copies originally sent to the late Professor G.H. Hardy in return for the original notebooks, which are now in the possession of the University of Madras. These copies were presumably given by Professor Hardy to Professor Watson because of Watson's interest in Ramanujan's work. Professor Watson did more than any other mathematician to provide proofs of the numerous results stated without proof by Ramanujan. Evidently Professor Watson and Dr B.M. Wilson divided the labour of proving Ramanujan's results between them, Wilson taking the earlier chapters and Watson the later. It was intended that this work of collaboration should be published. After Dr Wilson's death his work was presumably passed on to Professor Watson. From 1940 onwards, Watson appears to have lost interest in the notebooks and I doubt whether he did any work on them from that date.

It should be mentioned that a facsimile edition in two volumes of Ramanujan's original notebooks was published in India in 1957, and I enclose a review of it which I wrote for the Mathematical Gazette. The facsimiles have been published without any editorial comment. Volume 1 contains the material of A plus about 5 extra pages. Volume 2 contains the material in B omitting the 'loose papers' and Ramanujan's three reports, and also contains a smaller third notebook of about 30 pages. Watson does not seem to have possessed a copy of this third notebook.

If some mathematician were willing and able to edit and complete the work of Watson and Wilson on the notebooks, the publication of such a work would be of the greatest importance and interest. Such a task would occupy a long period, however, and I doubt whether an active mathematician would be willing to spare the time, as it would be almost a full-time job. Whether such a person will come forward or not, it would be of great value if it were known that these papers could be consulted in your library, and I hope very much that your College will feel able to accept them. I am sure that Ramanujan, Hardy, Watson and Wilson would all have been glad to know that the manuscripts would find a resting place in Trinity College.

Yours sincerely,
R.A. Rankin

Commentary. In the absence of the Librarian, the letter was acknowledged by the Sub-Librarian A. Halcrow. After information that the College was glad to accept the manuscripts was received, items A and B 1–4 were dispatched from Glasgow to Trinity. The remaining manuscripts were held by Mrs Watson in Leamington Spa and she sent those listed as GNW 1–3 to Cambridge in November 1965, but, for some reason, did not dispatch the remaining papers mentioned in the letter. However, after a visit to Leamington Spa, these were retrieved by Rankin and sent to Trinity in 1969, as stated in the Commentary to the following letter.

In May 1977, Berndt began the task of editing the notebooks, urged in the last paragraph of the letter. At this time (1995), four of five expected volumes have been completed [**32**], [**35**], [**37**], [**40**].

J.M. Whittaker to G.E. Andrews   **15 August 1979**

11B Endcliffe Crescent
Sheffield S10 3EB

Dear Professor Andrews

Looking through recent periodicals in the University Library yesterday I came across your paper on Ramanujan's notebook (of which I should be grateful for an offprint if you can spare one). You may be interested in the circumstances of its recovery.

When the Royal Society asked me to write G.N. Watson's obituary memoir I wrote to his widow to ask if I could examine his papers. She kindly invited me to lunch and afterwards her son took me upstairs to see them. They covered the floor of a fair sized room to a depth of about a foot, all jumbled together, and were to be incinerated in a few days. One could only make lucky dips and, as Watson never threw away anything, the result might be a sheet of mathematics but more probably a receipted bill or a draft of his income tax return for 1923. By an extraordinary stroke of luck one of my dips brought up the Ramanujan material which Hardy must have passed on to him when he proposed to edit the earlier notebooks. Rankin and I thought that Trinity was the right place for it rather than India which had done nothing for him.

Yours sincerely
J.M. Whittaker

Commentary. John (Jack) Macnaghten Whittaker (1905–1984) was a son of Sir Edmund Whittaker (1873–1956), who was Professor of Mathematics at Edinburgh University from 1912 to 1946, and had previously been Royal Astronomer of Ireland. Jack, like his father, was a Trinity man and a Wrangler. He had the added distinction of winning both a Smith's Prize and an Adams Prize at Cambridge. His field of interest was complex function theory, and, in particular, interpolatary function theory. In 1933 he accepted the position of Professor of Mathematics at Liverpool and left in 1952 on his appointment as Vice-Chancellor of Sheffield University. He retired from this post in 1965. The offprint requested by him is the paper [7].

Whittaker's visit to Watson's widow occurred on 20 November 1965, and its outcome was the memoir [295]. Earlier, on 12 July 1965, Rankin paid a similar visit (see page 84 of [233]) to obtain material for an obituary notice [230] that he had been asked to write for the London Mathematical Society. His experience was similar, and he excavated a large amount of work related to Ramanujan. Since Hardy, Watson, and Ramanujan had all been Fellows of Trinity, he suggested to Mrs Watson that the documents should be offered to Trinity College Library, and she readily agreed.

Shortly afterwards she dispatched a considerable batch of material to Rankin in Glasgow; the Librarian of Trinity College was contacted (see the previous letter of 1 September 1965) and given a detailed description of the material so far discovered. After Jack Whittaker's visit to Mrs Watson, a further batch of papers (including the 'Lost Notebook') was sent to Trinity in December 1968. A final collection of papers related to Ramanujan (consisting mainly of B.M. Wilson's work) was collected in April 1969 by Rankin from Mrs Watson in Leamington Spa and later sent to Trinity.

In his letter Whittaker states that Watson's papers were shortly to be incinerated, and they were certainly ultimately destroyed. However, his son George Watson, who was a graduate of Birmingham University in mathematics, closely examined the papers, and it seems likely that nothing of interest escaped his attention. In this connection it may be of interest to remark that other interesting manuscripts, not connected with Ramanujan, were discovered in the same room, namely, Watson's projected and incomplete revision of Whittaker and Watson's *Modern Analysis* [294] and his monograph on *Three Decades of Midland Railway Locomotives*. Finally, to correct the record, it should be mentioned that Whittaker's statement that India had done nothing for Ramanujan cannot be substantiated; it is clear that his Indian colleagues had done all that was within their power to assist him.

Born on 4 December 1938, George E. Andrews received his Ph.D. at the University of Pennsylvania in 1964 as the last student of Hans Rademacher. Andrews is generally recognized as the world's leading authority on partitions and is the author of [6], the foremost treatise on the subject. He has devoted much of his life to the study of Ramanujan's work, in particular, mock theta functions and other q-series, both primary topics in Ramanujan's lost notebook [227]. Much of Whittaker's letter is quoted by Andrews in his monograph [8].

A. Ranganathan to S. Chandrasekhar 12 January 1981
excerpt

Here is the piece from my grandfather's diary which was written on April 27, 1920.

"Ramanujan's death—on April 26, 1920—could have been avoided. If he had been allowed to follow my instructions, this double tragedy need not have taken place. The neglect of Ramanujan during his early phase—perhaps partly due to the ignorance of his contemporaries—as well as his relatives' (mother's and wife's) contributory (I almost feel like using the stronger word 'criminal') negligence—have contributed to this double tragedy—a tragedy which is too deep for tears. I cannot get over a particular experience for a physician. Tragically enough, Ramanujan himself told me that he shouldn't have returned to India. His death is the saddest event of my professional career. It is not for me to assess Ramanujan's mathematical genius. But at the human level, he was one of the noblest men I have met in my life—Shy, reserved and endowed with an infinite capacity to bear the agonies of the mind and spirit with fortitude."

... she [S. Janaki] did not get what she expected from her illustrious husband—companionship (not in the <u>intellectual</u> sense) and a certain degree of middle class comfort. Her life was lonely, frustrated and poverty-striken. Similarly Ramanujan's mother expected Ramanujan to provide her with a comfortable home. And to cite an example which acted as an irritant, Ramanujan's letter to the Registrar of the Madras University to distribute the moneys due to him between the family and certain poor students, had aggravated the domestic situation.

Commentary. In a letter of 20 February 1991 to one of us (BCB), S. Chandrasekhar remarked, " ... some women (elderly at the time I knew them) who according to them were 'friends' or neighbors of Ramanujan's wife and mother during Ramanujan's last illness, did say openly (at the time) that there were violent quarrels between the mother and the daughter-in-law with respect to the provisions in Ramanujan's will. And according to what they said, they were more concerned with the will and less with Ramanujan's health." A Brahmin widow in India did not have the possibility of remarriage and was normally relegated to a meager existence, especially if she was childless, unless her husband had provided for her before his death. Thus, it was natural for Janaki to be concerned about her future while her husband's health continued to wane. The views of Ramanujan's neighbors and that of Ramanujan's harassed physician, pained by the death of his most illustrious patient, toward Janaki must be mollified in light of the hostile future that she faced.

A. Ranganathan to S. Chandrasekhar 26 March 1981
excerpt

I think it would be most appropriate indeed to include the extract from my grandfather's diary among other papers which you have deposited in the archives of the Royal Society. And now, I shall try to answer the points raised by you in your letter. My grandfather was Dr. P.S. Chandrasekar, M.D. (1869–1956), Retired Director of the Government Tuberculosis Hospital, Madras ... Although Dr. P.S. Chandrasekar was a professor of Hygiene and Physiology at the Madras Medical College for a number of years, he had specialized in the study and practice of tuberculosis. Subsequently the Madras Government had appointed him as the first Director of the newly established Government Tuberculosis Hospital. He was also a member of the Syndicate as well as a member of the Presidency College Committee which selected students for higher studies in the United Kingdom ... .

My grandfather became Ramanujan's physician as a matter of course. Apart from being the Director of the Madras Government Tuberculosis Hospital, he had an all-India reputation as a T.B. specialist. (Here it is sad to reflect on the fact that despite enjoying a fairly successful record as a physician, he encountered two tragedies—the Ramanujan tragedy and his eldest daughter who died of tuberculosis.) And the offer was made in the characteristic British style—the Royal Society (which meant Prof. Hardy of course) and the Secretary of State's office in London had communicated with the Madras Government which deputed its Surgeon–General to formally request my grandfather to take up Ramanujan as his patient.

Commentary. P.S. Chandrasekhar was born on 12 August 1869 and died in Madras on 25 September 1956. He was the author of the book *Tuberculosis in the Madras Presidency*.

G.C. McVittie to Bruce Berndt 6 August 1987
excerpt

In my student days in Edinburgh the professor of mathematics was E.T. Whittaker who, as a young man, had been a Fellow of Trinity, Cambridge. He continued to be deeply interested in Cambridge gossip throughout his life and it was from him that I heard of Ramanujan. The story was that Ramanujan cooked his own Indian food in his College rooms in special copper vessels which became corroded through being improperly washed. The corrosion eventually poisoned him.

Commentary. In Ramanujan's time, cooking was done primarily in aluminum, copper, or brass vessels. Many foods, in particular, rasam, a thin lentil soup and a staple with every meal for Iyengars, are considered to taste better when cooked in a brass or copper vessel coated with an inside lining of tin. Because tin and lead are similar in appearance, tin is often mistaken for lead. Thus, the rumor that Ramanujan died of lead poisoning, which evidently was also asserted by the famous physicist P.A.M. Dirac [**49**, p. 78], does not appear to have any substance. Moreover, there is no medical diagnosis of Ramanujan which mentions any such possibility.

George C. McVittie was born on 5 June 1904 in Turkey of British parents and died in Canterbury, England, in March 1988 at the age of 83. He received a B.A. from Cambridge University and a Ph.D. in astronomy and cosmology from Edinburgh University in 1930. For his service during World War II, he was made an Officer of the Order of the British Empire. From 1952 to 1972 he was Head of the Department of Astronomy at the University of Illinois at Urbana–Champaign.

Much of the correspondence related in this book concerns Ramanujan's fatal illness of which no firm diagnosis has been made. Very recently, D.A.B. Young [**298**] has carefully examined all available descriptions of Ramanujan's symptoms and medical history. Although the documents in the Tamil Nadu Archives were not available to him, Young's diagnosis, it seems to us, is the most informed and scientific one that has been made. We quote extensively from Young's paper [**298**] and strongly suggest that readers examine it. Since readers may not be familiar with medical terminology, we have appended definitions in parentheses.

After discussing Ramanujan's symptoms, including intermittent fever, weight loss, and an increase of white blood cells, Young mentions four diseases that show these symptoms. Using further evidence, Young discounts three of them, namely, subacute bacterial endocarditis, Hodgkin's disease, and metastatic cancer of the liver.

"The fourth disease is hepatic (of or affecting the liver) amoebiasis (parasitic infection of intestines or liver) and the fact that it has been arrived at here by a process of elimination should not disguise the high probability that it could have been the cause of intermittent fever in someone of Ramanujan's background. For a well-known medical textbook has said: 'Always suspect hepatic amoebiasis in a patient with obscure pyrexia (intermittent fever) coming from a tropical country'. Other symptoms would be pain with tenderness in the right hypochondrium (either of the upper lateral abdominal regions containing the lower ribs) or the epigastrium (upper middle portion of the stomach), enlargement of the liver, and weakness. Progressive emaciation leading to cachexia (a generally weakened, emaciated condition of the body) is characteristic of the disease. If hepatic amoebiasis was suspected, the effect of emetine for 8–10 days on the patient's symptoms was the easiest way to confirm the diagnosis; and it remains so today, although metronidazole would now be preferred to emetine. It should be noted that although hepatic amoebiasis was not diagnosed in Ramanujan's case, it is contained in the comprehensive diagnosis of blood poisoning, for the 'obscure source' of toxaemia would be the amoebic abscess in the liver. Lastly, remembering Dr Shaw's diagnosis (cancer of the liver), it is of interest that in the differential diagnosis of hepatic amoebiasis, carcinoma (cancer) of the liver may cause great difficulties."

Young then argues that Ramanujan's medical history in India favors this diagnosis. "Amoebiasis is a protozoal infection of the large intestine that gives rise to dysentery. The disease is widespread in India, particularly in and around the large coastal cities of Calcutta, Bombay, and Madras. The prevalence of the disease, though very high, is imprecisely known since most of those infected are symptomless carriers ... ." Young then recounts that shortly after Ramanujan left home in 1906 to attend college in Madras, "he contracted a bad bout of dysentery that forced him to return home for three months. This was probably amoebic dysentery, the most common form in India at that time, where the onset is usually gradual and the symptoms progressive but generally limited to abdominal discomfort and some diarrhea .... Amoebiasis, unless adequately treated, is a permanent infection, although many patients may go for long periods with no overt signs of the disease. Relapses occur when the host-parasite relationship is disturbed." Such a relapse likely occurred in 1909 when Ramanujan became acutely ill and had to return home for rest and care. "Later that same year (1909), while still at home with his family, he developed a hydrocele (a collection of waterly fluid in the cavity of the body, especially the scrotum), which was operated on in January 1910." From Ramanujan's and Dr Shaw's descriptions, Young concludes that Ramanujan had a scrotal amoeboma rather than a hydrocele.

In his conclusion, Young writes "Hepatic amoebiasis was regarded in 1918 as a tropical disease ('tropical liver abscess'), and this would have had important implications for successful diagnosis, especially in provincial medical centres. Furthermore, the specialists called in were experts in either tuberculosis or gastric medicine. Another major difficulty is that a patient with this disease would not, unless specifically asked, recall as relevant that he had had two episodes of dysentery 11 and 8 years before. Finally, there is the very good reason that, because of the great variability in physical findings, the diagnosis was difficult in 1918 and remains so today; hepatic amoebiasis 'presents a severe challenge to the diagnostic skills of the clinician ... '."

Young also remarks that a diagnosis of cancer was not made until Ramanujan returned to India, but, in fact, this diagnosis was made in England, but eventually discarded.

A. Ranganathan to S. Chandrasekhar   **15 January 1992**

Locksley Hall
852, Poonamallee High Road,
KILPAUK, MADRAS-600 010.
INDIA

Dear Doctor,

I was delighted to receive your letter of January 1, 1992 last evening. It gives me great pleasure to reciprocate your New Year greetings. I am sure you'd have got my New Year card, by now. May I thank you once again for your several acts of graciousness over the years?

Let me give you the details of what my grandfather told me about Ramanujan's "unscrupulous relatives". The unscrupulousness of these relatives—the word 'unscrupulous' in this context is pretty mild—can be perceived at two levels, a) financial and b) intellectual.

As I had indicated earlier, my grandfather was something more than Ramanujan's physician. He was an admirer and a friend of Ramanujan; a) almost immediately after Ramanujan's death, my grandfather Dr. P.S. Chandra Sekar got in touch with a few Government and university officials. For he wanted Mrs. Ramanujan to live in a state of a reasonable comfort. Dr. P.S. C.S. did succeed in getting a substantial sum of money. However, he made a mistake in handing over the money to Ramanujan's mother (Komalatammal). Subsequently, my grandfather's bitter comment was: "Not even a single rupee reached the poor lady" (Mrs. Ramanujan); b) Ramanujan had instructed (Ramanujan realized that his end was near) my grandfather to send his four 'volumes'—for some strange reason, Dr. P.S. C.S. had always referred to Ramanujan's 'Notebooks' as Ramanujan's 'volumes'—to Prof. Hardy. And to his utter shock, my grandfather could locate only 3 'volumes' (i.e., a week after Ramanujan had passed away). ... My grandfather was convinced that a murky transaction had taken place. Ultimately, my grandfather could send only 3 'volumes' of Ramanujan's papers to Prof. Hardy. ...

I am taking the liberty of enclosing a typed copy of my reviews of Wali's biography and your volume of essays. My reviews will appear in a Delhi-based journal shortly.

With my warmest regards and New Year greetings to you and Mrs. Chandrasekhar.

Yours sincerely,
A. Ranganathan

Commentary. In the penultimate paragraph, Ranganathan refers to Wali's biography of Chandrasekhar [273].

A. Ranganathan to S. Chandrasekhar   15 March 1992

Locksley Hall
852, Poonamallee High Road,
KILPAUK, MADRAS-600 010.
INDIA

Dear Doctor,

Many thanks for your gracious reply of February 25, 1992. Let me answer your question. Yes, indeed, my grandfather did discuss this point with Prof. Hardy. In fact, my grandfather had visited England a few years after Ramanujan's death. And he met Hardy at his college in Oxford.

CHAPTER 8. RAMANUJAN'S PAPERS AND MANUSCRIPTS

As I had written to you in my letter of January 15, 1992, my grandfather referred to Ramanujan's <u>Notebooks</u> as Ramanujan's <u>Volumes</u>. Of course, this is not an important point. However it is clear that a Ramanujan manuscript (whether it is termed a Notebook or a Volume or a set of sheets of paper) is missing.

I agree with you that this loss is most disturbing. For it was done during his last phase of creativity in Madras. My grandfather had informed Hardy about this volume in his earlier letters as well as during a subsequent conversation. He had also alerted Dr. S. R. Ranganathan about it. He made a mistake in not informing you about it. And I am inclined to think that it could be located in the home of a third or a fourth generation relative of Ramanujan—it is just a possibility.

Yours sincerely,
A. Ranganathan

Chapter 9
Family Histories

Commentary. The following short history of Ramanujan's life can be found in the National Archives in Delhi. The beginning portion of the article of 5 April 1919 in the *Madras Times* is based on this family history. R.A. Askey [16] has also reproduced this family history and offered comments on it. In particular, note that, according to this account, Ramanujan began his study of hypergeometric series in about 1905, which is considerably earlier than had been formerly thought.

Mr. S. Ramanujan was born at Erode in December 1887 and is a native of Kumbakonam in Tanjore District. He was the eldest son of brahmin parents of the Vaishnava Sect. His Father not literate in English; his Mother knew something of Hindu astronomy. Very poor parents, two younger brothers and 3 sisters died, brothers 14 and 8 years. 1900 in his 12th year he began to work beyond what his teacher was teaching. Arithmetic & Geometric series. Trigonometry (copied Loney's II Part) sine, cosine series and not the ratios. 1902 learnt from another how to solve cubic equations. Solved biquadratic equation independently by resolving into two quadratic factors. 1903 Attacked solving quintic equation and failed. Beginning of 1904 began to investigate the series $1 \frac{1}{2} \frac{1}{3} \ldots$ and calculated Euler's constant in a regular series and also to 15 places of decimals, and also Bernoulli's numbers without knowing that these existed. Developed summation of series and subsequently learnt that this was only integral calculus. Then learnt what was differentiation and integration. In December passed Madras University Matriculation and joined Kumbakonam College intermediate, underwent 1st year's course. Took a good deal of interest in Mathematics at the sacrifice of other subjects. Failed to secure promotion to the 2nd year college course. There was a temporary unsoundness of mind for a period of six months. But during this period he was sane enough to work Mathematics alone. During this period he investigated much of what is known as Hypergeometric series. At the end of 1905 and the whole of 1906 he investigated relations of many integrals and series (1905 Waltair etc. places. 750 miles) and then subsequently

POSTAGE STAMP COMMEMORATING THE 75TH ANNIVERSARY OF RAMANUJAN'S BIRTH

learnt these related to what is now called Elliptic Functions. In 1906 joined Pachaiyappa's College and read the first year College course for 3 months and then subsequently fell seriously ill and thence discontinued studies. In 1907 December appeared privately for F.A. and had the great distinction of failing in all the subjects due to the derangement of brain which occasionally continued till the end of 1907. Developed continued fractions and investigated divergent series in 1908.

(Mother's side Sanskrit scholars. Have received presents from Kings) Betrothal in 1908. 1909. Underwent severe surgical operation. Was completely prostrated for nearly a year. 1910. Developed relations among Elliptic Modular equations. 1911. Went in search of Mathematicians to appreciate his work. Not met with much encouragement. Saw almost all Mathematicians V. Ramaswami Aiyar among others, Professor Ross of Christian College, Professor Middlemast of Presidency College. R. Ramachandra Rao, Collector of Nellore took some interest. He paid his boarding fees in Madras for nearly a period of one year. March 1912 joined the Madras Port Trust. May 1913 got a scholarship.

Commentary. Another family record was compiled by Ramanujan's brother S. Lakshmi Narasimhan. This is handwritten on over 40 pages, but many pages have very little writing on them. Very little is readable, but we have been able to decipher some of it. Page numbers are written in square brackets.

Ramanujan Family Record

Ramanujan born at Erode 22.12.1887 [17]
Father, K. Srinivasa Aiyangar, born 1863 [17]

CHAPTER 9. FAMILY HISTORIES

Mother, Komalattammal, born 28.1.1868 [17]
Son, Sadagopan, born at Erode 26.6.1889, died 9.9.1889 [17]
Daughter born Erode 1.11.1891, died 2.2.1892 [17]
[Maternal] Grandfather died 27.12.1894 at Kumbakonam [18]
[Paternal] Grandfather died 17.10.1895 at Madras [18]
Brother S. Lakshmi Narasimhan born 18.4.1898 at. K.M.U. [Kumbakonam?] [20, 37]
Brother S. Tirunarayanan born 19.4.1905 at K.M.U. [20]
Leg operation April, May 1909 [21]
Arrived London 14.4.1914. [31]
To Cambridge 18.4.14. [31]
Trinity College 30.4.14. [31]
Smallpox December 1889 [37, 38]
Marriage 14.7.1909 [39]
Sick testicles January 1910 [39]
Expressed intention of going to England 27.1.14 [39]
[There are several references to family quarrels.]
At end of 1910 in Madras and introduced to Professor E.B. Ross who was able to see real greatness in his talents but was unable to go through and understand his works. [42]
At Ross's suggestion Hardy written to. [43]
Early in 1914 or at the end of 1913 when Prof. Neville was there he got an offer from the Madras University which he took to K.N. [K. Narasimha Iyengar] and K.N. persuaded him to accept the offer promising to get his mother's consent. K.N. went to his mother whom he knew from his boyhood and prevailed upon her to give her consent to S.R. going to England. On his accepting the offer S.R. was consulting K.N. for everything relating to his visit to England who was in close touch with him till his departure to England. [44]

Commentary. Edward Burn Ross (1881–1947) was a graduate of Edinburgh University. He was admitted to Trinity College, Cambridge, in 1903, and was listed as 7th Wrangler in 1904. He obtained First Class Honours in Part II of the Tripos in 1906, and then went to India in the following year as Professor of Mathematics at Madras Christian College. He retired because of ill health in 1932 and returned to Edinburgh. He was acquainted with Hardy, and is mentioned several times in S.R. Ranganathan's book [**229**]. It is clear that he had earned the trust of Ramanujan.

Provenance of the Letters

A   Tamil Nadu Archives
B   B.C. Berndt Collection
C   S. Chandrasekhar Collection
D   National Archives, Delhi
G   Nicolas Griffin Collection
GEA   G.E. Andrews Collection
K   Kumbakonam Town High School
M   Mittag–Leffler Institute, Djursholm
N   Narosa edition of *Lost Notebook*
P   Ramanujan's *Collected Papers* [227]
R   S.R. Ranganathan's book [229]
RAR   R.A. Rankin Collection
S   P.K. Srinivasan's books [263] and personal collection
T   Trinity College Library, Cambridge, pressmark Add. ms.a.94. Different items are indicated by superscripts.
U   Cambridge University Library, pressmark Add.
V   K. Srinivasa Rao Collection of Letters to E. Vinayaka Row

Document	Date	Source
E.W. Middlemast—Letter of Recommendation	21 Sep 1911	S,49; D
Ramanujan to Madras Port Trust Office	9 Feb 1912	S,31; D
Memo from Madras Port Trust Office	Feb 1912	D
Chief Accountant to Ramanujan	Feb 1912	D
C.L.T. Griffith to Sir Francis Spring	12 Nov 1912	S,50; D
Alfred Bourne to Sir Francis Spring	14 Nov 1912	S,51; D
W. Graham to C.L.T. Griffith	27 Nov 1912	S,52; D
C.L.T. Griffith to Sir Francis Spring	28 Nov 1912	D
M.J.M. Hill to C.L.T. Griffith	3 Dec 1912	S,53; D
M.J.M. Hill to C.L.T. Griffith	7 Dec 1912	D
C.L.T. Griffith to Sir Francis Spring	5 Jan 1913	D
Ramanujan to G.H. Hardy	16 Jan 1913	P,xxiii; S,44

B. Russell to Ottoline Morrell	2 Feb 1913	G
Hardy to Ramanujan	8 Feb 1913	D
Gilbert T. Walker to F. Dewsbury	26 Feb 1913	S,55
Ramanujan to Hardy	27 Feb 1913	U,7011; P,xxvii
J.E. Littlewood to Hardy	? Mar 1913	T, $1^{(2)}$; N,380
F. Dewsbury to B. Hanumantha Rao	10 Mar 1913	A
A. Davies to B. Hanumantha Rao	10 Mar 1913	A
Sec. for Indian students to Sec. Ad. Comm.	3 Feb 1913	A
B. Hanumantha Rao to S.N. Aiyar	12 Mar 1913	D
Memo from Sir Francis Spring	? Mar 1913	D
B. Hanumantha Rao to F. Dewsbury	25 Mar 1913	A
Hardy to Ramanujan	26 Mar 1913	S,56
F. Dewsbury to Educational Dept.	5 Apr 1913	A
Memo, files of Educational Dept.	5 Apr 1913	A
Memo from S.N. Aiyar	11 Apr 1913	D
Ramanujan to F. Dewsbury	12 Apr 1913	D
Ramanujan to Hardy	17 Apr 1913	U,7011; S,45; P,xxix
Arthur Davies to S.N. Aiyar	7 Jun 1913	D
S.N. Aiyar to Sir Francis Spring	31 Oct 1913	D
Hardy to Ramanujan	24 Dec 1913	S,57
Ramanujan to Hardy	22 Jan 1914	U,7011
E.H. Neville to F. Dewsbury	28 Jan 1914	S,59
R. Littlehailes to F. Dewsbury	29 Jan 1914	S,61
Sir Francis Spring to C.B. Cottrell	5 Feb 1914	S,64; D
C.B. Cottrell to Sir Francis Spring	5 Feb 1914	D
Doc. of Educational Dept.	12 Feb 1914	A
Order from Educ. Dept.	12 Feb 1914	A
Notes connected with order above	12 Feb 1914	A
C. Mallet to Hardy	11 Feb 1914	
Hardy to E.H. Neville	12 Feb 1914	
Ramanujan to R. Krishna Rao	30 Mar 1914	S,3
Ramanujan to R. Krishna Rao	11 Jun 1914	S,5
Ramanujan to R. Krishna Rao	7 Aug 1914	S,9
Ramanujan to R. Krishna Rao	13 Nov 1914	S,13
Ramanujan to E. Vinayaka Row	11 Jun 1914	V
Ramanujan to E. Vinayaka Row	24 Mar 1915	V
Ramanujan to E. Vinayaka Row	10 Sep 1915	V
Ramanujan to Komalattamal	11 Sep 1914	S,35
Ramanujan to Srinivasa Aiyangar	17 Nov 1914	S,43
Ramanujan to C.N. Ganapathy Iyer	17 Dec 1914	S

PROVENANCE OF LETTERS

Ramanujan to S.M. Subramanian	7 Jan 1915	S,21
Ramanujan to S.M. Subramanian	3 Jun 1915	S,28
Ramanujan to S.M. Subramanian	1 Jul 1915	S,29
Memo from S.N. Aiyar	Apr 1915	D
E.W. Barnes to F. Dewsbury	8 Nov 1915	S,66
Ramanujan to S.N. Aiyar	11 Nov 1915	S,32; D
Ramanujan to R. Ramachandra Rao	11 Nov 1915	D
R. Ramachandra Rao to S.N. Aiyar	6 Dec 1915	D
S.N. Aiyar to Sir Francis Spring	15 Dec 1915	D
Sir Francis Spring to F. Dewsbury	15 Dec 1915	S,67; D
F. Dewsbury to Sir Francis Spring	16 Dec 1915	D
S.N. Aiyar to F. Dewsbury	20 Dec 1915	D
F. Dewsbury to Sir Francis Spring	17 Jan 1916	D
Ramanujan to S.M. Subramanian	30 Mar 1916	S,30
Memo from S.N. Aiyar	? May 1916	D
W.E.H. Berwick to Ramanujan	24 Oct 1916	T, $5^{(1)}$
Ramanujan to Hardy	Dec 1916	N,132
F. Dewsbury to S.N. Aiyar	14 May 1917	D
S.N. Aiyar to F. Dewsbury	14 May 1917	D
F. Dewsbury to S.N. Aiyar	14 May 1917	D
S.N. Aiyar to F. Dewsbury	14 May 1917	D
F. Dewsbury to S.N. Aiyar	14 May 1917	D
S.N. Aiyar to F. Dewsbury	14 May 1917	D
Ramanujan to Hardy	May–June 1917	T, $17^{(11)}$
Hardy to G. Mittag-Leffler	20 Aug 1917	M
Hardy to S.M. Subramanian	20 Sep 1917	S,69
Hardy to G. Mittag-Leffler	12 Oct 1917	M
Hardy to Ramanujan	6 Feb 1918	T, $16^{(1)}$; N,384
Hardy to Sir J.J. Thomson	Feb 1918	U,7654
Ramanujan to Hardy	?1 Mar 1918	
Excerpt from Records of Roy. Soc.	18 Dec 1917	Roy. Soc.
F. Dewsbury to Hardy	5 Mar 1918	T, $1^{(13)}$
Sir Francis Spring to S.N. Aiyar	9 Mar 1918	D
S.N. Aiyar to Sir Francis Spring	12 Mar 1918	D
P.V. Seshu Aiyar to S.N. Aiyar	17 Mar 1918	D
Indian Math. Soc. to Hardy	22 Mar 1918	T, $1^{(29)}$
F. Dewsbury to Hardy	10 Apr 1918	T, $1^{(14)}$
F. Dewsbury to Hardy	16 Apr 1918	T, $1^{(15)}$
Ramanujan to Royal Society	17 May 1918	K
Ramanujan to A.S. Ramalingam	19 Jun 1918	T, $1^{(6)}$
A.S. Ramalingam to Hardy	23 Jun 1918	T, $1^{(7)}$
A.S. Ramalingam to Ramanujan	23 Jun 1918	T, $1^{(10)}$

Ramanujan to Hardy, Matlock House, Sunday	? Jun 1918	T, $2^{(3)}$; N,97
Ramanujan to Hardy	28 Jun 1918	T, $2^{(19)}$; N,113
Ramanujan to Hardy, Matlock House, Monday	Jun/Jul 1918	T, $2^{(14)}$; N,105
Ramanujan to Hardy, Matlock House, Tuesday	Jun/Jul 1918	T, $2^{(11)}$; N,116
Ramanujan to Hardy, Fitzroy House, Friday	?11 Oct 1918	
Ramanujan to Hardy, Fitzroy House, Monday	? Oct 1918	T, $2^{(22)}$; N,121
Ramanujan to Hardy, Fitzroy House, Saturday	? Oct 1918	T, $2^{(1)}$; N,93
Hardy to F. Dewsbury	26 Nov 1918	S,76
Ramanujan to F. Dewsbury	11 Jan 1919	S,46; R
F. Dewsbury to Educ. Dept.	13 Mar 1919	A
Medical Report	1 Aug 1919	A
P.A. Pires to Dist. Medical Officer	30 Aug 1919	A
S.N. Aiyar to Ramanujan	23 Oct 1919	D
F. Dewsbury to Hardy	22 Dec 1919	T, $1^{(16)}$
Ramanujan to Madras Port Trust	12 Jan 1920	D
Ramanujan to Hardy	12 Jan 1920	P,xxxi; S,45
S.N. Aiyar to Ramanujan	?15 Jan 1920	D
F. Dewsbury to Hardy	15 Jan 1920	T, $1^{(17)}$
J.J. Hensman to Hardy	29 Apr 1920	T, $1^{(18)}$
S. Lakshmi Narasimhan to Hardy	29 Apr 1920	T, $1^{(19)}$
Mr Ramanujan's Accounts	? May 1920	T, $1^{(20)}$
E.M. Clark (Barclays Bank) to Hardy	7 May 1920	T, $1^{(21)}$
Komalatammal to Gov. of Madras	14 May 1920	A
E.M. Clark to Hardy	17 May 1920	T, $1^{(22)}$
E.M. Clark to Hardy	19 May 1920	T, $1^{(23)}$
S. Janaki to F.F. Thomas	20 May 1920	A
E.M. Clark to Hardy	25 May 1920	T, $1^{(24)}$
Hardy to Sir J.J. Thomson	May 1920	U,7654
Hardy to F. Dewsbury	26 May 1920	S,78
J.J. Hensman to Educ. Dept.	26 May 1920	A
Komalatammal to Educ. Dept.	19 Jun 1920	A
Komalatammal to Gov. of Madras	26 Jul 1920	A
R. Littlehailes to Gov. of Madras	28 Jul 1920	A
K. Ramunni Menon to Madras St. Gov.	28 Jul 1920	T, $1^{(25)}$
Gov. memo	3 Sep 1920	A

Komalatammal to Sec. of Gov., Madras	29 Sep 1920	A
R. S. Iyangar to Regis., Univ. of Madras	28 Oct 1920	A
R. Ramachandra Rao to Hardy	3 Dec 1920	T, $12^{(22)}$
Sir Francis Spring to S.N. Aiyar	15 Feb 1921	D
Sir Francis Spring to Hardy	18 Feb 1921	T, $1^{(26)}$
S.N. Aiyar to Sir Francis Spring	10 Mar 1921	D
K. Ramunni Menon to Sec., Gov. of Madras	24 Mar 1921	A
H. Sharp to Sec., Gov. of Madras	4 May 1921	A
K. Ananda Rau to Hardy	26 May 1921	T, $12^{(23)}$
S. Janaki to Sec., Gov. of Madras	5 Aug 1921	A
R.H. Fowler to Hardy	9 Jun 1922	T, $12^{(34)}$
J.H. Jeans to Hardy	15 Jun 1922	T, $12^{(33)}$
F. Dewsbury to Hardy	13 Jul 1922	T, $12^{(40)}$
F. Dewsbury to Hardy	2 Aug 1923	T, $12^{(25)}$; N,415
F. Dewsbury to Hardy	6 Aug 1923	T, $12^{(29)}$; N,417
F. Dewsbury to Hardy	30 Aug 1923	T, $12^{(28)}$; N,419
Hardy to B.M. Wilson	10 Jun 1925	T, $12^{(41)}$
P.V. Seshu Aiyar to B.M. Wilson	8 Jul 1925	T, $12^{(49)}$
B.M. Wilson to P.V. Seshu Aiyar	18 Aug 1925	K
P.V. Seshu Aiyar to B.M. Wilson	22 Apr 1926	T, $12^{(58)}$
Komalattamal to Hardy	25 Aug 1927	T, $1^{(27)}$
E.H. Neville to B.M. Wilson	28 Dec 1927	T, $12^{(70)}$
T. Vijayaraghavan to B.M. Wilson	4 Jan 1928	T, $12^{(72)}$; N,409
Hardy to G.N. Watson	early 1928	T, $12^{(76)}$; N,393
G.N. Watson to B.M. Wilson	28 Jun 1929	T,Add.ms.b.102
Hardy to G.N. Watson	Apr 1930	T, $18^{(1)}$; N,390
G.N. Watson to B.M. Wilson	1 Oct 1930	T,Add.ms.b.102
K. Ananda Rau to Hardy	7 Apr 1936	T, $1^{(32)}$; N,400
Hardy to S. Chandrasekhar	19 Feb 1936	C
S. Chandrasekhar to Hardy	4 Aug 1937	T, $8^{(1)}$
Hardy to S. Chandrasekhar	15 Dec 1937	C
D.H. Lehmer to Hardy	2 Nov 1937	T, $6^{(1)}$; N,404
S. Chandrasekhar to Hardy	28 Dec 1937	T, $8^{(2)}$
Hardy to G.N. Watson	late 1930s	T, $3^{(1)}$; N,392
G.N. Watson to H.A. Heilbronn	19 Mar 1939	T, $4^{(1)}$; N,402
Hardy to A.F. Scholfield	23 Jul 1940	U,7011(C)
R.A. Rankin to Trinity College	1 Sep 1965	RAR
J.M. Whittaker to G.E. Andrews	15 Aug 1979	GEA
A. Ranganathan to S. Chandrasekhar	12 Jan 1981	C

A. Ranganathan to S. Chandrasekhar	26 Mar 1981	C
G.C. McVittie to B.C. Berndt	6 Aug 1987	B
A. Ranganathan to S. Chandrasekhar	15 Jan 1992	C
A. Ranganathan to S. Chandrasekhar	15 Mar 1992	C

References

1. P. V. Seshu Aiyar, "The late Mr. S. Ramanujan, B.A., F.R.S.," *J. Indian Math. Soc.* **12** (1920), 81–86.
2. S. N. Aiyar, "The distribution of primes," *J. Indian Math. Soc.* **5** (1913), 60–61.
3. K. Ananda Rau, "On the convergence and summability of Dirichlet's series," *Proc. London Math. Soc.* (2) **34** (1932), 414–440.
4. G. E. Andrews, "On the theorems of Watson and Dragonette for Ramanujan's mock theta functions," *Amer. J. Math.* **88** (1966), 454–490.
5. G. E. Andrews, "On q-difference equations for certain well-poised basic hypergeometric series," *Quart. J. Math. Oxford* **19** (1968), 433–447.
6. G. E. Andrews, *The Theory of Partitions*, Addison-Wesley, Reading, 1976.
7. G. E. Andrews, "An introduction to Ramanujan's 'lost' notebook," *Amer. Math. Monthly* **86** (1979), 89-108.
8. G. E. Andrews, *q-series: Their Development and Application in Analysis, Number Theory, Combinatorics, Physics, and Computer Algebra*, CBMS Regional Conf. Ser. in Math. No. 66, Amer. Math. Soc., Providence, 1986.
9. G. E. Andrews, "On the proofs of the Rogers-Ramanujan identities," in *q-Series and Partitions* (D. Stanton, ed.), Springer-Verlag, New York, 1989, pp. 1–14.
10. G. E. Andrews, "Mock theta functions," in *Theta Functions Bowdoin, Part 2, Proceedings of Symposia in Pure Mathematics*, vol. 49, American Mathematical Society, Providence, Rhode Island, 1989, pp. 283–298.
11. G. E. Andrews, R. A. Askey, B. C. Berndt, K. G. Ramanathan, and R. A. Rankin, eds., *Ramanujan Revisited*, Academic Press, Boston, 1988.
12. G. E. Andrews, B. C. Berndt, L. Jacobsen, and R. L. Lamphere, "Variations on the Rogers-Ramanujan continued fraction in Ramanujan's notebooks," in *Number Theory, Madras, 1987* (K. Alladi, ed.), Lecture Notes in Math. No. 1395, Springer-Verlag, Berlin, 1989, pp. 73–83.
13. G. E. Andrews, B. C. Berndt, L. Jacobsen, and R. L. Lamphere, *The continued fractions found in the unorganized portions of Ramanujan's notebooks*, Mem. Amer. Math. Soc. **99** (1992), no. 477.
14. R. A. Askey, "Beta integrals in Ramanujan's papers, his unpublished work and further examples," in *Ramanujan Revisited* (G. E. Andrews et al., eds.), Academic Press, Boston, 1988, pp. 561–590.
15. R. A. Askey, "Beta integrals and q-extensions," *Proc. Ramanujan Centennial International Conf., 15-18 December 1987, Annamalainagar*, The Ramanujan Mathematical Soc., Madras, 1988, pp. 85–102.

16. R. A. Askey, "Ramanujan and hypergeometric and basic hypergeometric series," *Russian Math. Surveys* **45** (1990), 37–86, *Ramanujan International Symposium on Analysis* (Pune, 1987) (N. K. Thakare, ed.), Macmillan of India, New Delhi, 1989, pp. 1–83.

17. Association of Mathematics Teachers of India, *Ramanujan Centenary Year*, XXII *Annual Conference 3–6 Dec.* 1987, Town Higher Secondary School, Kumbakonam, Tamilnadu, 1987.

18. A. O. L. Atkin, "Proof of a conjecture of Ramanujan," *Glasgow Math. J.* **8** (1967), 14–32.

19. A. J. Ayer, *Russell*, Fontana/Collins, London, 1972.

20. W. N. Bailey, "Ernest William Barnes," *J. London Math. Soc.* **29** (1954), 498–503.

21. W. N. Bailey, *Generalized Hypergeometric Series*, Stechert–Hafner, New York, 1964.

22. H. F. Baker, "Percy Alexander MacMahon," *J. London Math. Soc.* **5** (1930), 307–318.

23. R. Balasubramanian, "The circle method and its implications," *J. Indian Inst. Sci. (Special issue)* (1987), 39–44.

24. R. Balasubramanian and J.-M. Deshouillers, "Problème de Waring pour les bicarrés. I. Schéma de la solution," *C. R. Acad. Sci. (Paris), Sér. I Math.* **303** (1986), 85–88.

25. R. Balasubramanian and J.-M. Deshouillers, "Problème de Waring pour les bicarrés. II. Résultats auxiliaires pour le théorème asymptotique," *C. R. Acad. Sci. (Paris), Sér. I Math.* **303** (1986), 161–163.

26. E. W. Barnes, "The asymptotic expansion of integral functions defined by generalized hypergeometric series," *Proc. London Math. Soc.* (2) **5** (1907), 59–116.

27. G. Bauer, "Von den Coefficienten der Reihen von Kugelfunctionen einer Variabeln," *J. Reine Angew. Math.* **56** (1859), 101–121.

28. G. Bauer, "Von einem Kettenbruche Euler's und einem Theorem von Wallis," *Abh. Bayer. Akad. Wiss.* **11** (1872), 96–116.

29. B. C. Berndt, "Analytic Eisenstein series, theta-functions, and series relations in the spirit of Ramanujan," *J. Reine Angew. Math.* **303/304** (1978), 332–365.

30. B. C. Berndt, "The quarterly reports of S. Ramanujan," *Amer. Math. Monthly* **90** (1983), 505–516.

31. B. C. Berndt, "Ramanujan's quarterly reports," *Bull. London Math. Soc.* **16** (1984), 449–489.

32. B. C. Berndt, *Ramanujan's Notebooks, Part* I, Springer-Verlag, New York, 1985.

33. B. C. Berndt, "Ramanujan's modular equations," in *Ramanujan Revisited* (G. A. Andrews et al., eds.), Academic Press, Boston, 1988, pp. 313-333.

34. B. C. Berndt, "Introduction to Ramanujan's modular equations," in *Proc. Ramanujan Centennial Conf., 15-18 December* 1987, *Annamalainagar*, The Ramanujan Mathematical Soc., Madras, 1988, pp. 15-20.

35. B. C. Berndt, *Ramanujan's Notebooks, Part* II, Springer-Verlag, New York, 1989.

36. B. C. Berndt, "Srinivasa Ramanujan," *The American Scholar* **58** (1989), 234–244.

37. B. C. Berndt, *Ramanujan's Notebooks, Part* III, Springer-Verlag, New York, 1991.

38. B. C. Berndt, "Hans Rademacher (1892–1969)," *Acta Arith.* **61**, (1992), 209–231; "The Rademacher Legacy to Mathematics", *Contemp. Math.*, Vol. 166 (G. E. Andrews, D. M. Bressoud, and L. A. Parson, eds.), Amer. Math. Soc., Providence, RI, 1994, pp. xii-xxxvi.

39. B. C. Berndt, "On a certain theta-function in a letter of Ramanujan from Fitzroy House," *Ganita* **43** (1992), 33–43.

40. B. C. Berndt, *Ramanujan's Notebooks, Part* IV, Springer-Verlag, New York, 1994.

41. B. C. Berndt, *Ramanujan's Notebooks, Part* V, Springer-Verlag, New York (to appear).

42. B. C. Berndt, S. Bhargava, and F. G. Garvan, "Ramanujan's theories of elliptic functions to alternative bases," *Trans. Amer. Math. Soc.* **347** (1995), 4163–4244.

43. B. C. Berndt and H. H. Chan, "Some values for the Rogers–Ramanujan continued fraction," *Canad. J. Math.* **47** (1995), 897–914.

44. B. C. Berndt and H. H. Chan, "Ramanujan's explicit values for the classical theta-function," *Mathematika* **42** (1995), 278–294.

45. R. Bharathi, ed., *Prof. Srinivasa Ramanujan Commemoration Volume*, Jupiter Press, Madras, 1974.

46. P. Bialek, *Ramanujan's formulas for the coefficients in the power series expansions of certain modular forms*, Ph.D. thesis, University of Illinois at Urbana–Champaign, 1995.

47. B. J. Birch, "A look back at Ramanujan's notebooks," *Math. Proc. Cambridge Philos. Soc.* **78** (1975), 73–79.

48. B. Bollobás, *Littlewood's Miscellany*, Cambridge University Press, Cambridge, 1986.

49. B. Bollobás, "Ramanujan—a glimpse of his life and his mathematics," *The Cambridge Review* (1988), 76–80.

50. B. Bollobás, "Ramanujan—a glimpse of his life and his mathematics," *Eureka* **48** (1988), 81–98.

51. J. M. Borwein and P. B. Borwein, "A cubic counterpart of Jacobi's identity and the AGM," *Trans. Amer. Math. Soc.* **323** (1991), 691–701.

52. J. M. Borwein, P. B. Borwein, and F. G. Garvan, "Some cubic modular identities of Ramanujan," *Trans. Amer. Math. Soc.* **343** (1994), 35–47.

53. J. Brillhart, "Derrick Henry Lehmer 1905-1991," *Notices Amer. Math. Soc.* **40** (1993), 31–32.

54. T. J. I'A. Bromwich, *An Introduction to the Theory of Infinite Series*, 2nd. ed., Macmillan, London, 1926.

55. C. E. Buckland, *Dictionary of Indian Biography*, Swan Sonnenschein, London, 1906.

56. D. Buell, *Binary Quadratic Forms*, Springer-Verlag, New York, 1989.

57. J. C. Burkill, *Dictionary of Scientific Biography*, Charles Scribner's Sons, New York, 1971.

58. J. C. Burkill, "John Edensor Littlewood," *Bull. London Math. Soc.* **11** (1979), 59–103.

59. G. S. Carr, *A Synopsis of Elementary Results Pure Mathematics*, Francis Hodgson, London, 1880; 1886; *Formulas and Theorems in Pure Mathematics*, 2nd ed., Chelsea, New York, 1970.

60. A. Cauchy, *Oeuvres, Série* II, t. VII, Gauthier-Villars, Paris, 1889.

61. A. Cayley, *An Elementary Treatise on Elliptic Functions*, 2nd ed., Dover, New York, 1961.

62. H. H. Chan, "On Ramanujan's cubic continued fraction," *Acta Arith.* **73** (1995), 343–355.

63. S. Chandrasekhar, "On Ramanujan," in *Ramanujan Revisited* (G. A. Andrews, et al., eds.), Academic Press, Boston, 1988, pp. 1–6.

64. K. Chandrasekharan, "T. Vijayaraghavan," *Math. Student* **24** (1956), 251–267.

65. P. L. Chebyshev, "Lettre de M. le professeur Tchébychev à M. Fuss sur un nouveau théoreme relatif aux nombres premiers contenus dans les formes $4n+1$ et $4n+3$," *Bull. Cl. Phys.—Math. de l'Acad. Imp. Sci., St. Petersburg* **11** (1853), 208.

66. P. L. Chebyshev, *Oeuvres*, t. 1, Chelsea, New York, 1961, pp. 697–698.

67. H. Cohen, "q-identities for Maass waveforms," *Invent. Math.* **91** (1988), 409–422.

68. H. Davenport, "On Waring's problem for fourth powers," *Ann. of Math.* **40** (1939), 731–747.

69. H. Davenport, "T. Vijayaraghavan," *J. London Math Soc.* **33** (1958), 252–255.

70. H. Davenport, *Collected Papers*, vol. III, Academic Press, London, 1977.

71. J.-M. Deshouillers and F. Dress, "Sommes de diviseurs et structure multiplicative des entiers," *Acta Arith.* **49** (1988), 341–375.

72. J.-M. Deshouillers and F. Dress, "Sums of biquadrates: on the representation of large integers," *Ann. della Scuola Norm. Sup. Pisa* **19** (1992), 113–153.

73. J.-M. Deshouillers and F. Dress, "Sur la majoration des sommes de Weyl biquadratiques," *Ann. della Scuola Norm. Sup. Pisa* **19** (1992), 291–305.

74. J.-M. Deshouillers and F. Dress, "Numerical results for sums of 5 and 7 biquadrates and consequences for sums of 19 biquadrates," *Math. Comp.* **61** (1993), 195–207.

75. P. G. L. Dirichlet, "Recherches sur diverses applications de l'analyse infinitésimale à la théorie des nombres," *J. Reine Angew. Math.* **21** (1840), 1–12.

76. L. Dragonette, "Some asymptotic formulae for the mock theta series of Ramanujan," *Trans. Amer. Math. Soc.* **72** (1952), 474–500.

77. J. Dutka, "Wallis's product, Brouncker's continued fraction, and Leibniz's series," *Arch. History Exact Sci.* **26** (1982), 115–126.

78. R. J. Evans and D. Stanton, "Asymptotic formulas for zero-balanced hypergeometric series," *SIAM J. Math. Anal.* **15** (1984), 1010–1020.

79. L. N. G. Filon, "Micaiah John Muller Hill (obituary notice)," *J. London Math. Soc.* **4** (1929), 313–318.

80. A. Frater, *Chasing the Monsoon*, Alfred A. Knopf, New York, 1991.

81. J. W. L. Glaisher, "On the series which represent the twelve elliptic and the four zeta functions," *Mess. Math.* **18** (1889), 1–84.

82. L. A. Goldberg, *Transformations of Theta-functions and Analogues of Dedekind Sums*, Ph.D. thesis, University of Illinois, Urbana-Champaign, 1981.

83. J. P. Gram, "Undersøgelser angaaende Maengden af Primtal under en given Graense," *K. Videnskab. Selsk. Skr.* (6) **2** (1884), 183–308.

84. A. G. Greenhill, *The Applications of Elliptic Functions*, Macmillan, London, 1892.

85. N. Griffin, *The Selected Letters of Bertrand Russell, Vol. 1, The Private Years 1884–1914*, Allen Lane, London, 1992.

86. H. Gupta, *Tables of Partitions*, Indian Mathematical Society, Madras, 1939.

87. J. L. Hafner, "New omega theorems for two classical lattice point problems," *Invent. Math.* **63** (1981), 181–186.

88. G. H. Hardy, "Note on the function $\int_x^\infty e^{-\frac{1}{2}(x^2-t^2)}dt$," *Quart. J. Math.* **35** (1904), 193–207.

89. G. H. Hardy, *A Course of Pure Mathematics*, Cambridge University Press, Cambridge, 1908.

90. G. H. Hardy, *Orders of Infinity*, Cambridge Univ. Press, London, 1910.

91. G. H. Hardy, "Proof of a formula of Mr. Ramanujan," *Mess. Math.* **44** (1915), 18–21.

92. G. H. Hardy, "On Dirichlet's divisor problem," *Proc. London Math. Soc.* (2) **15** (1916), 1–25.

93. G. H. Hardy, "On the representation of a number as the sum of any number of squares, and in particular of five or seven," *Proc. London Math. Soc.* (2) **17** (1918), xxii–xxiv.

94. G. H. Hardy, "On the representation of a number as the sum of any number of squares, and in particular of five or seven," *Proc. Nat. Acad. Sci.* **4** (1918), 340–344.

95. G. H. Hardy, "Note LII, On some definite integrals considered by Mellin," *Mess. Math.* **49** (1919), 85–91.

96. G. H. Hardy, "On the representation of a number as the sum of any number of squares, and in particular of five," *Trans. Amer. Math. Soc.* **21** (1920), 255–284.

97. G. H. Hardy, "On two theorems of F. Carlson and S. Wigert," *Acta Math.* **42** (1920), 327–339.

98. G. H. Hardy, "Srinivasa Ramanujan," *Proc. London Math. Soc.* (2) **19** (1921), xl–lviii.

99. G. H. Hardy, "A chapter from Ramanujan's note-book," *Proc. Cambridge Philos. Soc.* (2) **21** (1923), 492–503.

100. G. H. Hardy, "Some formulae of Ramanujan," *Proc. London Math. Soc.* (2) **22** (1924), xii–xiii.

101. G. H. Hardy, "A formula of Ramanujan," *J. London Math. Soc.* **3** (1928), 238–240.

102. G. H. Hardy, "A formula of Ramanujan in the theory of primes," *J. London Math. Soc.* **12** (1937), 94–98.

103. G. H. Hardy, *Divergent Series*, Clarendon Press, Oxford, 1949.

104. G. H. Hardy, *A Mathematician's Apology*, Cambridge University Press, Cambridge, 1967.

105. G. H. Hardy, *Collected Papers*, Vol. 1, Clarendon Press, Oxford, 1966.

106. G. H. Hardy, *Collected Papers*, Vol. 2, Clarendon Press, Oxford, 1967.

107. G. H. Hardy, *Collected Papers*, Vol. 4, Clarendon Press, Oxford, 1969.

108. G. H. Hardy, *Collected Papers*, Vol. 5, Clarendon Press, Oxford, 1972.

109. G. H. Hardy, *Collected Papers*, Vol. 6, Clarendon Press, Oxford, 1974.

110. G. H. Hardy, *Ramanujan*, Cambridge University Press, Cambridge, 1940, reprinted by Chelsea, New York, 1978.

111. G. H. Hardy, *Bertrand Russell and Trinity*, Cambridge University Press, Cambridge, 1942, 1970.

112. G. H. Hardy, *Collected Papers*, Vol. 7, Clarendon Press, Oxford, 1979.

113. G. H. Hardy and H. Heilbronn, "Edmund Landau," *J. London Math. Soc.* **13** (1938), 302–310.

114. G. H. Hardy and J. E. Littlewood, "Contributions to the arithmetic theory of series," *Proc. London Math. Soc.* (2) **11** (1913), 411–478.

115. G. H. Hardy and J. E. Littlewood, "Some problems of Diophantine approximation. I. The fractional part of $n^k\theta$," *Acta Math.* **37** (1914), 155–191.

116. G. H. Hardy and J. E. Littlewood, "Contributions to the theory of the Riemann zeta-function and theory of the distribution of primes," *Acta Math.* **41** (1918), 119–196.

117. G. H. Hardy and J. E. Littlewood, "A new solution of Waring's problem," *Quart. J. Math.* **48** (1920), 272–293.

118. G. H. Hardy and J. E. Littlewood, "Some problems of 'Partitio Numerorum': I. A new solution of Waring's problem," *Nachr. K. Gesell. Wiss. Göttingen, Math.-phys. Kl.* (1920), 33–54.

119. G. H. Hardy and J. E. Littlewood, "Some problems of 'Partitio Numerorum': II. Proof that every large number is the sum of at most 21 biquadrates," *Math. Zeit.* **9** (1921), 14–27.

120. G. H. Hardy and J. E. Littlewood, "Some problems of Diophantine approximation: The lattice-points of a right-angled triangle," *Proc. London Math. Soc.* (2) **20** (1922), 15–26.

121. G. H. Hardy and J. E. Littlewood, "Some problems of Diophantine approximation: The lattice-points of a right-angled triangle (Second Memoir)," *Abh. Math. Sem. Univ. Hamburg* **1** (1922), 212–249.

122. G. H. Hardy and J. E. Littlewood, "Note on the theory of series (XX): On Lambert Series," *Proc. London Math. Soc.* (2) **41** (1936), 257–270.

123. G. H. Hardy and S. Ramanujan, "Une formule asymptotique pour le nombre des partitions de n," *C. R. Acad. Sci. (Paris)* **164** (1917), 35–38.

124. G. H. Hardy and S. Ramanujan, "Asymptotic formulae in combinatory analysis," *Proc. London Math. Soc.* (2) **17** (1918), 75–118.

125. G. H. Hardy and S. Ramanujan, "On the coefficients in the expansions of certain modular functions," *Proc. Royal Soc. A* **95** (1918), 144–155.

126. G. H. Hardy and E. M. Wright, *An Introduction to the Theory of Numbers*, fifth ed., Clarendon Press, Oxford, 1979.

127. U. S. Haslam-Jones, "Bertram Martin Wilson," *Proc. Edinburgh Math. Soc.* **4** (1936), 268–269.

128. M. Hausman and H. N. Shapiro, "On Ramanujan's right triangle conjecture," *Comm. Pure Appl. Math.* **42** (1989), 885–889.

129. E. Hecke, "Über Modulfunktionen und die Dirichletschen Reihen mit Eulerscher Produktentwicklung I," *Math. Ann.* **114** (1937), 1–28.

130. E. Hecke, *Mathematische Werke*, Vandenhoeck & Ruprecht, Göttingen, 1970.

131. H. A. Heilbronn, *The Collected Papers of Hans Arnold Heilbronn* (E. J. Kani and R. A. Smith, eds.), Wiley, New York, 1988.

132. D. Hickerson, "A proof of the mock theta conjectures," *Invent. Math.* **94** (1988), 639–660.

133. M. J. M. Hill, "On a formula for the sum of a finite number of terms of the hypergeometric series when the fourth element is equal to unity," *Proc. London Math. Soc.* (2) **5** (1907), 335–341.

134. M. J. M. Hill, "On a formula for the sum of a finite number of terms of the hypergeometric series when the fourth element is unity (second communication)," *Proc. London Math. Soc.* (2) **6** (1908), 339–348.

135. M. D. Hirschhorn, "A continued fraction," *Duke Math. J.* **41** (1974), 27–33.

136. M. D. Hirschhorn, "Ramanujan's contributions to continued fractions," *Toils and Triumphs of Srinivasa Ramanujan the Man and the Mathematician* (W. H. Abdi, ed.), National Publishing House, Jaipur, 1992, pp. 236–246.

137. M. N. Huxley, "Exponential sums and lattice points II," *Proc. London Math. Soc.* **66** (1993), 279–301.

138. A. Ivić, *The Riemann Zeta-function*, John Wiley, New York, 1985.

139. H. Iwaniec and C. J. Mozzochi, "On the divisor and circle problems," *J. Number Theory* **29** (1988), 60–93.

140. C. G. J. Jacobi, "De fractione continua, in quam integrale $\int_x^\infty e^{-xx}dx$ evolvere licet," *J. Reine Angew. Math.* **12** (1834), 346–347.

141. L. Jacobsen, "Domains of validity for some of Ramanujan's continued fraction formulas," *J. Math. Anal. Appl.* **143** (1989), 412–437.

142. R. Kanigel, *The Man Who Knew Infinity*, Charles Scribner's, New York, 1991.

143. P. Kaplan and K. S. Williams, "The Chowla–Selberg formula for non-fundamental discriminants," preprint.

144. J. N. Kapur, ed., *Some Eminent Indian Mathematicians of the Twentieth Century*, vol. 2, Mathematical Sciences Trust Society, Kanpur, 1989.

145. F. Klein and R. Fricke, *Vorlesungen über die Theorie der Elliptischen Modulfunctionen*, B. G. Teubner, Leipzig, 1892.

146. M. I. Knopp, *Modular Functions in Analytic Number Theory*, Markham, Chicago, 1970, reprinted by Chelsea, New York, 1993.

147. E. Krätzel, *Lattice Points*, Kluwer, Dordrecht, 1988.

148. C. Krishnamachari, "Certain definite integrals and series connected with Bernoulli's numbers," *J. Indian Math. Soc.* **12** (1920), 14–31.

149. E. Landau, "Über die Einteilung der positiven ganzen Zahlen in vier Klassen nach der Mindeszahl der zu ihrer additiven Zusammensetzung erforderlichen Quadrate," *Archiv Math. Phys.* (3) **13** (1908), 305–312.

150. E. Landau, *Primzahlen*, Chelsea, New York, 1953.

151. E. Landau, *Collected Papers*, Vol. 4, Thales Verlag, Essen, 1985.

152. P. S. Laplace, *Traité de Mécanique Céleste 4*, J. B. M. Duprat, Paris, 1805.

153. D. H. Lehmer, "The mechanical combination of linear forms," *Amer. Math. Monthly* **35** (1928), 114–121.

154. D. H. Lehmer, "Hunting big game in the theory of numbers," *Scripta Math.* **1** (1932–33), 229–235.

155. D. H. Lehmer, "A photo-electric number sieve," *Amer. Math. Monthly* **40** (1933), 401–406.

156. D. H. Lehmer, "On the series for the partition function," *Trans. Amer. Math. Soc.* **43** (1938), 271–295.

157. D. H. Lehmer, "A history of the sieve process," *A History of Computing in the Twentieth Century*, Academic Press, New York, 1980, pp. 445–456.

158. D. H. Lehmer, "Exploitation of parallelism in number theoretic and combinatorial computations," *Proc. Sixth Manitoba Conf. on Numerical Math.*, Utilitas Math. Publ. Inc., Winnipeg, 1976, pp. 95–111.

159. J. Lehner, "Ramanujan's identities involving the partition function for the moduli 11^a," *Amer. J. Math.* **65** (1943), 492–520.

160. M. Lerch, "Sur la fonction $\zeta(s)$ pour valeurs impaires d'argument," *J. Sci. Math. Astron., pub. pelo Dr. F. Gomes Teixeira, Coimbra* **14** (1901), 65–69.

161. J. E. Littlewood, "Sur la distribution des nombres premiers," *C. R. Acad. Sci. (Paris)* **158** (1914), 263–266.

162. J. E. Littlewood, "Review of Collected Papers of Srinivasa Ramanujan," *Math. Gaz.* **14** (1929), 427–428.

163. J. E. Littlewood, *A Mathematician's Miscellany*, Methuen & Co., London, 1953, revised edition, edited by B. Bollobás, Cambridge University Press, Cambridge, 1986.

164. S. L. Loney, *Plane Trigonometry*, Parts I, II, Cambridge Univ. Press, Cambridge, 1893.

165. S. L. Malurkar, "On the application of Herr Mellin's integrals to some series," *J. Indian Math. Soc.* **16** (1925/26), 130–138.

166. L. J. Mordell, "On the representation of numbers as a sum of $2r$ squares," *Quart. J. Math.* **48** (1917), 93–104.

167. L. J. Mordell, "On the representations of a number as a sum of an odd number of squares," *Trans. Cambridge Philos. Soc.* **22** (1919), 361–372.

168. L. J. Mordell, "On Mr. Ramanujan's empirical expansions of modular functions," *Proc. Cambridge Philos. Soc.* **19** (1920), 117–124.

169. M. T. Naraniengar, "V. Ramaswami Aiyar, the founder of the Society," *Math. Student* **3** (1935), 111–119.

170. S. Narayanan, "Tributes and Reminiscences," *Math. Student* **9** (1941), 36.

171. E. H. Neville, "Srinivasa Ramanujan," *Nature* **149** (1942), 292–294.

172. E. H. Neville, *Jacobian Elliptic Functions*, second edition, Oxford University Press, Oxford, 1951.

173. J.-L. Nicolas, "On highly composite numbers," in *Ramanujan Revisited* (G. E. Andrews, et al., eds.), Academic Press, Boston, 1988, pp. 215–244.

174. J.-L. Nicolas, "Appendix: On composite numbers," *Number Theory, Madras 1987* (K. Alladi, ed.), Lecture Notes in Math. No. 1395, Springer-Verlag, Berlin, 1989, pp. 18–20.

175. J.-L. Nicolas and G. Robin, "Highly composite numbers by Srinivasa Ramanujan," The Ramanujan J. **1** (1997).

176. N. Nielsen, *Theorie des Integrallogarithmus*, Chelsea, New York, 1965.

177. I. Niven, H. S. Zuckerman, and H. L. Montgomery, *An Introduction to the Theory of Numbers*, fifth ed., Wiley, New York, 1991.

178. F. W. J. Olver, *Asymptotics and Special Functions*, Academic Press, New York, 1974.

179. A. M. Ostrowski, "Bemerkungen zur Theorie der Diophantischen Approximationen," *Abh. Math. Sem. Univ. Hamburg* **1** (1922), 77–98; 250–251.

180. O. Perron, "Über die Preeceschen Kettenbrüche," *Sitz. Bayer. Akad. Wiss. München Math. Phys. Kl.* (1953), 21–56.

181. O. Perron, *Die Lehre von den Kettenbrüchen*, Band 2, dritte Auf., B. G. Teubner, Stuttgart, 1957.

182. C. T. Preece, "Theorems stated by Ramanujan (I): Theorems on integrals," *J. London Math. Soc.* **3** (1928), 212–216.

183. C. T. Preece, "Theorems stated by Ramanujan (III): Theorems on transformation of series and integrals," *J. London Math. Soc.* **3** (1928), 274–282.

184. C. T. Preece, "Theorems stated by Ramanujan (VI): Theorems on continued fractions," *J. London Math. Soc.* **4** (1929), 34–39.

185. C. T. Preece, "Theorems stated by Ramanujan (X)," *J. London Math. Soc.* **6** (1931), 22–32.

186. C. T. Preece, "Theorems stated by Ramanujan (XIII)," *J. London Math. Soc.* **6** (1931), 95–99.

187. H. Rademacher, "On the partition function $p(n)$," *Proc. London Math. Soc.* (2) **43** (1937), 241–254.

188. H. Rademacher, "On the Selberg formula for $A_k(n)$," *J. Indian Math. Soc.* **21** (1957), 41–55.

189. H. Rademacher, *Collected Papers*, Vol. 2, MIT Press, Cambridge, 1974.

190. H. Rademacher and A. Whiteman, "Theorems on Dedekind sums," *Amer. J. Math.* **63** (1941), 377–407.

191. H. Rademacher and H. S. Zuckerman, "On the Fourier coefficients of certain modular forms of positive dimension," *Ann. of Math.* **39** (1938), 433–462.

192. S. Raghavan, "On Ramanujan and Dirichlet series with Euler products," *Glasgow Math. J.* **25** (1984), 203–206.

193. C. T. Rajagopal, "K. Ananda Rau—A sketch," *Publ. Ramanujan Inst.* **1** (1968–69), 1–10.

194. C. T. Rajagopal, "K. Ananda Rau," *J. London Math. Soc.* **44** (1969), 1–6.

195. S. Ram, *Srinivasa Ramanujan*, National Book Trust, India, New Delhi, 1972.

196. K. Ramachandra, "Srinivasa Ramanujan (the inventor of the circle method)," *J. Math. Phys. Sci.* **21** (1987), 545–564; *Hardy-Ramanujan J.* **10** (1987), 9–24.

197. K. G. Ramanathan, "Remarks on some series considered by Ramanujan," *J. Indian Math. Soc.* **46** (1982), 107–136, published in 1985.

198. K. G. Ramanathan, "On Ramanujan's continued fraction," *Acta Arith.* **43** (1984), 209–226.

199. K. G. Ramanathan, "On the Rogers-Ramanujan continued fraction," *Proc. Indian Acad. Sci. (Math. Sci.)* **93** (1984), 67–77.

200. K. G. Ramanathan, "Ramanujan's continued fraction," *Indian J. Pure Appl. Math.* **16** (1985), 695–724.

201. K. G. Ramanathan, "Some applications of Kronecker's limit formula," *J. Indian Math. Soc.* **52** (1987), 71–89.

202. K. G. Ramanathan, "Hypergeometric series and continued fractions," *Proc. Indian Acad. Sci. (Math. Sci.)* **97** (1987), 277–296.

203. K. G. Ramanathan, "On some theorems stated by Ramanujan," *Number Theory and Related Topics*, Oxford Univ. Press, Bombay, 1989, pp. 151–160.

204. K. G. Ramanathan, "Ramanujan's modular equations," *Acta Arith.* **53** (1990), 403–420.

205. S. Ramanujan, "Some properties of Bernoulli's numbers," *J. Indian Math. Soc.* **3** (1911), 219–234.

206. S. Ramanujan, "Question 386," *J. Indian Math. Soc.* **4** (1912), 120.

207. S. Ramanujan, "Question 294," *J. Indian Math. Soc.* **4** (1912), 151–152.

208. S. Ramanujan, "Question 295," *J. Indian Math. Soc.* **5** (1913), 65.

209. S. Ramanujan, "Irregular numbers," *J. Indian Math. Soc.* **5** (1913), 105–106.

210. S. Ramanujan, "Modular equations and approximations to π," *Quart. J. Math. Oxford* **45** (1914), 350–372.

211. S. Ramanujan, "On the sum of the square roots of the first n natural numbers," *J. Indian Math. Soc.* **7** (1915), 173–175.

212. S. Ramanujan, "Some definite integrals," *Mess. Math.* **44** (1915), 10–18.

213. S. Ramanujan, "On the integral $\int_0^x (\tan^{-1} t)/t \, dt$," *J. Indian Math. Soc.* **7** (1915), 93–96.

214. S. Ramanujan, "Some definite integrals connected with Gauss's sums," *Mess. Math.* **44** (1915), 75–85.

215. S. Ramanujan, "Summation of certain series," *Mess. Math.* **44** (1915), 157–160.

216. S. Ramanujan, "New expressions for Riemann's functions $\xi(s)$ and $\Xi(t)$," *Quart. J. Math.* **46** (1915), 253–260.

217. S. Ramanujan, "Highly composite numbers," *Proc. London Math. Soc.* (2) **14** (1915), 347–409.

218. S. Ramanujan, "On certain infinite series," *Mess. Math.* **45** (1916), 11–15.

219. S. Ramanujan, "On certain arithmetical functions," *Trans. Cambridge Philos. Soc.* **22** (1916), 159–184.

220. S. Ramanujan, "A series for Euler's constant γ," *Mess. Math.* **46** (1917), 73–80.

221. S. Ramanujan, "On certain trigonometrical sums and their applications in the theory of numbers," *Trans. Cambridge Philos. Soc.* **22** (1918), 259–276.

222. S. Ramanujan, "Some properties of $p(n)$, the number of partitions of n," *Proc. Cambridge Philos. Soc.* **19** (1919), 207–210.

223. S. Ramanujan, "Proof of certain identities in combinatory analysis," *Proc. Cambridge Philos. Soc.* **19** (1919), 214–216.

224. S. Ramanujan, "Congruence properties of partitions," *Math. Z.* **9** (1921), 147–153.

225. S. Ramanujan, *Notebooks* (2 volumes), Tata Institute of Fundamental Research, Bombay, 1957.

226. S. Ramanujan, *Collected Papers*, Cambridge University Press, Cambridge, 1927, 2nd ed., Chelsea, New York, 1962.

227. S. Ramanujan, *The Lost Notebook and Other Unpublished Papers*, Narosa, New Delhi, 1988.

228. S. S. Rangachari, "Ramanujan and Dirichlet series with Euler products," *Proc. Indian Acad. Sci. (Math. Sci.)* **91** (1982), 1–15.

229. S. R. Ranganathan, *Ramanujan The Man and the Mathematician*, Asia Publishing House, Bombay, 1967.

230. R. A. Rankin, "George Neville Watson," *J. London Math. Soc.* **41** (1966), 551–565.

231. R. A. Rankin, "Ramanujan's unpublished work on congruences," *Modular Functions of One Variable, V*, Lecture Notes in Math. No. 601, Springer–Verlag, Berlin, 1977, pp. 3–15.

232. R. A. Rankin, "Gertrude Katherine Stanley," *Bull. London Math. Soc.* **14** (1982), 554–555.

233. R. A. Rankin, "Ramanujan's manuscripts and notebooks," *Bull. London Math. Soc.* **14** (1982), 81–97.

234. R. A. Rankin, "Ramanujan as a patient," *Proc. Indian Acad. Sci. (Math. Sci.)* **93** (1984), 79–100.

235. R. A. Rankin, "Ramanujan's manuscripts and notebooks, II," *Bull. London Math. Soc.* **21** (1989), 351–365.

236. R. A. Rankin, "Fourier coefficients of cusp forms," *Math. Proc. Cambridge Philos. Soc.* **100** (1986), 5–29.

237. A. N. Rao, "P.V. Seshu Aiyar," *Math. Student* **4** (1936), 116–118.

238. A. N. Rao, "Dewan Bahadur R. Ramachandra Rao," *Math. Student* **4** (1936), 168–169.

239. A. N. Rao, "Rao Bahadur S. Narayana Ayyar," *Math. Student* **4** (1936), 169.

240. A. N. Rao and V. G. Iyer, "R. Vaidyanathaswamy (1894–1960)," *Math. Student* **29** (1961), 1–14.

241. K. Srinivasa Rao, "Mrs. Janakiammal Ramanujan," *The Hindu, Madras; The Indian Express; The Statesman, Delhi; The Statesman, Calcutta* (14 April 1994).

242. M. B. Rao and M. V. Ayyar, "On some infinite series and products, Part I," *J. Indian Math. Soc.* **15** (1923/24), 150–162.

243. R. Ramachandra Rao, "In Memoriam S. Ramanujan, B.A., F.R.S.," *J. Indian Math. Soc.* **12** (1920), 87–90.

244. B. Riemann, "Ueber die Anzahl der Primzahlen unter einer gegebenen Grösse," *Monatsber. König. Preuss. Acad. Wiss. Berlin* (1859), 671–680.

245. B. Riemann, *Mathematische Werke*, zweite Auf., Dover, New York, 1953, pp. 145-153.

246. L. J. Rogers, "Second memoir on the expansion of certain infinite products," *Proc. London Math. Soc.* **25** (1894), 318–343.

247. L. J. Rogers, "On a type of modular relation," *Proc. London Math. Soc.* (2) **19** (1920), 387–397.

248. J. M. Rushforth, *Congruence Properties of the Partition Function and Associated Functions*, Ph.D. Thesis, University of Birmingham, 1950.

249. J. M. Rushforth, "Congruence properties of the partition function and associated functions," *Proc. Cambridge Philos. Soc.* **48** (1952), 402–413.

250. B. W. A. Russell and A. N. Whitehead, *Principia Mathematica*, Cambridge University Press, Cambridge, 1910, 1912, 1913.

251. I. Schur, "Ein Beiträg zur additiven Zahlentheorie und zur Theorie der Kettenbrüche," *S.-B. Preuss. Akad. Wiss. Phys.-Math. Kl.* (1926), 302–321.

252. I. Schur, *Gesammelte Abhandlungen*, Band 2, Springer–Verlag, Berlin, 1973.

253. J.-A. Séguier, *La Théorie des Formes Quadratiques et de la Multiplication Complexe*, Gauthier-Villars, Paris, 1894.

254. A. Selberg, "Über die Mock-Thetafunktionen siebenter Ordnung," *Arch. Math. og Naturvidenskab* **41** (1938), 1–15.

255. A. Selberg, *Collected Papers*, Vol. 1, Springer-Verlag, Berlin, 1989, pp. 695–706.

256. A. Selberg and S. Chowla, "On Epstein's zeta-function (I)," *Proc. Nat. Acad. Sci. USA* **35** (1949), 367-370.

257. A. Selberg and S. Chowla, "On Epstein's zeta-function," *J. Reine Angew. Math.* **227** (1967), 86-110.

258. S. P. Sen, *Dictionary of National Biography* (4 volumes), Institute of Historical Studies, Calcutta, 1972–1974.

259. N. M. Shah and B. M. Wilson, "On an empirical formula connected with Goldbach's theorem," *Proc. Cambridge Philos. Soc.* **19** (1919), 238–244.

260. E. Shils, "Reflections on tradition, centre and periphery and the universal validity of science: the significance of the life of S. Ramanujan," *Minerva* **29** (1991), 393–419.

261. R. Sitaramachandrarao, "Some formulae of S. Ramanujan III," *J. Madras Univ., Ser. B* **51** (1988), 67–70.

262. S. Skewes, "On the difference $\pi(x) - \operatorname{li} x$ (I)," *Proc. London Math. Soc.* **8** (1933), 277–283.

263. P. K. Srinivasan, *Ramanujan: An Inspiration*, Vols. 1, 2, Muthialpet High School, Madras, 1968.

264. G. Szegö, "Über einige von S. Ramanujan gestellte Aufgaben," *J. London Math. Soc.* **3** (1928), 225–232.

265. G. Szegö, *Collected Papers*, Vol. 2, Birkhäuser, Boston, 1982.

266. J. Tannery and J. Molk, *Éléments de la Théorie des Fonctions Elliptiques* (4 volumes), Gauthier–Villars, Paris, 1893–1902.

267. E. C. Titchmarsh, *Theory of Fourier Integrals*, second ed., Oxford University Press, Oxford, 1948.

268. H. W. Turnbull, "Bertram Martin Wilson," *Proc. Roy. Soc. Edinburgh* **55** (1936), 176–177.

269. R. C. Vaughan, *The Hardy–Littlewood Circle Method*, Cambridge University Press, Cambridge, 1981.

270. K. Venkatachaliengar, *Development of Elliptic Functions According to Ramanujan*, Tech. Report 2, Madurai Kamaraj University, Madurai, 1988.

271. T. Vijayaraghavan and G. N. Watson, "Solution to Question 784," *J. Indian Math. Soc.* **19** (1931), 12–23.

272. S. S. Wagstaff, "Ramanujan's paper on Bernoulli numbers," *J. Indian Math. Soc.* **45** (1981), 49–65.

273. K. C. Wali, *Chandra*, The University of Chicago Press, Chicago, 1991.

274. G. N. Watson, *A Treatise on the Theory of Bessel Functions*, Cambridge University Press, Cambridge, 1922.

275. G. N. Watson, "Theorems stated by Ramanujan (II): Theorems on summation of series," *J. London Math. Soc.* **3** (1928), 216–225.

276. G. N. Watson, "Theorems stated by Ramanujan (IV): Theorems on approximate integration and summation of series," *J. London Math. Soc.* **3** (1928), 282–289.

277. G. N. Watson, "Solution to Question 722," *J. Indian Math. Soc.* **18** (1929), 113–117.

278. G. N. Watson, "Theorems stated by Ramanujan (V): Approximations connected with e^x," *Proc. London Math. Soc.* (2) **29** (1929), 293–308.

279. G. N. Watson, "Theorems stated by Ramanujan (VII): Theorems on continued fractions," *J. London Math. Soc.* **4** (1929), 39–48.

280. G. N. Watson, "Theorems stated by Ramanujan (VIII): Theorems on divergent series," *J. London Math. Soc.* **4** (1929), 82–86.

281. G. N. Watson, "Theorems stated by Ramanujan (IX): Two continued fractions," *J. London Math. Soc.* **4** (1929), 231–237.

282. G. N. Watson, "Theorems stated by Ramanujan (XI)," *J. London Math. Soc.* **6** (1931), 59–65.

283. G. N. Watson, "Theorems stated by Ramanujan (XII): A singular modulus," *J. London Math. Soc.* **6** (1931), 65–70.

284. G. N. Watson, "Theorems stated by Ramanujan (XIV): A singular modulus," *J. London Math. Soc.* **6** (1931), 126–132.

285. G. N. Watson, "Ramanujan's note books," *J. London Math. Soc.* **6** (1931), 137–153.

286. G. N. Watson, "Some singular moduli (I)," *Quart. J. Math.* **3** (1932), 81–98.

287. G. N. Watson, "Some singular moduli (II)," *Quart. J. Math.* **3** (1932), 189–212.

288. G. N. Watson, "Scraps from some mathematical note books," *Math. Gazette* **18** (1934), 5–18.

289. G. N. Watson, "The final problem: an account of the mock theta functions," *J. London Math. Soc.* **11** (1936), 3–15.

290. G. N. Watson, "The mock theta functions (2)," *Proc. London Math. Soc.* (2) **42** (1937), 274–304.

291. G. N. Watson, "Ramanujans Vermutung über Zerfällungsanzahlen," *J. Reine Angew. Math.* **179** (1938), 97–128.

292. H. Weber, "Zur Theorie der elliptischen Functionen," *Acta Math.* **11** (1887–88), 333–390.

293. H. Weber, *Lehrbuch der Algebra, Dritter Band,* 2nd. ed., Friedrich Vieweg und Sohn, Braunschweig, 1908, reprinted by Chelsea, New York, 1961.

294. E. T. Whittaker and G. N. Watson, *A Course of Modern Analysis*, Cambridge University Press, Cambridge, 1915.

295. J. M. Whittaker, "George Neville Watson," *Biographical Memoirs of Fellows of the Royal Society* **12** (1966), 521–530.

296. S. Wigert, "Sur l'order de grandeur du nombre des diviseurs d'un entier," *Arkiv f. Mat. Astron. Fys.* **3** 18 (1906–07), 1–9.

297. K. S. Williams and Zhang Nan-Yue, "The Chowla–Selberg formula for genera," preprint.

298. D. A. B. Young, "Ramanujan's illness," *Notes Rec. Royal Soc. London* **48** (1994), 107–119.

299. L.-C. Zhang, "Ramanujan's continued fractions for products of gamma functions," *J. Math. Anal. Appl.* **174** (1993), 22–52.

300. I. J. Zucker, "The evaluation in terms of Γ-functions of the periods of elliptic curves admitting complex multiplication," *Math. Proc. Cambridge Philos. Soc.* **82** (1977), 111–118.

301. H. S. Zuckerman, "On the coefficients of certain modular forms belonging to subgroups of the modular group," *Trans. Amer. Math. Soc.* **45** (1939), 298–321.

302. H. S. Zuckerman, "On the expansion of certain modular forms of positive dimension," *Amer. J. Math.* **62** (1940), 127–152.

Index

Abel, N., 17
Abel–Plana summation formula, 40–41
Aberdeen University, 288
Accountant General, Madras, 1
Acta Mathematica, 146
Ananda Rau, K., 106, 110–112, 118, 259–261, 264, 266, 289, 290
Anantharaman, M., 128, 136
Andhra University, 277
Andrews, G.E., 42, 43, 65, 66, 224, 267, 304, 305
Apollonius Circle, 113
Arber, E.A.N., 154
Askey, R.A., 37, 65, 313
Association of Mathematics Teachers of India, 271
Atkin, A.O.L., 194, 297
Ayer, A.J., 45
Ayyar, M.V., 283

Bahadur, 106
Bailey, W.N., 40, 44, 66, 129
Baker, H.F., 2, 155
Balasubramanian, R., 42, 199
Bapatla, 125, 126
Barclays Bank, 234, 235, 237, 241
Barnes, E.W., 50, 77, 82, 105, 129, 132, 133
Barrow, I., 7
Bauer, G., 38, 43
Berndt, B.C., 255, 285, 290, 304, 306, 307, passim
Bernoulli numbers, 34, 38–39, 49, 54, 63, 206, 313
Berry, A., 49, 77, 82
Berwick, W.E.H., 138–140, 286
Bessel functions, 283
Bharathi, R., 154
Bhargava, S., 67

Bhujanga Rao, Dr., 214
Bialek, P., 191
Binet, A., 38
Birch, B.J., 194
Birmingham University, 282, 284, 288, 289, 299
Bollobás, B., 174
Bolton, Dr., 233, 234
Bombay, 126
Borwein, J.M. and P.B., 67, 199
Bose, J.C., 161
Bourne, A., 13, 157, 207
Boyd, J.W.H., 173
Bradley, F.H., 115
Bromwich, T.J.I'A., 16–19, 46, 53, 155, 208
Brouncker, Lord, 43
Buell, D., 140
Burkill, J.C., 30, 51

Cambridge University Press, 263
Cardiff, University College of, 286
Carlson–Wigert, 282, 284, 286, 287
Carr, G.S., 1, 5
Cartwright, M.L., 51
Cathedral Road, Madras, 214
Cauchy, A.L., 38, 39, 106
Cauchy–Poisson formula, 279–281
Cauchy's theory of residues, 46
Cauvery River, 215, 229
Cayley, A., 289, 290
Cazalet, V.A., 237, 238
Chan, H.H., 64, 66, 141
Chandrakantha, Smt., 113
Chandrasekhar, P.S., 306, 307, 310, 311
Chandrasekhar, S., 193, 255, 261, 290–294, 298, 306, 307, 309–311
Chandrasekharan, K., 261
Chasles, M., 230
Chebychev, P.L., 34
Chetpet, 219, 224, 231
Chettiar, T.A., 252
Chetty, P. Adinarayana, 227–230
Choultry, 245
Chowla, S., 140, 297
Chowry-Muthu, Dr., 173, 233, 234
circle method, 42

Clark, E.M., 234, 235, 237, 241, 242
Cohen, H., 224
Coimbatore, 215, 216
Colinette Road, Putney, 174, 201, 234
Cooum River, 52
Cottrell, C.B., 98, 100
Cours d'Analyse, 31
Course of Pure Mathematics, 32
Cranleigh, 30
Cromwell Road, London, 109, 110, 114, 162, 164, 173
Cuddalore, 164
Cumbum, 73
Cursetjee, A., 161

Darling, H.B.C., 298
Davenport, H., 198
Davies, A., 11, 70, 76, 86, 87, 99, 204, 209–211
Dedekind eta–function, 42, 194
definite integrals, 24–26, 28, 47, 57–59, 116, 120
de la Valleé–Poussin, C.J., 87
Deshmukh, C.D., 230
Deshouillers, J.-M., 199
Dewan, 106
Dewsbury, F., 51, 53, 70, 76, 78, 80, 94, 96, 100, 129, 132–136, 142–144, 156, 159, 160, 193, 199, 201, 202, 218, 224, 242, 263–266, 274
diets, 224
Dirac, P.A.M., 308
Dirichlet L-function, 40, 281
Dirichlet, P.G.L., 199
Dirichlet series, 194, 260, 261, 299
Dougall, J., 38
Dragonette, L., 224
Dress, F., 199
Dundee, 268
Durairajan, Mr., 124
Dutka, J., 43

Eddington, A.S., 154
Edinburgh University, 315
Educational Times, 75
Eisenstein series, 176–186, 188–191, 194, 195, 299, 300
Ekanatha Rao, S., 113
elliptic functions, 19, 96, 116, 182, 206, 289, 290, 314
elliptic integrals, 140, 141

Erode, 1, 215, 216, 313–315
Euler–Maclaurin summation formula, 33, 46
Euler products, 194, 300, 301
Everest, Mount, 33

Fawcett, P., 8
Field, R.E., 171, 173
First Arts Examination, 1, 76, 144, 206, 314
Fitzroy House, 3, 174, 192–199
Forsyth, A.R., 155
Fort St George, 15, 78, 261
Fourier integral, 286
Fourier transform, 280
Fowler, R.H., 262
Francis, W., 102, 103
Frater, A., 53
Freeman–Thomas, Freeman, 238, 241

Ganapathy Iyer, C.N., 120–123
Ganapathy, M.M., 128, 136
Garvan, F.G., 67
Gauss, C.F., 123, 139
Gibraltar, 238
Glaisher, J.W.L., 38
Glasgow University, 302, 305
Godavery, 120
Goldberg, L.A., 42
Gometra, 231
Göttingen, 243
Government College of Kumbakonam, 1, 5, 93, 98, 158, 247, 269, 289, 290, 313
Govindarajan, Mr., 124
Grace, J.H., 155
Graham, J.F., 13, 14, 207, 213
Gram, J.P., 34, 63
Greenhill, A.G., 138, 140, 289, 290
Griffin, N., 45
Griffith, C.L.T., 2, 12–17, 19, 207
Guindy, 214, 215, 241
Gupta, H., 297

Hadamard, J., 87
Hafner, J.L., 36
Halcrow, A., 304

Hall, S., 30, 142
Hanumantha Rao, B., 70, 73, 76, 209
Hardy, G.E., 142
Hardy, G.H., biography, 30–33, passim
Hardy, I., 30
Harley Street, London, 215
Harry Ransom Humanities Center, 45
Hausman, M., 35
Hecke, E., 194, 301
Hecke zeta-function, 187, 188
Heffer, W., 229, 230
Heilbronn, H., 243, 299–301
Hensman, J.J., 231, 244
hepatic amoebiasis, 3, 308, 309
Hermite, C., 138–140
Hickerson, D., 224
highly composite numbers, 138, 286, 287
Hill Grove, 173, 174, 234
Hill, M.J.M., 2, 13, 15–19, 49, 208, 209
Hindu High School, Kumbakonam, 73, 292
Hirschhorn, M., 65
Hobson, E.W., 2, 32, 155
Holland, T., 157
Horne, W.O., 103
Huband, G.D., 238
Hutton's Mathematical Tables, 232, 233
Huxley, M.N., 36
hypergeometric functions, 40, 59, 66, 67

Indian Mathematical Club, 74, 75, 106
Indian Mathematical Society, 74, 158
Ingham, A.E., 32
Ivić, A., 36
Iwaniec, H., 36

Jacobi, C.G.J., 41, 69, 106
Jacobsen, L., 43, 65, 66
Janaki, S. (Mrs. Ramanujan), 1, 10, 238–241, 248, 252, 255–257, 261, 292, 293, 306, 310
  annuity to, 249, 252, 253, 255–257
Jeans, J.H., 31, 153, 154, 262
Journal of the Indian Mathematical Society, 1, 17, 34, 39–41, 70, 75, 116, 117, 128, 253, 265, 270, 272, 288, 289
Journal of the London Mathematical Society, 288, 289

Kanigel, R., 30, 93, 193, 216, 230
Kaplan, P., 141
K.C.I.E., 135
Kesava Pai, Dr., 214
Kiepert, L., 138, 140
Kilpauk, 309, 310
Kincaid, F., 153, 168, 169, 173, 233, 234
Klein, F., 138, 139
Knopp, M.I., 193, 194, 297
Knorr, W.R., 230
Kodaikanal, 218
Kodumudi, 214, 215, 229, 230
Komalatammal, 1, 10, 118–119, 157, 235, 236, 244–247, 249–251, 273–274, 310, 315
Krätzel, E., 36
Kreczmer, W., 298
Krishnamachari, C.?, 127, 128
Krishna Rao, R., 1, 106–107, 109–112, 260
Krishnaswamy Iyer, A., 113
Kronecker, L., 138
Kumbakonam, 1, 106, 126, 158, 205, 217, 219, 229, 232, 245, 247, 250, 272, 291, 313, 315
Kumbakonam High School, 1, 4, 97, 102, 113, 246, 248

Lagrange inversion formula, 37
Lahore, 291
Lakshmi Narasimhan, S., 231–233, 244, 246, 247, 273, 274, 314, 315
Lambert series, 290
Lamphere, R., 43, 65, 66
Landau, E., 32, 36, 42, 176, 182, 199, 243, 244, 300
Laplace, P.S., 41
Larmor, J., 155
Laurence, R.V., 143
Lehigh University, 295
Lehmer, D.H., 194, 196, 295–297
Lehner, J., 299
Lerch, M., 38
Littlehailes, R., 13, 14, 87, 96, 98, 99, 101, 102, 122, 204, 209, 210, 247, 248, 274
Littlewood, J.E., 3, 32, 34, 35, 42, 44, 47, 48, 50, 51, 68, 70, 76, 77, 81, 82, 89, 91, 105, 112, 121, 123, 152, 154, 155, 192, 198, 211, 242, 259, 260, 282, 290, 291
Livens, G.H., 285, 286
Liverpool University, 268, 289

Lloyd George, David, 100
London Mathematical Society, 18, 70, 112, 113, 122, 131, 140, 151, 224, 260, 287, 299, 302, 305
Loney, S.L., 206, 313
'lost notebook', 182, 191, 223, 224, 261, 267, 283, 293, 305
Love, A.E.H., 31

MacDonald, Ramsay, 129
MacMahon, P.A., 141, 142, 154, 155, 192, 297
Macphail, E.M., 264, 265
Madgshon, J.B., 173
Madhava Rao, T., 106, 113
Madras Central Station, 229
Madras Mail, 105, 106, 156, 203, 227–230, 252
Madras Port Trust, 2, 7–12, 21, 51, 53, 71, 73, 74, 78–81, 90, 97, 122, 128, 129, 143, 204, 207, 209, 213, 217, 219, 255, 314
Madras Times, 203–213, 253, 254, 313
Madras University, 2, 10, 80, 95, 97–99, 102, 105, 113, 128, 131, 133–135, 156, 183, 201, 203–205, 210, 213, 232, 234, 238, 239, 252, 253, 257, 259, 263–265, 271, 273, 285, 286, 313, 315
Madurai, 73
MahaRaja Rajashry, 11, 86
Mallet, C.E., 90, 103–105, 209
Malurkar, S.L., 39, 283
Mambalam, 238, 241, 245, 261
Mathematical Association, 96
mathematical tripos, 7, 8
Mathematician's Apology, 30, 32
Mathews, G.B., 138, 139
Matlock House, 3, 153, 154, 161, 162, 166, 172–175, 185, 188, 189, 191, 193, 233
McLean, W., 53
McVittie, G.C., 307, 308
Medical Register, 219
Mellin's integral, 278, 280–282, 284, 286
Mexborough, 168, 174
Middlemast, E.W., 7, 11, 13, 15, 122, 206–208, 314
Minakshisundaram, S., 261
Minkowski, H., 243
Mittag-Leffler, G., 146, 269
Möbius function, 63
mock theta functions, 220–224, 260
modular equation, 60–62, 67, 68, 116, 138, 139, 290
Molk, J., 70

Montagu, Mr., 147
Mordell, L.J., 49, 151, 152, 194, 287, 298, 299, 301
Morrell, Lady Ottoline, 44, 45
Mozzochi, C.J., 36
Mussolini, Benito, 33
Mutlin, Dr., 233
Mysore, 132

Nan-Yue, Z., 141
Narasimha Iyengar, K., 315
Narayana Aiyar, S., 2, 11, 34, 49, 73–75, 79, 80, 86, 87, 93, 128, 130, 131, 133–135, 138, 143, 144, 156, 157, 207–209, 211–213, 217, 219, 253–255
Narayanan, S., 96
Narayanan, W., 257
Narayaniengar, M.T., 96
National Archives, Delhi, 130, 131, 133, 313
Nature, 153, 274, 275
Nehru, Jawaharlal, 255
Nellore, 1, 106, 131, 206, 209
nested radicals, 275–278
Neville, E.H., 2, 89–91, 94–99, 101, 103–105, 109–111, 114, 121, 204, 210, 211, 233, 274, 275, 301, 315
Neville, R., 110
New India, 213, 214
Newton, I., 7, 44, 103, 213
Nicholson, J.W., 155
Nicolas, J.-L., 287
Nilgris, 233
Nungambakkam, 52, 96
Nursing Hostel, Thompson's Lane, Cambridge, 145, 147

Offord, A.C., 51
Old College, Madras, 14
Olver, F.W.J., 41
Ootacamund, 131–132, 214, 216, 233, 247
Orders of Infinity, 21, 123
Ostrowski, A.M., 35

Pachaiyappa's College, 1, 6, 14, 93, 113, 114, 118, 314
Paley, R.E.A.C., 51
Palghat, 158
panchangam, 126
Parseval's theorem, 40
Partitio Numerorum, 259, 260

partition function $p(n)$,
  asymptotic formula, 41, 42, 141, 142, 145, 146, 295, 296
  congruences, 192–196, 295–299
Patrachariar, K., 122, 123
Pentland, Lord, 100, 211
Perron, O., 43, 66
Phillips, S.M., 233, 234
Pires, P.A., 216
Poisson summation formula, 40, 283
Polya, G., 32
Preece, C.T., 37, 38, 40, 43, 65–67, 288
Presidency College, Madras, 14, 75, 96, 101, 122, 157, 257, 259–261, 264, 269, 273, 289, 292, 307, 314
Principia Mathematica, 45
Proceedings of the London Mathematical Society, 140, 253, 255, 260
Purvis, J.E., 154
Putney, 3

Rademacher, H., 42, 297, 305
Rajagopal, C.T., 261
Ram, L., 162, 166–168, 171–174, 233, 234
Ramachandra, K., 42
Ramachandra Rao, R., 1, 2, 6, 19, 106, 118, 131, 147, 200, 206, 208–209, 214–215, 218, 232, 252–253, 255, 259, 260, 264, 265, 270, 291, 314
Ramalingam, A.S., 161–174, 234
Raman, C.V., 122, 161, 291, 292
Ramanathan, K.G., 43, 64, 66, 68, 188
Ramaniah, R., 158
Ramanujachari, C., 103
Ramanujaha, 101
Ramanujan,
  cause of death, 308, 309
  Collected Papers, 37, 43, 63, 67, 68, 80, 223, 242, 253, 263, 265, 272, 288, 291, 292, 298, 301
  copyright of papers, 249
  degree from Cambridge, 131, 136–137, 138
  diets during his last months, 224–225
  Fellow of the Royal Society, 152–157, 159, 161, 192
  Fellow of Trinity College, 3, 192, 193, 201, 234, 242, 243
  medical accounts, 233–234
  medical reports, 214–216
  notebooks, 259, 264–266, 269, 270, 272, 285, 288, 310, 311
  passport photo, 292–294, 298
  postage stamp, 314

quarterly reports, 80, 265
scholarship to Cambridge, 96, 98, 100–103, 128, 131–136, 142, 143, 159, 160, 202
scholarship to University of Madras, 78–81, 86
tau-function, 195, 283
theory of prime numbers, 22, 23, 33, 34, 48, 49, 54–56, 63, 64, 68, 69, 76, 77, 81–89, 91–93

Ramanujan Institute, Madras, 277
Ramanujan Museum, Royapuram, 122
Ramaswami Aiyar, V., 75, 96, 206, 314
Ramaswami Iyer, C.P., 113
Ramunni Menon, K., 249, 256, 257
Ranganathan, A., 306, 307, 309–311
Ranganathan, S.R., 201, 230, 311, 315
Rangaswami, T.V., 203, 224
Rankin, R.A., 194, 267, 287, 294, 302–305
Rao, M.B., 283
Reading, 96, 274
Riemann, B., 34, 123
Riemann hypothesis, 50
Riemann zeta-function, 17, 44, 63, 68, 77, 116, 188, 278–281
Riesz, M., 32, 260
Roberts, S.C., 294
Robin, G., 287
Rogers, L.J., 43, 64, 220, 298
Rogers–Ramanujan continued fraction, 29, 43, 57, 64, 66, 77, 87, 90
Rogers–Ramanujan identities, 43, 65, 244
Rogosinski, W.W., 32
Ross, E.B., 206, 314, 315
Royal Society, 3, 32, 51
Royapuram, 245
Rushforth, J.M., 196
Russell, B., 44, 45, 291
Russell, Dr., 218, 219
Russell, R., 138

Sadagopan, Ms., 315
Saidapet, 238, 241
St Andrews University, 269, 286
Sankara Rao, Mr., 118
Sarangapani Temple, 217, 218
Sarker, R.P., 53
Sastri, P.P.S., 230
Satagopan, T.A., 80, 265, 302

Satyamangalam, 75
Scholfield, A.F., 43, 244, 301, 302
Schröter, H., 138, 140
Schur, I., 243, 244
Schwarz, W., 243
Séguier, J.-A., 139
Selberg, A., 42, 140, 224, 297
Senate House, Cambridge, 137
Seshu Aiyar, P.V., 106, 112, 118, 157–159, 235, 236, 244, 245, 253, 255, 259, 264–272, 291
Shah, N.M., 271, 272
Shapiro, H.N., 35
Sharp, H., 257, 258
Shaw, H. Batty, 152, 154, 308, 309
Shils, E., 51
Sikorski, General, 238
Simla, 99
Sinclair, J., 100, 211
Singaravelu Mudaliyar, S.P., 117, 118
singular moduli, 60, 62, 67, 68, 288, 289
singular series, 147–152
Sitapati Ayyar, Mr., 214, 215
Sitaramachandrarao, R., 39
Sivasailam, A., 205
Skewes, S., 123
Snow, C.P., 30
Sohncke, L.A., 138, 140
Soldner, J., 63
Spring, Sir F., 2, 12, 13, 15, 19, 75, 79, 80, 87, 98–100, 130, 133–136, 156, 157, 162, 207–209, 211–213, 253–255
Srinivasa Aiyangar, K., 1, 119, 120, 143, 144, 205, 247, 255, 314
Srinivasa Iyangar, R., 245, 252, 261
Srinivasa Rao, K., 113, 203, 224, 255
Srinivasan, P.K., 52, 78, 90, 106, 119, 122, 230, 271
Srinivasan, P.S., 271
Stanley, A.E., 154
Stanley, G.K., 286–288
Subbanarayanan, N., 75
Subramanian, L.V., 158
Subramanian, S.M., 116, 123–128, 136, 146–147
Surya Narayana, T., 116–118
Suryanarayanan, S.S., 113–115, 117, 147
Swaminathan, C.G., 122
Szegö, G., 41

Tamil Nadu Archives, 70, 76, 100, 215, 249, 250, 308
Tanjore, 113
Tannery, P., 70
Thacker, M.S., 255
Thomson, J.J., 152–154, 242
Tirunarayanan, S., 119, 233, 246, 248, 273, 315
Tiruva Aeswaranpet, 73
Tiruvanakoil, 291
Titchmarsh, E.C., 30, 32
Trinity College, Cambridge, 2, 3, 30, 31, 45, 46
Trinity College, Dublin, 13
Triplicane, 73, 157, 235, 244, 273

Vacation Lectures Fund, 101, 103, 132, 202, 256
Vaidyanathaswamy, R., 284–286
Vaughan, R.C., 42, 199
Venkatachaliengar, K., 67
Venkata Vilas, 214
Vijayaraghavan, T., 275–278, 288, 289
Vinayaka Row, E., 113–117, 147
Vinogradov, I.M., 42
Vizagapatnam, 1
Von Neumann, J., 297
Vythynathaswamy, R., 260

Wagstaff, S.S., 17
Wali, K.C., 292, 310
Walker, Sir G., 2, 51–53, 75, 76, 81, 86, 157, 204, 209
Walker's quasi–biennial oscillation, 52
Waring's problem, 152, 196–199, 259
Watson, E.G., 267, 302, 304, 305
Watson, G.N., 36, 38–41, 43, 44, 49, 64, 66–68, 80, 194, 196, 224, 263, 265, 267, 269, 278, 282–289, 295, 297–305
Watson, G. T., 286, 305
Webb, R.R., 31
Weber, H., 67, 289
Wells, 3
Westfield College, London, 287
Whipple, F.J.W., 66
Whitehead, A.N., 45, 155
Whittaker, E.T., 155, 283, 305, 307
Whittaker, J.M., 304, 305
Williams, K.S., 141
Wilson, B.M., 267–272, 274, 275, 284, 285, 288, 289, 302–304

Winchester College, 30
Wingate, W.W., 233, 234
Wittgenstein, L.J.J., 44
Wrangler, 8
Wrenn Bannett, 106, 107
Wright, E.M., 32, 35, 36, 288
Wynn, W.H., 233, 234

Young, D.A.B., 308, 309
Young, W.H., 155

Zhang, L.-C., 43
Zucker, I.J., 141
Zuckerman, H.S., 42, 191

American Mathematical Society

Journal of the Ramanujan Mathematical Society

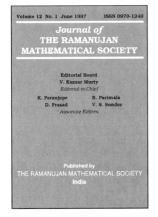

Editorial Board:

V. Kumar Murty, *University of Toronto, ON, Canada,* **Kapil Paranjape,** *Institute of Mathematical Sciences, Madras, India,* **R. Parimala,** *Tata Institute of Fundamental Research, Bombay, India,* **Dipendra Prasad,** *Mehta Research Institute, Allahabad, India,* **V. S. Sunder,** *Institute of Mathematical Sciences, Madras, India*

In 1997, the *Journal of the Ramanujan Mathematical Society* takes on a new look. With a freshly constituted Editorial Board containing some of the best young mathematicians from India, the journal is sure to be of significant interest to a wide spectrum of the mathematical public. The journal is dedicated to publishing high-quality original papers in all areas of mathematics. One volume of two numbers is published each year. Backlog will be kept to a minimum so as to ensure timely publication.

Distributed worldwide by the AMS (outside of India).

ISSN 0970-1249; 1997 Subscription: All Individuals $25, List $90; Order code 97JRMSRMSAD

All prices subject to change. *Prepayment required.* Order from: **American Mathematical Society,** P. O. Box 5904, Boston, MA 02206-5904. For credit card orders, fax (401) 331-3842 or call toll free 800-321-4AMS (4267) in the U. S. and Canada, (401) 455-4000 worldwide. Or place your order through the AMS bookstore at http://www.ams.org/bookstore/. Residents of Canada, please include 7% GST.